Spies and Shuttles

Smithsonian
National Air and Space Museum

UNIVERSITY PRESS OF FLORIDA

Florida A&M University, Tallahassee
Florida Atlantic University, Boca Raton
Florida Gulf Coast University, Ft. Myers
Florida International University, Miami
Florida State University, Tallahassee
New College of Florida, Sarasota
University of Central Florida, Orlando
University of Florida, Gainesville
University of North Florida, Jacksonville
University of South Florida, Tampa
University of West Florida, Pensacola

SPIES AND SHUTTLES

ASA's Secret Relationships with the DoD and CIA

James E. David

Smithsonian National Air and Space Museum,
Washington, D.C., in association with
University Press of Florida
Gainesville · Tallahassee · Tampa · Boca Raton
Pensacola · Orlando · Miami · Jacksonville · Ft. Myers · Sarasota

Published in the United States of America

First cloth printing, 2015
First paperback printing, 2024

29 28 27 26 25 24   6 5 4 3 2 1

Library of Congress Cataloging-in-Publication Data

David, James E., 1951– author.
Spies and shuttles : NASA's secret relationships with the DOD and CIA / James David.
pages cm
Includes bibliographical references and index.
ISBN 978-0-8130-4999-1 (cloth) | ISBN 978-0-8130-8091-8 (pbk.) | ISBN 978-0-8130-4765-2 (ebook)
1. Astronautics—United States—History. 2. Astronautics, Military—Government policy—United States. 3. United States. National Aeronautics and Space Administration—History. 4. United States. Department of Defense—History. 5. United States. Central Intelligence Agency—History. 6. National security—United States. I. Title.
TL789.8.U5D26 2015
327.1273—dc23   2014030770

The University Press of Florida is the scholarly publishing agency for the State University System of Florida, comprising Florida A&M University, Florida Atlantic University, Florida Gulf Coast University, Florida International University, Florida State University, New College of Florida, University of Central Florida, University of Florida, University of North Florida, University of South Florida, and University of West Florida.

University Press of Florida
2046 NE Waldo Road
Suite 2100
Gainesville, FL 32609
http://upress.ufl.edu

For Jeanne and Cristina

Contents

Illustrations

Acknowledgments

Many people have helped me greatly in the research and writing of this book. At the top of the list is Michael Neufeld, my colleague at the National Air and Space Museum. His encouragement, guidance, and support throughout the project have been of immense value. Jeffrey Richelson, senior fellow at the National Security Archive, provided excellent comments on all the draft chapters.

The staffs at the many declassification offices to which I submitted numerous requests worked long and hard locating and reviewing the records that form the basis of much of this book. Two merit particular mention: the Washington Headquarters Services, Records and Declassification Division of the Office of the Secretary of Defense and the Information Access and Release Team of the National Reconnaissance Office.

Judith Barnes and the Transfer and Disposition Office at the Washington National Records Center in Suitland, Maryland, helped me during my many visits in the huge task of determining exactly which records remained at that facility and which had been already transferred to the National Archives in College Park, Maryland. David Fort, deputy director of the Freedom of Information Act (FOIA) and Mandatory Declassification Review (MDR) Division at College Park, processed many collections of records at my request and enabled me to review records I otherwise would not have been able to access.

Meredith Babb, Nevil Parker, Elaine Durham Otto, and the other staff at the University Press of Florida have been superb. Their guidance and contributions have greatly improved this book.

Lastly, I would like to thank my wife, Jeanne Delasko, for her love and support during the many years I worked on the book.

Abbreviations

AAP	Apollo Applications Program
BoB	Bureau of the Budget
CIA	Central Intelligence Agency
CREST-NARA	CIA Records Search Tool, National Archives and Records Administration
DCI	Director of Central Intelligence
DDR&E	Director of Defense Research and Engineering
DDS&T	Deputy Director for Science and Technology
DMSP	Defense Meteorological Satellite Program
DoD	Department of Defense
ELV	Expendable Launch Vehicle
ESSA	Environmental Science Services Administration
FMSAC	Foreign Missile and Space Analysis Center
MSFN	Manned Space Flight Network
MSFPC	Manned Space Flight Policy Committee
NACA	National Advisory Committee on Aeronautics
NASA	National Aeronautics and Space Administration

NGSP	National Geodetic Satellite Program
NOAA	National Oceanic and Atmospheric Administration
NORAD	North American Air Defense Command
NRO	National Reconnaissance Office
NSAM	National Security Action Memorandum
PSAC	President's Science Advisory Committee
SACC	Survey Applications Coordinating Committee
SAR	Synthetic Aperture Radar
STADAN	Satellite Tracking and Data Acquisition Network

Introduction

The Soviet Union's successful launch of *Sputnik 1* in October 1957 and that of *Sputnik 2* the following month led to intensive debate in the United States on how to respond. President Dwight Eisenhower and his administration initially downplayed the significance and tried to reassure the nation that America was not militarily inferior to the USSR. However, many politicians, journalists, scientists, and others argued that the Soviet accomplishment was a great propaganda victory and proved the Cold War adversary was ahead in critical military technologies such as the intercontinental ballistic missile. They believed that America was now in a race for control of space and must win it whatever the cost.[1]

Eisenhower directed the acceleration of the nation's missile programs and took some other limited steps, but he continued to adamantly oppose any large increases in spending. These actions were taken against the backdrop of America's attempt to launch its first satellite in December, which resulted in national humiliation when the launch vehicle exploded on the pad. The nation finally succeeded in placing a spacecraft in orbit on 31 January 1958, but this did not quiet the administration's critics.[2]

Pressure grew to place the space program in one agency, and the military services and others soon set forth competing proposals. Eisenhower directed the recently created President's Science Advisory Committee (PSAC) to assess what the space program should be and what organization should lead it. In its late March 1958 report, PSAC initially noted that there were four factors driving the United States into space: the urge to explore the unknown, defense objectives, national prestige, and scientific observation and experimentation. PSAC rejected Air Force and Army arguments for control of the space program and recommended establishment of a new civilian agency to run the non-defense part. Eisenhower endorsed the report, stating: "These opportunities reinforce my conviction that we and other nations have a great responsibility to promote the peaceful use of space and to utilize the new knowledge obtainable from space science and technology for the benefit of all mankind."[3]

Eisenhower quickly submitted the National Aeronautics and Space Act of 1958 to Congress. The new agency was designated the National Aeronautics and Space Administration (NASA), and it would take over the personnel and facilities of the National Advisory Committee on Aeronautics formed in 1915: the Langley Aeronautical Laboratory in Virginia, Ames Aeronautical Laboratory in California, Lewis Flight Propulsion Laboratory in Ohio, High Speed Flight Station at Edwards AFB in California, and Wallops Island launch complex in Virginia.[4]

It was very important for propaganda purposes to contrast NASA and its aeronautical and space programs with the Soviet Union's secretive and military-oriented efforts. The legislation stated that the U.S. policy was "activities in space should be devoted to peaceful purposes for the benefit of all mankind." It assigned NASA responsibility for all U.S. government aeronautical and space research and operations, except those associated with weapons systems, military operations, and the defense of the nation, which were assigned to the Department of Defense (DoD). Several goals were set forth, including the expansion of human knowledge, the building and operation of vehicles to carry instruments and living organisms through space, the continuation of the role of the United States as a leader in aeronautical and space science and technology and their application to peaceful activities, and international cooperation in the pursuit of peaceful applications. NASA was to make available to defense agencies "discoveries that have military value or significance," and defense agencies were to give "information as to discoveries which have value or significance" to NASA. The legislation directed that information obtained or developed by NASA "shall be made available for public inspection," unless federal law required it to be withheld or it was classified. Congress added provisions establishing a Civil-Military Liaison Committee to promote technological cooperation between the two programs and a National Aeronautics and Space Council to develop space strategy. Eisenhower signed the bill in July soon after Congress passed it, and NASA began operations on 1 October 1958.[5]

These mandates were guiding principles for NASA's officials and the nation's political leaders, but NASA could not and did not always follow them for several reasons. First, NASA and the national security community had a common interest in a wide range of aeronautical and space technologies and frequently needed to utilize each other's hardware and facilities to accomplish their missions. Second, there were certain NASA programs designed for civilian purposes that the national security community had to monitor and restrict to eliminate any threat to classified programs. Third, NASA and the national security community relied on each other for data and expertise concerning foreign aeronautical and space programs. Fourth, many of NASA's top officials had previously held government or industry positions deeply involved in defense or intelligence matters and supported a strong national defense. Within NASA's restraints, they were willing to contribute its immense resources and expertise to achieving it.

In the aeronautical field, the DoD maintained the close relationships it had developed with the former National Advisory Committee on Aeronautics research centers and continued using them to help build virtually all of its aircraft and some of the weapons carried on them. Although much of this support was overt, it received little publicity and virtually no criticism for blurring the boundary between the civilian and national security sectors directed by NASA's founding legislation.

The situation with NASA's spaceflight programs and related activities was radically different. During the intense competition in space with the Soviets in the 1960s, it was very careful not to appear to be engaged in any secretive or military-oriented efforts. NASA and the nation promoted virtually all of its spaceflight programs, from applications satellites to human spaceflight, as demonstrations of America's use of space for peaceful and scientific purposes. These efforts received extensive worldwide publicity and earned considerable goodwill by engaging international partners and acquiring and disseminating a wide range of valuable scientific data to nations around the world. These restraints became less important after the United States won the race to the Moon, but they did not disappear.

NASA utilized DoD boosters, upper stages, launch complexes, and command and control networks in many of its spaceflight programs. This support did not change their peaceful and scientific character and was done openly. NASA employed classified technologies or their equivalent for use in some of its astronomy and early lunar photography programs to acquire the requisite high-quality imagery. This too did not alter the nature of the programs, but the hardware's true origins and acquisition procedures remained hidden pursuant to classified agreements reached with the defense and intelligence agencies. Restrictions imposed by these agencies on certain NASA activities, particularly on the employment of certain technologies to image the Earth and on the acquisition and dissemination of photography of the planet beginning in the 1960s, also did not affect the character of the underlying programs. However, the existence of the restrictions and the interaction concerning them was classified and, of course, they limited the scientific data NASA collected.

The national security community at the same time frequently utilized NASA launch vehicles, satellites, other spaceflight equipment, and command and control networks to meet their requirements. Although the existence of this support was usually public knowledge, the details frequently were not. During the 1960s, the DoD received a wide range of mostly unclassified assistance for its Dyna-Soar, Blue Gemini, and Manned Orbiting Laboratory human spaceflight projects, which were designed to deliver nuclear weapons or conduct reconnaissance. The most notable example of dependence was the Shuttle program which blurred the boundary between the civilian and national security space sectors more than any other. To ensure the critical political support of the defense and intelligence agencies, NASA developed the Shuttle in an unprecedented and unequal partnership with them to serve as a cheaper and more reliable launch vehicle for all U.S. government payloads. It readily accepted their requirements for a larger orbiter and more exacting performance specifications. NASA agreed to pay most of the program's costs and to national security missions having priority over civilian missions. For the first time ever, NASA openly and repeatedly flew classified experiments and payloads, provided secure command and control, and withheld considerable mission information from the public.

The defense and intelligence agencies acquired important data from NASA's applications satellites. A joint geodetic satellite program conducted in the 1960s and managed by NASA, as well as NASA's more advanced geodetic satellites the following decade built in close cooperation with these agencies, provided critical information for improving the accuracy of long-range missiles. Because of national security concerns, for the first time in an applications satellite program NASA agreed in the mid-1970s to limit the dissemination of some of the data acquired. This restriction remained in place until the DoD reversed its position and President Jimmy Carter ordered all of the data freely distributed. A NASA-managed joint meteorological satellite program in the early 1960s sought to satisfy all tactical and strategic requirements, but it proved unable to do so and was quickly terminated. Nevertheless, NASA worked very closely with the defense and intelligence agencies in building its subsequent weather satellites that helped support tactical operations in particular. The Central Intelligence Agency (CIA) and others covertly acquired imagery from NASA's land remote sensing satellites for intelligence and mapping purposes beginning in the 1970s.

NASA's command and control networks provided unclassified and classified support to DoD's missile tests and on-orbit satellites during the 1960s and 1970s. Although there were no limitations on the use of the networks' domestic ground stations for these purposes, there were some with respect to most of the foreign ground stations, and NASA had to act carefully.

There was also considerable mutual support in areas related to their spaceflight programs, almost all of which were classified. From its earliest days NASA received large amounts of finished intelligence on the Soviet space program from the CIA and helped it and others analyze Soviet aeronautical and space programs. It worked closely with the North American Air Defense Command in the space surveillance field. During the 1960s, NASA collaborated with the CIA and others in retrieving space debris that had returned to Earth and formulating uniform policies on this process. It cooperated with the CIA and others in developing and implementing cover stories for intelligence projects.

The extensive literature on NASA covers the mostly open and unclassified interaction between it and the national security agencies in several areas—the massive support it gave to building DoD aircraft;[6] its use of DoD boosters, upper stages, launch complexes, command and control networks, and ships and aircraft in the early human spaceflight programs;[7] and its support to the Air Force's Dyna-Soar, Blue Gemini, and Manned Orbiting Laboratory human spaceflight projects.[8] In contrast, only a small number of works address the hidden and largely classified interaction in the spaceflight and related fields.

NASA's employment of classified cameras in the Lunar Orbiter program was first disclosed in *SAMOS to the Moon: The Clandestine Transfer of Reconnaissance Technology between Federal Agencies*, written by historian Cargill Hall at the National Reconnaissance Office (NRO). Dwayne Day's "Mapping the Dark Side of the World, Part 2" and the author's "Astronaut Photography and the Intelligence Community: Who Saw What?" discuss the use of classified cameras in the Apollo and Skylab programs. Vance Mitchell, another NRO historian, describes the planned use of the GAMBIT 1 imagery intelligence satellite camera in the Apollo program in "Showing the Way: NASA, the NRO, and the Apollo Lunar Reconnaissance Program, 1963–1967." The early procedures established to give access to selected classified overhead imagery to cleared federal civilian agency personnel for civilian applications and their limited success are examined in the author's "The Intelligence Agencies Help Find Whales: Civilian Use of Classified Overhead Photography under Project Argo." The author's "Was It Really 'Space Junk'? U.S. Intelligence Interest in Space Debris That Returned to Earth" discusses NASA's participation with the national security agencies during the 1960s in retrieving U.S. and foreign space fragments and in formulating a uniform policy on the issue. Pamela Mack's *Viewing the Earth: The Social Construction of the Landsat Satellite System* is the most complete account of building the first civilian remote sensing satellite. Written long before any records were declassified, it describes the national security community's general opposition to the program but does not give many details of the restrictions imposed on NASA or the mechanisms instituted to enforce them.

Robert Smith's *The Space Telescope: A Study of NASA, Science, Technology, and Politics* is the comprehensive history of the *Hubble Space Telescope*. Published before the launch of the spacecraft and declassification of any of the imagery intelligence satellite programs, it discusses generally how these programs developed the technologies that enabled the telescope's optical system to be built and some of the security restrictions imposed on NASA. Eric Chaisson, a scientist at the Space Telescope Science Institute and a holder of high-level security clearances, wrote *The Hubble Wars: Astrophysics Meets Astropolitics in the Two-Billion-Dollar Struggle over the Hubble Space Telescope* shortly after the first Shuttle mission to repair the telescope. Acknowledging that he was writing so as not to disclose any classified information, Chaisson notes that much of the design and fabrication of the telescope derived from classified satellites, and he briefly describes many areas in which NASA and the national security agencies did or did not cooperate.

There have been several articles on the U.S. intelligence agencies targeting the Soviet lunar landing program during the 1960s and supplying finished intelligence on it to NASA. The most comprehensive are Dwayne Day and Asif Siddiqi's "The Moon in the Crosshairs: CIA Intelligence on the Soviet Manned Lunar Programme, Part 1—Launch Complex J" and "Part 2—The J Vehicle." Michael Beschloss's book *MAYDAY: Eisenhower, Khrushchev, and the U-2 Affair* covers many aspects of NASA's participation in the U-2 cover story.

Several works address the extensive and complex interaction in the Shuttle program. T. A. Heppenheimer's *The Space Shuttle Decision: NASA's Search for a Reusable Launch Vehicle* and *History of the Space Shuttle, Volume 2, Development of the Shuttle, 1972–1981* are the most detailed histories of the program up to the first flight. The first volume contains an extensive discussion of the background of the decision to build the Shuttle, the critical need to obtain the support of the national security agencies for the project, and how their requirements determined the orbiters' size, configuration, and performance specifications. The second volume covers the continuing political battles surrounding the Shuttle, the overt agreements between NASA and the national security agencies on its development and operation, the many technical problems encountered during its development, and the resulting delays and performance shortfall. L. Parker Temple III, an Air Force officer working on space programs during the Shuttle's development, examines the Shuttle and other programs in *Shades of Gray: National Security and the Evolution of Reconnaissance*. It examines many of the same issues as Heppenheimer does and some important new ones as well—the controversies within the national security establishment over the Shuttle during the Ford and Carter administrations, the technical and political difficulties in transitioning national security payloads to the Shuttle, the long and successful fight of the defense and intelligence agencies beginning in 1983 to retain expendable launch vehicles for some payloads, and the plans for a secure command and control system.

Hans Mark was NRO director and then secretary of the Air Force during the Carter administration and NASA's deputy administrator from 1981 to 1984. As a result, he is probably as knowledgeable as anybody concerning the interaction between NASA and the defense and intelligence agencies on the Shuttle during this period. However, his book *The Space Station: A Personal Journey* only discusses his role in trying to accelerate transition to the Shuttle generally and the key support he and Secretary of Defense Harold Brown gave the program at the White House during the Carter administration. Peter Hays's "NASA and the Department of Defense: Enduring Themes in Three Key Areas," Dwayne Day's "Invitation to Struggle: The History of Civilian-Military Relations in Space," and David Spires's *Orbital Futures: Selected Documents in Air Force Space History* are good overviews of the interaction concerning the Shuttle.

This work demonstrates that the relations were far greater and deeper than previously covered in the open literature and that there were never completely separate and distinct civilian and national security space programs. NASA actively contributed to meeting national security requirements, received considerable support from the defense and intelligence agencies to meet its civilian objectives, and at times was subject to restrictions on the technologies it could use or the scientific data it could collect or disseminate because of national security concerns.

It is organized chronologically, covering the period 1958 through the end of the DoD classified missions on the Shuttle in the early 1990s. This does not signify that the mutual support and close working relationships ceased at this point, but rather that there is little information available on them afterward.

Chapter 1 gives a brief overview of NASA's creation and early leaders and then examines the close collaboration and liaison that developed in several areas between 1958 and 1961. These include NASA's receipt of high-level intelligence on the Soviet space program from the CIA, the first steps taken to share its expertise in analyzing it, participation in the U-2 cover story and the resulting fallout, cooperation with the DoD in the acquisition and dissemination of space surveillance data, collaboration on the design and operation of command and control networks, and the initial support its networks gave to DoD satellites in orbit. It also details the national security concerns regarding the first-generation Tiros weather satellites and the procedures implemented to address them until it was quickly determined that the concerns were unfounded.

Chapter 2 begins with a summary of NASA's tremendous growth in size and stature during the Apollo era and its new leaders. It then focuses on the expansion of the relationship regarding foreign intelligence during the 1960s and reviews the increasing amounts of intelligence the CIA provided, the occasional conflicts over the conclusions contained therein on the state of the Soviet manned lunar landing program and the CIA's refusal to change them, NASA's largely unsuccessful attempts to use intelligence at the White House and Congress to increase its budget, and the influence of intelligence on the Soviet manned circumlunar program on changing Apollo 8 from an Earth-orbital to a lunar-orbital mission. The chapter ends with a discussion of NASA's greatly expanded contributions to the analysis of foreign space and aeronautical programs.

Additional activities during the 1960s are examined in chapter 3. These include NASA's participation in cover stories beyond the U-2, continued cooperation with the DoD on the acquisition and dissemination of space surveillance data, increasing unclassified and classified support to DoD missile tests and on-orbit satellites, NASA's largely successful opposition to proposals to combine the civilian and DoD command and control networks, and early collaboration on developing data relay satellites. It also examines the cooperation with the CIA and others in retrieving space debris that had returned to Earth and in establishing uniform procedures for this process, the partial success of the joint National Geodetic Satellite Program managed by NASA in meeting civilian and national security requirements, and the conflicts and compromises regarding the classification of data from that program. The chapter concludes with a discussion of the failure and early termination of the joint National Operational Meteorological Satellite System managed by NASA to satisfy civilian and national security needs, the development of the DoD's highly classified weather satellites to meet tactical and strategic requirements, NASA's close coordination with the defense and intelligence agencies in developing its second- and third-generation polar-orbiting weather satellites, and their increasing contributions to satisfying tactical requirements.

Chapter 4 focuses on NASA's plans beginning in the 1960s to routinely conduct land remote sensing from space and the political and technological threats they posed to the National Reconnaissance Program. It discusses the restrictions the defense and intelligence agencies imposed on NASA's use of image-forming sensors and other sensitive technologies beginning in 1965 to alleviate these concerns, the joint bodies established to review all of NASA's activities in this area and their impact on them, the delays imposed on NASA's development of a robotic land remote sensing satellite, and NASA's role in the creation and operation of Project Argo under which a limited number of federal civilian scientists gained access to classified overhead photography to compensate in part for the lower-quality imagery NASA was restricted to collecting.

The involvement of the national security agencies in NASA's human spaceflight experiments from the Mercury through Apollo-Soyuz programs and its monitoring of and support to the lunar and astronomical programs from the 1960s through the *Hubble Space Telescope* are examined in chapter 5. It covers the DoD's unsuccessful attempts in the early 1960s to wrest control of the human spaceflight program from NASA, its limited participation in Mercury experiments, the sponsorship of numerous Gemini experiments designed primarily to evaluate the usefulness of humans performing reconnaissance from space, the conflicts with NASA over the classification of data from some, and the compromises NASA made to accommodate them. The chapter also discusses the procedures established by the national security agencies early in the Gemini program to conduct post-mission review of all photographic experiments to ensure no imagery of sensitive domestic or foreign sites was released, the additional pre-mission review of all photographic experiments instituted early in the Apollo program for the same purpose, and the waiver granted NASA by the national security agencies and the White House to use a camera in Skylab exceeding the technical limitations established in 1965 and the accompanying strict conditions. It then examines the national security community's early concerns over NASA's proposed lunar photography programs and its unsuccessful attempt to conduct them for NASA, NASA's subsequent planned and actual use of classified cameras in them, and the classified agreements which governed their procurement and the information that could be released concerning their origins and use. The chapter concludes with a review of the national security concerns regarding NASA's space-based astronomical programs and its monitoring of them beginning in the late 1960s, the apparent relaxation of restrictions in 1969 on the use of sophisticated optical systems and pointing/stabilization systems in these programs, and the known and probable involvement with the development of the *Hubble Space Telescope*.

Chapter 6 gives a brief overview of NASA in the Shuttle era, including the succession of new leaders. It then examines the other post-Apollo projects that NASA initially preferred, the changed political climate that required it to abandon them and instead develop a vehicle to deliver practical and tangible benefits, the unprecedented partnership that NASA forged out of necessity with the national security agencies to design and operate the Shuttle as the exclusive launch vehicle for all U.S. government payloads, its acceptance of their requirements for a larger orbiter and more demanding performance specifications, the unequal division of responsibility it readily accepted for building and operating the vehicle, and the uncertainty surrounding the national security community's early transition plans. It then describes the unprecedented review of all the early proposed civilian experiments by the defense and intelligence agencies and the conditions imposed on some, their agreement with NASA on building a secure command and control system, the high-level interest concerning the Shuttle's survivability against anti-satellite weapons and other threats, and the apparent limited protection measures implemented to address them. It concludes with a discussion of the program's growing technical and financial problems during the Carter administration, their effect on the transition plans for specific military and intelligence space programs, and the critical political support of the national security community's civilian leadership at the White House, which saved the program from cancellation or drastic reduction during this period.

Chapter 7 discusses the early national security missions flown, the limited secure command and control system finally implemented for them, NASA's unprecedented operational security and information withholding policies instituted for the flights, the known and probable involvement of the defense and intelligence agencies in several civilian scientific experiments, and the Shuttle's failure to meet the planned flight rate or original performance specifications and the resulting impact on national security space programs. It next reviews the successful efforts of the defense and intelligence agencies in 1983 and 1984 to resume acquisition of a small number of expendable launch vehicles to launch the heaviest payloads, the effect of the *Challenger* accident in dramatically increasing their procurement of expendable launch vehicles to completely replace the Shuttle as the means to access space, and the final eight dedicated national security missions flown from 1988 to 1992. Lastly, the chapter evaluates the costs and benefits to the national security community of its participation in the program.

Chapter 8 reviews NASA's application satellites during the Shuttle era. It discusses the extensive liaison concerning the design and operation of NASA's fourth-generation polar-orbiting weather satellites, the continued support of the civilian satellites to satisfying some tactical requirements, their largely unsuccessful attempt to meet strategic requirements when all the DoD's weather satellites in orbit malfunctioned between 1980 and 1982, and the failed attempts to merge or privatize the DoD and civilian programs. The chapter then examines the close collaboration on developing and operating NASA's geodetic and oceanographic satellites to satisfy both civilian and defense needs, the national security community's approval of the use of the sensors used on the latter, the conflicts and compromises over the means to limit the acquisition and dissemination of selected geodetic and oceanographic data, the White House reviews of the restrictions regarding geodetic data and their eventual overturning by President Jimmy Carter due to the DoD's changed position on its sensitivity, and the ultimately unsuccessful plans to develop a single program under NASA management to meet all civilian and defense requirements in the geodetic and oceanographic fields. It finally details the evolution of NASA's land remote sensing satellites, the conflicts and compromises over NASA's attempt to conduct a worldwide crop survey with the second one, their increasing contributions to meeting national security requirements in the areas of estimating foreign agricultural production and mapping, and the short and unsuccessful effort to privatize them in the late 1980s.

There are gaps in the work due to the continued unavailability of many government records.[9] For example, there are a number of declassified records from the 1960s concerning NASA's command and control networks supporting DoD missile tests and satellites, cooperating with the CIA and others in recovering space debris, working with the national security agencies in developing data relay satellites, receiving intelligence from the CIA on the Soviet space program, and helping the CIA and others analyze the Soviet aeronautical and space programs. However, there are very few accessible documents on these topics from after this period. There are, in addition, virtually no declassified records from any period on important subjects such as joint management of frequencies used in space, the national security community's utilization of NASA's Tracking and Data Relay Satellite System, NASA's use of classified space surveillance data in mission planning and operations, its participation in arms control negotiations on anti-satellite weapons beginning in the Carter administration, or its problems in exporting sensitive space-related technologies such as Landsat ground stations. The release of records on these and other topics in the coming years will undoubtedly demonstrate that there was even closer cooperation and interaction between NASA and the national security community in the spaceflight and related fields.

Forging Close Ties in NASA's Early Years

President Eisenhower selected T. Keith Glennan as NASA's first administrator and Hugh Dryden as deputy administrator. Both were sworn in 19
August 1958. Glennan had been president of Case Institute of Technology since 1947. He had worked on classified projects at the U.S. Navy's
Underwater Sound Laboratories during World War II and, as a member
of the Atomic Energy Commission from 1950 to 1952, had also received
highly classified data on foreign atomic energy programs from U.S. intelligence agencies. Glennan shared Eisenhower's view that NASA should
engage in an orderly series of steps exploring space and should not enter into an expensive, unrestrained race with the Soviets.[1] He made it
a condition of his appointment that Dryden be deputy administrator.
Dryden, then director of the National Advisory Committee on Aeronautics (NACA), had worked closely with industry and the military services
developing aircraft, rockets, and missiles beginning in World War II. He
was familiar with the classified world from these projects as well as his
service on many postwar scientific advisory committees of the Air Force
and others. Additionally, Dryden had worked closely with the CIA in the
mid-1950s to test U-2 reconnaissance aircraft models in NACA's wind
tunnels and to develop its cover story as a NACA high-altitude weather
research vehicle.[2]

NASA did not have any space programs when it began operations on 1 October 1958, but this changed quickly. Eisenhower assigned the human spaceflight mission to NASA later that year, and it initiated Project Mercury. NASA quickly took over the Navy's Vanguard satellite project, and its personnel formed the core of the new Goddard Space Flight Center in Greenbelt, Maryland. It also soon assumed the Army and Air Force's lunar probe projects. With White House support, NASA overcame opposition from the Army and took over the Jet Propulsion Laboratory in Pasadena, California, at the end of 1958. The following year it assumed the Army's Tiros (Television InfraRed Observation Satellite) program to provide weather data from space and the Advanced Research Project Agency's Centaur upper stage rocket project. Over considerable Army opposition, Eisenhower supported NASA's takeover of the Army Ballistic Missile Agency's Development Operations Division (with the von Braun rocket- and missile-building group) in 1959. The transfer was completed the following year, and the over 4,000 personnel formed the core of the new Marshall Space Flight Center in Huntsville, Alabama.[3]

The defense and intelligence agencies, of course, were still engaged in many space activities despite the transfer of the above personnel and projects to NASA. Among others, these involved developing launch vehicles and launch complexes; spacecraft to conduct imagery and signals intelligence, early warning of missile attack, communications, and navigation; a command and control network to track and receive data from and send data to these satellites; a space surveillance network to detect and track Soviet satellites; and a worldwide network of ground stations, ships, and aircraft to detect Soviet launches and collect telemetry and any mission data transmitted by Soviet vehicles.

NASA's space programs soon dwarfed its aeronautical programs in size and scope. By the end of 1959, for example, 27 percent of its research and development was devoted to human spaceflight, 16 percent to lunar and planetary probes, 14 percent to launch vehicles, 11 percent to supersonic aircraft, 8 percent to robotic satellites, 7 percent to ballistic missiles, 6 percent to vertical take-off and landing and other special aircraft, 5 percent to hypersonic aircraft, 3 percent to maneuverable missiles, and 3 percent to subsonic aircraft.[4] NASA contracted out to industry and academic institutions for some of the work, but the majority was done in-house. Glennan described the close relationship between much of NASA's efforts and national security to Congress in April 1959: "It is well to keep in mind that a great deal of our work is intimately concerned with our national security. While we are charged with direction of all U.S. aeronautical and space research and development except for military projects, our charter makes it clear that we are to work very closely with the armed services and make available to them all developments of military interest."[5]

An organizational change occurred at NASA during its early years that played an important role in increasing its ties with the national security agencies. The Civil-Military Liaison Committee created under the National Aeronautics and Space Act of 1958 rarely met and was an ineffective NASA-DoD coordinating body. Glennan and James Douglas, deputy secretary of defense, sought to remedy this problem when they signed an agreement in September 1960 establishing the Aeronautics and Astronautics Coordinating Board. Its objectives were to avoid unnecessary duplication, coordinate activities of common interest, identify problems requiring solution by either party, and promote the exchange of information. Its cochairs were NASA's deputy administrator and the director of defense research and engineering (DDR&E) within the Office of the Secretary of Defense. With both NASA and DoD representatives, the initial panels formed were Manned Space Flight, Unmanned Spacecraft, Launch Vehicles, Space Flight Ground Support, Supporting Space Research and Technology, and Aeronautics.[6] It is important to note that the Aeronautics and Astronautics Coordinating Board oversaw NASA-DoD technological cooperation and coordination, not that between NASA and non-DoD agencies such as the CIA. Furthermore, the Board's available records indicate that when it worked with classified information, it only did so at the Confidential or Secret level. Matters classified Top Secret or at the codeword level were handled by other permanent or ad hoc groups.

Despite the opposition of many national security agencies to NASA's creation and assumption of certain projects and facilities, NASA and these organizations quickly forged close ties in a number of areas, of which one of the most important was foreign intelligence. It was imperative for NASA's leadership and other top government officials to obtain timely and accurate information on the Soviet space program, and the only source of this data was the U.S. intelligence community. NASA's top officials received the highest level reports and estimates from the CIA. Although there were extensive discussions on the contributions NASA could make in analyzing the USSR's efforts, the available evidence suggests that it did not provide any assistance until the following decade.

Another area of close interaction was NASA's active participation in the cover story for the U-2. After the shoot down of Gary Powers's aircraft in May 1960 and resulting embarrassment to NASA, it firmly rebuffed the CIA's pleas to continue its public role in any cover story for the U-2. Nevertheless, it continued participating in future cover stories by giving confidential advice on what they should be.

There was also considerable liaison concerning other elements of the civilian and national security space programs, including developing separate command and control networks, conducting and disseminating information on space surveillance, and supporting on-orbit DoD satellites. Despite their very different purposes, NASA and the defense and intelligence agencies recognized the need to work together to ensure the success of their separate efforts and to avoid costly and unnecessary duplication.

Extensive interaction took place concerning the Tiros weather satellites, the project NASA took over from the Army in 1959. Despite its promise of substantial practical benefits for many nations and enhanced international prestige, there were national security concerns over the possibility that the imagery would disclose information of intelligence value or would intensify the Soviet's propaganda campaign against reconnaissance from space. As a result, several changes were made in the program at the request of the CIA and others. They only remained in place briefly after it was determined that the photography contained nothing of intelligence interest and there was no international reaction against it.

NASA as a Consumer of Intelligence

An important requirement of NASA, as well as top policymakers across the government, was reliable data on current and future Soviet space programs. The USSR imposed a blanket of secrecy over these efforts, as they did with almost all facets of their national life. In contrast to the United States, they did not announce their launches beforehand, and no international press covered them. The Soviets publicized successful missions, but often only after a long delay and in an incomplete fashion. They were purposefully vague about future plans and never publicly acknowledged failures.

As a result, U.S. intelligence agencies were the only source of timely and accurate information. Their effort in this area utilized all sources—

imagery intelligence, signals intelligence, human intelligence, open sources, technical exploitation of foreign hardware, space surveillance— of which the first two were by far the most important. Imagery intelligence revealed the existence, location, configuration, and dimensions of launch complexes; the configuration and dimensions of rockets and missiles at these sites; rocket and missile fabrication and testing facilities; tracking, data acquisition, and command and control networks; and other key components of foreign space programs. CIA U-2 aircraft provided the only systematic photographic coverage of the USSR in the late 1950s, flying 23 missions before the shoot down of Gary Power's plane in May 1960 permanently ended them.[7]

Imagery intelligence from space began in August 1960 with the first successful mission of a joint Air Force/CIA CORONA satellite. For several years thereafter, its coverage remained sporadic and the resolution relatively poor. By the mid-1960s, however, advances in cameras and other technologies resulted in much more frequent coverage and greatly improved resolution. Additionally, the GAMBIT 1 started flying in 1963. It eventually achieved a maximum resolution of two feet, which exceeded CORONA's by several times. The follow-on GAMBIT 3 system, which became operational in 1966, achieved a maximum resolution of better than two feet. These satellites did not provide the broad area coverage that CORONA did, but their superior resolution provided technical intelligence that it could not.[8]

Signals intelligence consisted of electronic intelligence and communications intelligence. The former targeted radars, telemetry from rockets, missiles, and spacecraft, command and control signals sent to these vehicles, and mission data transmitted by spacecraft back to Earth. Relying primarily on it, U.S. analysts attempted to determine the range, accuracy, payload capacity, and other performance characteristics of Soviet rockets and missiles. Similarly, they used it to try to establish the missions and performance of the USSR's spacecraft and to evaluate claims concerning their accomplishments.[9] A worldwide network of DoD and CIA ground stations, ships, and aircraft collected electronic intelligence.[10]

Although collection efforts against the Soviet space program had been under way since before *Sputnik 1*, in 1960 the U.S. Intelligence Board formally established as a first priority objective obtaining data on Soviet satellites carrying weapons or conducting military or intelligence missions. Reflecting the importance of the competition with the USSR in the human spaceflight area, at the same time it established as a second priority objective acquiring data on "Biological probes and satellites, Manned space vehicles, Lunar and planetary probes (manned and unmanned)." NASA's associate administrator, Robert Seamans, wrote the director of the National Security Agency in early 1963 asking if the downlinks from Soviet lunar spacecraft could be intercepted and analyzed because of their potential value to the Apollo program. He enclosed a list of NASA's requirements and, among other things, stated that the proposed 85-foot diameter antenna at the intercept site in Asmara, Ethiopia, would be ideal for this purpose. Codenamed STONEHOUSE, the antenna was installed and was operational from 1965 to 1975. This was the only known instance in which NASA passed on specific collection requirements to the intelligence agencies.[11] In 1960, GRAB 1 was the first satellite to acquire electronic intelligence from space when it intercepted Soviet air defense and antiballistic missile radars. The follow-on POPPY system targeted these radars from 1962 to 1977.[12] All the other satellite programs that collected electronic intelligence remain classified. The role of communications intelligence in assessing foreign space programs remains totally classified and its contribution cannot be ascertained, but it must have been substantial.

Although various organizations processed imagery and signals intelligence, in the early years of the space age the Office of Scientific Intelligence under the CIA's Directorate of Intelligence had the primary responsibility for analyzing and reporting on Soviet missile and space programs using all sources. Information from this and other organizations was used to prepare intelligence assessments and briefings for the president and other top policymakers. Among the most important assessments on a topic were *National Intelligence Estimates* (usually prepared periodically) and *Special National Intelligence Estimates* (commonly a one-time publication). The CIA's Office of National Estimates wrote the initial drafts, which were then reviewed and voted on by the U.S. Intelligence Board. Any dissents were incorporated therein.[13]

The existence, technologies, and product of the various imagery and signals intelligence systems were highly classified. In every case, they were protected at one of three levels (Confidential, Secret, or Top Secret) within the National Security Information classification system.[14] Additional special security control systems guarded much of the data as well—TALENT the U-2's cameras, signals intelligence equipment, and product; CANES the existence of and data acquired by the GRAB electronic intelligence satellites; TALENT-KEYHOLE the fact of imagery intelligence from space and its product; BYEMAN the technical details of and procurement procedures for overhead collection platforms; Special Intelligence the decrypted texts or the substance of encrypted foreign communications.[15] Access to data covered by any of these special security control systems (often referred to as Sensitive Compartmented Information or codeword material) was more restricted than National Security Information and required additional security clearances and a demonstrated need-to-know. Much of the intelligence on foreign space programs fell into this category.

Even before NASA started operations, the CIA contacted it concerning the sharing of intelligence on Soviet space programs. Gen. Charles Cabell, USAF (Ret.), acting director of central intelligence, wrote Glennan on 10 September 1958:

> You are undoubtedly aware of the deep concern of this Agency with the challenge to the United States posed by Soviet advance[s] in space technology. Our work in this area of intelligence may be of material assistance to you in furthering the aims of NASA. With this in mind, it occurs to me that you may desire an oral briefing from our Office of Scientific Intelligence. This Office monitors and reports on the entire spectrum of Soviet science and devotes a sizeable effort to assessing the Soviet space program. I would be pleased to ask Dr. Herbert Scoville, Jr., the Assistant Director for Scientific Intelligence, to arrange an appropriate briefing for you at your convenience. Additionally, when you feel the time to be appropriate, we can arrange to make available to you on a continuing basis intelligence reports prepared by us which may bear on your problem.[16]

Although NASA's reply is completely redacted except for the letterhead, an attached CIA "Action Sheet" states that NASA was then receiving all relevant Office of Scientific Intelligence reports.[17]

Glennan and Dryden also received intelligence on the Soviet space program at the new National Aeronautics and Space Council.[18] Both attended the second meeting in October 1958 at which the CIA's assistant director for scientific intelligence gave a briefing on the "historical antecedents, the indicated scope of the USSR program, their emphasis on biological experiments, and their immediate capabilities."[19] The CIA occasionally gave intelligence briefings to the Council in subsequent years.[20]

From its first days, NASA also sought access to *National Intelligence Estimates (NIEs)* and *Special National Intelligence Estimates (SNIEs)*. The CIA's Office of National Estimates wrote the U.S. Intelligence Board in late November 1958 that NASA wished to obtain these publications in the areas of "guided missiles, space vehicles, atomic energy, and over-all Soviet military and scientific policies and capabilities." It recommended that 10 be provided, including the August 1958 *NIE 11-5-58: Soviet Capabilities in Guided Missiles and Space Vehicles*, the first on the Soviet space program since *Sputnik 1*. The board quickly approved, and the documents were sent directly to Glennan. The contents of *NIE 11-5-58* and *Annex A*, which discussed probable guided missile and spacecraft development programs, were Top Secret. *Annex B*, which described test ranges and their activities, and *Annex C*, which dealt with the USSR's nuclear warheads and was classified Top Secret/Restricted Data, remain classified today.[21]

Only 5 of the more than 50 pages of *NIE 11-5-58* and *Annex A* were devoted to the Soviet space program. It noted that the USSR would likely pursue the full range of spaceflight goals, including a human spaceflight program for "scientific and/or military purposes," but because of competition with other projects, "We cannot at this time determine which specific space flight activities enjoy the highest priority and will be pursued first." At the earliest, the time frame for "surveillance satellites, recoverable aeromedical satellites, lunar probes and impacts, lunar satellites and planetary probes to Mars and Venus" was 1958–59; "soft landings by lunar rockets and recoverable manned earth satellites" in 1959–60; "a manned glide-type high altitude research vehicle" in 1960–61; heavy earth satellites and manned circumlunar flights in 1961–62; manned lunar landings after 1965.[22]

NASA and the CIA increased contact in 1959 concerning intelligence on the Soviet missile and space programs. In March, the director of central intelligence (DCI), Allen Dulles, wrote Glennan that the U.S. Intelligence Board was establishing "priority national intelligence requirements for foreign space programs and activities" and was "exploring means and procedures by which such intelligence requirements and intelligence support needs in the general field of space can best be met—including particularly guidance and support to space surveillance." After noting the existing relationship between NASA and the Office of Scientific Intelligence, he concluded:

In view of the interests of the National Aeronautics and Space Administration in these matters, I wish to assure you that the intelligence community stands ready to provide whatever appropriate guidance or support you believe might be useful to NASA. In any event, the intelligence community will undoubtedly be calling on NASA from time to time for support on intelligence problems related to space surveillance.[23]

Glennan replied quickly, thanking him for the offer of assistance and assuring him that NASA would make "full use of all available sources of technical information." He also informed Dulles that he had appointed Harold Lawrence as the "focal point in all NASA relations with the intelligence community."[24] (Lawrence was assistant director of the Office of International Programs until he left NASA in 1960.) During April 1959, the CIA briefed Lawrence on U.S. intelligence collection operations, including CHALICE (CIA U-2 operations), the feasibility study for GUSTO (the CIA's A-12 reconnaissance aircraft, the U-2's planned successor), and CORONA.[25] Later that year, the CIA gave an extensive presentation on CORONA and ARGON (the mapping camera that flew on several CORONA satellites in the early 1960s) to Dryden and Richard Horner, a NASA associate administrator.[26]

Office of Naval Intelligence personnel gave a presentation to Glennan, Dryden, and four unknown NASA employees in March 1959 on Project TATTLETALE, the name of the Naval Research Laboratory's project at the time to build the GRAB electronic intelligence satellite.[27]

NASA received the Top Secret *NIE 11-5-59: Soviet Capabilities in Guided Missiles and Space Vehicles* and its *Annex A* three months after its publication in November 1959.[28] It gave more details on the different possible future space missions than the previous year's estimate and pushed back the earliest possible dates for some, including a manned circumlunar flight in 1964–65 and a manned lunar landing in about 1970.[29]

NASA obtained 15 additional estimates in October 1960, including *NIE 11-5-60: Soviet Capabilities in Guided Missiles and Space Vehicles* and its *Annexes A, B*, and *C*. Published five months earlier, the estimate and its *Annexes A* and *B* were Top Secret. *Annex C* was Top Secret/Codeword and because NASA had no facility for storing codeword information, Glennan read it and immediately returned it to the CIA courier. It remains classified today.[30] In the half page devoted to the space program,

NIE 11-5-60 initially noted that although the USSR had achieved several "spectacular firsts," the number of launchings had been less than anticipated, and there was no evidence of a "systematic program designed to achieve maximum progress toward clearly defined scientific goals." As a result, the intelligence agencies were not able "to predict with confidence the future course of the Soviet program." Nevertheless, they stated that the Soviets could achieve in the next year an unmanned lunar satellite or soft lunar landing, a Mars or Venus probe, and the orbiting and recovery of capsules with animals and perhaps a human. The expected first dates for manned lunar missions remained unchanged from *NIE 11-5-59:* a circumlunar flight in 1964–65 and a landing in about 1970.[31]

NASA's Use of Intelligence

There is no evidence that NASA's leadership utilized the intelligence it received at the White House or Congress in budget negotiations or for programming purposes, as it would the following decade. The CIA, of course, was already providing the White House with all the important data on the Soviet space program in briefings and publications such as the *President's Intelligence Checklist/President's Daily Brief, Current Intelligence Bulletin, National Intelligence Estimates*, and *Special National Intelligence Estimates*. With respect to Congress, the CIA regularly briefed the CIA Subcommittees of the House and Senate Armed Services Committees and Appropriations Committees on its budget, administration, operations, and foreign intelligence assessments. The CIA also had regular contact with NASA's authorization committees, the Senate Committee on Aeronautical and Space Sciences, and the House Committee on Science and Astronautics. From 1958 to 1961, the DCI, often accompanied by the assistant director for scientific intelligence and others, gave at least three presentations to the former and four to the latter on Soviet missile and space programs and their comparison to U.S. efforts.[32]

NASA as an Analyst of Intelligence

Along with becoming a regular consumer of intelligence, NASA also desired to contribute to its analysis and production. Shortly after its creation, Glennan and the CIA's assistant director for scientific intelligence wanted to establish liaison between lower levels of their organizations. A January 1959 meeting on this issue was attended by personnel from the

Office of Scientific Intelligence and NASA's Franklin Phillips, an assistant to Dr. Dryden; Henry Billingsley, head of the Office of International Cooperation; Robert Littell; and R. L. Bell, director of security. The CIA officials explained the functions and contributions of their agency, the U.S. Intelligence Board, and its Guided Missiles and Astronautics Intelligence Committee and Joint Atomic Energy Intelligence Committee.[33] When the NASA officials asked about NASA's possible contributions to these bodies and establishing an intelligence unit of its own, they were informed that the issue was under discussion.[34]

An internal Office of Scientific Intelligence memo several weeks later recommended that NASA create a small intelligence office because it specialized in "fields which have many intelligence and security implications," the leadership needed to know what was going on in the National Security Council and U.S. Intelligence Board, and its "intelligence activities must extend into their liaison relationships with other countries, especially with the United Kingdom and Canada." (The fourth and last reason in the memo is completely redacted.) The memo advocated that to ensure the administrator of NASA was kept current on national intelligence activities, NASA personnel should probably have observer status at the U.S. Intelligence Board, full membership on its Guided Missiles and Astronautics Intelligence Committee, and observer status on its Joint Atomic Energy Intelligence Committee and Scientific Intelligence Committee.[35] It is not certain what further discussions, if any, were held with NASA on these issues. However, for reasons probably related to the potential damage to its public image if it were revealed that it was participating in intelligence matters, NASA did not establish an intelligence office, and none of its personnel sat on the U.S. Intelligence Board or one of its committees for many years. However, NASA would begin contributing to the analysis of intelligence in earnest early the following decade through other channels.

NASA's Participation in Cover Stories

The CIA enlisted the Air Force's Air Weather Service and Dryden and a few other top NACA officials in devising a cover story for the U-2 as it was nearing operational status in early 1956. They agreed that the public would be told that NACA operated the aircraft for upper atmosphere research, with the assistance of the Air Weather Service. On 5 May 1956, NACA issued a press release to that effect, adding that the current research flights over the United States would be expanded to include flights overseas using Air Force bases. Another NACA press release two months later stated that the data acquired thus far was very valuable and that they had started making flights from England. In February 1957, NACA released a photo of a U-2 with NACA markings and NACA tail number. In fact, the U-2s were carrying some weather instrumentation on some domestic and overseas flights, and the data was released to the public. NACA published a research memorandum in March 1957 entitled "Preliminary Measurements of Atmospheric Turbulence at High Altitudes as Determined from Acceleration Measurements on Lockheed U-2 Airplane." Air Weather Service personnel presented a paper at the March 1959 American Meteorological Conference on typhoon-eye cloud patterns photographed from a U-2.[36]

NASA continued participating in the U-2 cover story, but its efforts after the shoot down of Gary Powers's aircraft on 1 May 1960 ended up embarrassing it and straining its relationship with the CIA. After Powers did not return on schedule and attempts to communicate with him were unsuccessful, the CIA convened a meeting on the afternoon of 1 May. Among those in attendance were Richard Bissell, deputy director (plans); his deputy, Richard Helms; Hugh Cumming, director of the State Department's Bureau of Intelligence and Research; Walter Bonney, NASA's head of press relations; and various U-2 project personnel. They agreed that the commander of the plane's detachment in Turkey should issue a press release stating that a NASA high-altitude weather research aircraft was missing, it had last been heard from near Lake Van in Turkey, and the pilot had last reported having oxygen problems. The commander issued the release the following day.[37]

Premier Nikita Khrushchev made the first public Soviet statement on 5 May when he stated that a U.S. airplane had been shot down deep inside the USSR. President Eisenhower and several key advisers met later that day and agreed that Acting Secretary of State Douglas Dillon would draft a public statement and the State Department would be the sole source of information. The White House press secretary, Jim Hagerty, read a short statement to reporters several hours later indicating that Eisenhower had ordered NASA and the State Department to conduct an inquiry and that they would release the results. Bissell had been in contact with Glennan several times during the previous days and had given him cover information in question-and-answer form, but both Glennan and key White House advisers thought it best to issue a statement based on that information instead of waiting for the results of the inquiry. When the press contacted Bonney after Hagerty's actions, Bonney had such a statement prepared and read it to reporters. It noted that one of NASA U-2s had been missing since the 1st and the pilot had reported that he was having oxygen problems near Lake Van. The statement mentioned a fictitious route, flight time, and altitude for the plane. In response to questions, Bonney admitted that the aircraft carried cameras but that they were only for photographing clouds and that all of its U-2s had been grounded for equipment checks.[38]

The cover story fell apart on 7 May when Khrushchev announced that Powers was alive and gave detailed information on him and his mission. Top policymakers met several times in the coming days to decide the course of action after Khrushchev's startling disclosures, but no NASA personnel were present. That same day a U-2 with NASA markings was displayed for the press at Edwards Air Force Base in California. It appears that the CIA and Lockheed arranged this, and NASA was not involved in any way. Later on the 7th, the State Department issued a release disclosing that the inquiry previously ordered by Eisenhower had determined there had been no authorization for any flight described by Khrushchev. However, in an attempt to obtain data concealed behind the Iron Curtain, "a flight over Soviet territory was probably taken by an unarmed civilian U-2 plane." This did not quiet the growing domestic and foreign criticism. The State Department told the press on 9 May that to protect against surprise attack, unarmed civilian aircraft had performed aerial surveillance along the periphery of and occasionally over the USSR under general presidential direction. The Soviets publicly displayed parts of Powers's plane on the 11th. Eisenhower spoke for the first time the following day, essentially saying that in the absence of inspection regimes, such activities were needed to deter war. He left for the long-awaited summit meeting with Khrushchev in Paris on the 14th. Although Eisenhower stated publicly that all flights over the USSR would cease, his refusal to meet Khrushchev's demand for an apology caused the early end of the summit and cancellation of a subsequent visit to the Soviet Union.[39]

During the late summer and early fall of 1960, the CIA tried to get NASA's permission to continue using it as a cover in the U-2 program. Bissell met several times with Glennan, but he refused to approve any continued participation. NASA's long involvement in the U-2 cover story had finally ended. At Bissell's urging, Gen. Cabell wrote a note to Bonney shortly before his departure from NASA in October 1960 thanking him "for the service you have rendered to major programs of concern both to this Agency and to the Nation's security and well-being."[40]

Cooperation in Command and Control and Space Surveillance

Neither NASA nor the national security agencies could successfully conduct any space missions without a worldwide network of ground stations to track their spacecraft, receive telemetry and other data from them, and send commands to them. Radars and optical devices performed the tracking, while large antennas received signals from and sent them to the satellites. Communications circuits tied all the stations with a single site that managed the operations. A separate space surveillance network with radars and optical instruments was needed to track foreign satellites and space debris such as inactive satellites, boosters, upper stages, and their fragments. The stations in this network were also connected with a single operations center.

NASA assumed the Naval Research Laboratory's Minitrack stations built to support the Vanguard program, most of which were along a north-south axis in the Western Hemisphere between the latitudes of 35 degrees. These stations could only track and communicate with U.S. satellites operating on a specified frequency and could not perform these functions with any in high-inclination orbits. The Goddard Space Flight Center's Computing Center controlled Minitrack and analyzed the data. NASA also took over responsibility for the contract under which the Smithsonian Institution operated 12 Baker-Nunn cameras around the world. Although providing more precise tracking data than Minitrack and able to track vehicles that did not carry beacons, the cameras could only do so for spacecraft in low-inclination orbits and in clear weather at dawn or dusk. Additionally, there was a long delay in processing their data. The Smithsonian Astrophysical Observatory Computer and Operations Control Center controlled the Baker-Nunn cameras and analyzed the data for NASA. The Center also received reports of visual observations made by volunteers around the world under Operation Moonwatch. NASA also acquired the Army's large antenna at Goldstone in the California desert for tracking lunar and deep-space probes.[41]

Although the DoD had not launched any national security payloads by the time NASA was created, it did have the Army's Microlock stations in the United States, Europe, Africa, and Singapore that had tracked and communicated with the Vanguard and Explorer satellites. With respect to space surveillance, the Air Force radar at Millstone Hill in Massachusetts had tracked *Sputnik 1* and was undergoing improvements. Some segments of the Navy's radar fence across the southern United States began operations in 1958. Acquiring information from these and other DoD and civilian assets, the Air Force Spacetrack Center acted as the central data and cataloging center.[42]

The joint NASA–Advanced Research Projects Agency Committee on United States Satellite and Space Vehicle Tracking and Surveillance Requirements made a presentation to Glennan and various DoD officials in January 1959. Two days later, Glennan and Secretary of Defense Neil McElroy signed an unclassified agreement that recognized the need for separate NASA and DoD tracking, data acquisition, and command and control networks. (There were several reasons for this, including the fact that many NASA facilities would be located in foreign countries who would object to their use for supporting military spacecraft.) Under the agreement, NASA was to establish stations similar to Goldstone in Australia and South Africa to complete a network for lunar and deep-space probes, build a network to support Project Mercury, and add Minitrack stations and upgrade them so they could service polar-orbiting spacecraft. NASA and DoD were to continue operating their separate data analysis centers, but there was to be a free exchange of data and assistance between them. Similarly, the worldwide communications circuits connecting the NASA and DoD stations to one another and the data analysis centers were to be shared.[43]

The agreement served as the blueprint for expansion and what became NASA's three tracking, data acquisition, and command and control networks. For lunar and planetary probes, NASA added large antennas in Australia and Spain. Along with Goldstone, these formed what was designated the Deep Space Network in 1962 under the Jet Propulsion Laboratory's direction. To serve the Mercury program, NASA originally established more than 15 stations around the world in what soon became the Manned Space Flight Network operated by the Goddard Space Flight Center. Stations were added and dropped for the Gemini and Apollo programs on account of different mission profiles. Substantial assistance was given by DoD ground stations, aircraft, and ships during all three human spaceflight programs. Lastly, NASA improved existing Minitrack stations, added new ones, and installed powerful radars at locations around the world to form what was designated the Satellite Tracking and

Data Acquisition Network in 1964 to serve robotic satellites in orbits of any inclination. It too was operated by Goddard. The Office of Tracking and Data Acquisition at NASA Headquarters had overall responsibility for development and operation of the three networks.[44]

The first known on-orbit support provided by a NASA network to a DoD spacecraft began on 20 June 1960 with the launch of the Naval Research Laboratory's GRAB 1 with its open solar radiation experiment and highly classified electronic intelligence package. NASA had agreed that its Minitrack stations in North and South America would provide tracking and telemetry support when Director of Naval Intelligence personnel briefed NASA's top leadership on the project in March 1959. It is not clear how long the assistance continued (the electronic intelligence package operated until 20 September 1960 and the solar radiation experiment until April 1961). NASA was probably requested to furnish similar support to each of the four subsequent GRABs, but there is only confirmation of this for the last two in 1962.[45]

The DoD also began a huge expansion of its ground support networks in years following the January 1959 agreement. To create the Air Force Satellite Control Facility network designed primarily to service high-inclination reconnaissance satellites, combined tracking, data acquisition, and command and control stations were initially built in Alaska, Hawaii, and at Sunnyvale and Vandenberg Air Force Base (Vandenberg) in California. It quickly added stations in New Hampshire, Greenland, and Guam.[46] The DoD ranges also increased their capabilities to support suborbital and orbital launches. The Atlantic Missile Range eventually included a network of at least 10 tracking, data acquisition, and command and control stations stretching from Florida through Puerto Rico and other islands and ending in South Africa. It also employed ships and aircraft to augment them. The Pacific Missile Range soon expanded into a network of more than 10 such stations stretching from California to Wake, Midway, and Eniwetok islands. It too employed ships and aircraft to assist them.[47]

The DoD similarly greatly expanded its space surveillance capabilities. Radars installed in Canada, the Aleutian Islands, Texas, Trinidad, Turkey, mainland Alaska, Greenland, and England were either dedicated to space surveillance or performed it as a secondary mission (the last three were part of the Ballistic Missile Early Warning System). The Air Force also acquired some Baker-Nunn cameras beginning in 1959 to assist in tracking. The Navy completed the radar fence across the southern

United States in 1959. All the dedicated space surveillance assets were designated the Space Detection and Tracking System and placed under the North American Air Defense Command (NORAD) in 1960.[48]

The DoD and NASA initially shared information concerning the objects they were tracking pursuant to the January 1959 agreement, and DoD released it to the public. However, after the first successful CORONA mission in August 1960, the DoD evidently stopped doing so on the basis that it did not want the orbital elements and other data of these highly classified spacecraft openly published.

A classified January 1961 agreement between DDR&E Herbert York and Dryden attempted to resolve the problem.[49] Among other things, it provided that NORAD had the responsibility to monitor and track U.S. launched military space vehicles, new foreign-launched objects, and space debris. NASA was responsible for doing the same for NASA and non-NASA launched objects of civilian or scientific value, DoD and non-NASA launched objects for which DoD has requested NASA support, and foreign-launched objects for which the government has requested NASA support. Goddard was regularly to furnish NORAD data on its cataloged objects, and NORAD was to reciprocate with unclassified data from its catalog for information and distribution to the public. NORAD was to provide classified data from its catalog to Goddard on a need-to-know basis. Goddard was to disseminate publicly all unclassified data on U.S. and foreign vehicles, and it soon began regularly doing so in its Satellite Situation Report.[50]

National Security Concerns Regarding NASA's Weather Satellite Program

Tiros was the first in a series of applications satellites that NASA developed over the decades to acquire valuable scientific data about the Earth for civilian purposes and that increased international cooperation in its space programs and generated considerable goodwill. The fact that the national security agencies transferred the program to NASA did not in any way signify a lack of interest in meteorological satellites on their part. The armed forces desired an operational system to improve the weather data received by military units operating worldwide, and the intelligence agencies wanted one to provide near-real time data over Sino-Soviet Bloc targets for controlling the imagery intelligence satellites being developed.[51]

Several bodies ensured close cooperation between NASA and other civilian and national security agencies in the meteorological field. The Joint Meteorological Group under the Joint Chiefs of Staff also had representatives of the U.S. Weather Bureau. It focused on technical cooperation and established the military requirements for weather data to be met by various collection systems. The Joint Meteorological Satellite Advisory Committee had NASA, DoD, and U.S. Weather Bureau personnel and worked primarily on technical issues. The Unmanned Spacecraft Panel of the Aeronautics and Astronautics Coordinating Board frequently provided policy-level coordination between NASA and the DoD.[52]

There were two critical national security issues involving Tiros and NASA's subsequent space-based Earth-imaging programs that were repeatedly addressed at the highest levels of the government. First, the USSR had commenced an intensive propaganda campaign against reconnaissance from space in the late 1950s. Articles in Soviet journals and statements at the United Nations and other forums charged that the acknowledged U.S. reconnaissance programs such as SAMOS constituted illegal espionage and were for the purpose of preparing a surprise attack. NASA's photography from space had to be carefully planned and conducted so that it would not reveal information of intelligence value of any nation, particularly those in the Sino-Soviet Bloc. Second, any NASA program could not disclose the existence, technologies, operational details, or product of the nation's highly classified imagery intelligence satellites and aircraft.[53]

NASA planned to fly a limited number of Tiros as research vehicles and, once the hardware and procedures were proven, develop an operational system designated Nimbus. It initially wanted to place *Tiros 1* in a near-polar orbit from Vandenberg in California in October 1959. The higher-resolution television camera would produce a maximum ground resolution of 500 to 600 feet from the planned orbital altitude of 400 miles. Although this was so poor that the imagery would likely have no intelligence value, the CIA still did not want the satellite imaging any part of the Sino-Soviet Bloc and recommended that it only photograph the Southern Hemisphere. If it were to image the Northern Hemisphere, the CIA believed priorities had to be established between it and the classified SAMOS and CORONA imagery intelligence programs and that the Tiros program should even be delayed until the end of the latter. Richard Bissell, deputy director (plans), recommended to the DCI that the subject be brought up before the U.S. Intelligence Board, and it was put on the agenda for the 21 July 1959 meeting.[54] What, if anything, the Board decided is unknown.

NASA soon postponed the initial flight to early 1960, changed the launch site to Cape Canaveral, and reduced the inclination to 50 degrees.[55] *Tiros 1* would thus only overfly the southernmost regions of the USSR on some passes. It is not clear whether technical reasons, pressure from the national security agencies, or both caused NASA to make these changes.

Glennan agreed in November 1959 to have the imagery from the higher-resolution camera screened before public release. The following month the U.S. Intelligence Board approved this action and declared that the imagery from the lower-resolution camera would be treated as unclassified and would be immediately distributed publicly.[56]

The Joint Chiefs of Staff established the military's requirements for weather data from satellites in early 1960. DDR&E Herbert York quickly forwarded them to Glennan. He wrote that the first and third objectives were common to civilian and military applications and believed they were being met for the most part in NASA's program. In contrast, the second objective was exclusively military, and York stated that they "should be incorporated into the meteorological satellite program at the appropriate time."[57]

NASA finally launched *Tiros 1* from Cape Canaveral on 1 April 1960 into an orbit with an inclination of 48 degrees. The narrow-angle television camera (which stopped operating after 12 days) covered an area about 65 miles on a side and achieved the highest resolution of 1,000 feet. Its wide-angle television camera covered an area 750 miles on a side and had a maximum resolution of 1.5 miles. Because of the spacecraft's low-inclination orbit and design, it was unable to photograph above 55 degrees latitude (north and south) and at most imaged only 25 percent of the Earth's surface over which it passed. Up to 32 photographs were stored for transmission to the ground terminals in New Jersey or Hawaii when the satellite was in their range. Couriers took the imagery from them to the Naval Photographic Interpretation Center in the Washington, D.C., area for processing and screening. As expected, the photography from both cameras was of such low resolution that no man-made objects could be detected on the ground. Glennan brought the initial images to President Eisenhower and others at the White House late on 1 April, but they were not released. After consultation with the DCI, NASA distributed selected pictures to the press the following day. A weather map of much of the Northern Hemisphere was subsequently prepared from several days' worth of imagery, and the CIA, State Department, and others approved its public release. About two weeks after the launch, the first photographs of the USSR were distributed to the press. After the first week, the screening process ended and NASA immediately released all the photography publicly after processing and analysis. There was virtually no negative international reaction to *Tiros 1*, which operated for three months.[58]

Bissell and Dryden discussed in September 1960 the possibility that subsequent Tiros could supply weather data for use in the CORONA program. Some CIA personnel advocated establishment of direct communication links between the Tiros ground stations and the ground network controlling CORONA. However, this never occurred because the low inclinations of the early Tiros, their limited photographic coverage, nonoptimal crossing times over the Sino-Soviet Bloc, and other technical factors prevented them from acquiring the meteorological data needed for strategic applications.[59]

President Eisenhower spoke before the United Nations in September about international cooperation in peaceful uses of outer space. As part of this effort, NASA offered *Tiros 2* orbital data to the Soviet Union and 20 other countries for their use with the photography over their locality. Seventeen nations accepted (the USSR was not one of them). In November 1960, NASA placed *Tiros 2* with the same two cameras and several infrared sensors in the identical 48 degree orbit. It operated until December of the following year.[60]

Summary

During its first years of operation, NASA established both overt and covert ties with the national security agencies in a number of areas. Much of the open and publicized interaction concerned NASA's use of DoD facilities and hardware such as ranges and boosters. The hidden and frequently classified relationships were more varied. NASA received invaluable assistance from the CIA in the form of finished intelligence on the USSR's space program. It provided critical support to the national security community, including participating in the U-2 cover story, furnishing on-orbit assistance to DoD satellites, and acquiring and disseminating space surveillance data. Additionally, the defense and intelligence agencies imposed restrictions on Tiros, NASA's first space-based imaging program. The resulting changes NASA had to make were minimal and short-lived, but this would not be the case in many subsequent Earth-observation programs.

NASA's interaction with the defense and intelligence agencies increased in both scope and complexity in the 1960s. The next chapter examines this in the areas of NASA's receipt, use, and contributions to the analysis of foreign intelligence.

CHAPTER 2

NASA, the CIA, and Foreign Intelligence during the Apollo Era

The Apollo era began in May 1961 when President John F. Kennedy announced to the world that the United States would land a man on the Moon by the end of the decade. It ended with the Apollo 17 mission in December 1972. America's decision to enter the race to the Moon had many repercussions for NASA, including greatly expanding its size and budget, making its human spaceflight programs the dominant and most visible, and increasing its interaction with the national security community.

NASA's leadership changed during this period and was instrumental in these growing ties. Kennedy selected James Webb in late January 1961 to be NASA's second administrator. Webb was an accomplished lawyer, businessman, and administrator who had extensive experience in both the private and public sectors. He served as director of the Bureau of the Budget from 1946 to 1950 and undersecretary of state from 1950 to 1952.[1] These two positions involved a large amount of contact with the national security agencies and required intimate knowledge of classified information and programs at the highest levels. Webb took office on 14 February. He retained Dryden as deputy administrator and Robert Seamans (who had come over from Radio Corporation of America's Missile Electronics and Controls Division in 1960) as associate administrator.

Among other duties, Seamans was the principal liaison with the defense and intelligence agencies.[2]

Dryden remained as deputy administrator until his death in 1965, at which time Seamans assumed the position. He continued being the chief liaison with the intelligence agencies.[3] Webb soon recruited Willis Shapley from the Bureau of the Budget to serve as associate deputy administrator. Shapley had been deputy chief of the Military Division and a special coordinator for space programs at the Bureau of the Budget and had intimate knowledge of all the unclassified and classified projects and activities in this area.[4]

Seamans resigned as deputy administrator in early 1968, and Webb selected Thomas Paine as his replacement. An engineer at General Electric since 1949, Paine had worked on a number of national security projects. Paine became acting administrator after Webb's unexpected resignation in late October 1968 and administrator after his confirmation early the next year.[5]

Webb brought several individuals into NASA's headquarters with extensive national security experience. In 1962, he selected Adm. Walter F. Boone, USN (Ret.), to be the first deputy associate administrator for defense affairs . Several retired military officers joined this office. Gen. Jacob Smart, USA (Ret.), became a special assistant to the administrator in 1966 and two years later assistant administrator for DoD and interagency affairs after Boone left NASA. Shortly after Seamans's resignation in early 1968, Webb appointed both Willis Shapley and Gen. Smart to be the chief liaison with the intelligence agencies. Gen. Charles Cabell, USAF (Ret.), became a special assistant to the administrator in 1965 after holding the position of deputy director of central intelligence from 1953 until 1962.[6]

The number of active duty personnel detailed to NASA headquarters and centers grew dramatically during the Apollo era. The best-known, of course, were the astronauts that came on three-year tours (which were inevitably renewed). After the detail of Brig. Gen. Samuel Phillips, USAF, as deputy director of the Apollo program in January 1964 (he became director later that year), NASA and the Air Force entered into several agreements that provided 128 officers from that service to work on the Gemini and Apollo projects. In total, NASA had over 330 active duty personnel from all three services working for it by the summer of 1966. This number slowly declined in the following years.[7]

One area in which the existing ties between NASA and the national security community became much deeper and more complex was NASA's growing demand for and use of intelligence on the USSR's space programs. Timely and accurate information on these activities—especially its manned lunar landing effort—became even more important to NASA and the nation's top policymakers. The United States had staked its prestige on placing the first humans on the Moon and would suffer a tremendous blow if the USSR won the competition.

With a growing number of Soviet space missions and more sophisticated collection and analytical capabilities, the intelligence agencies produced far more reports and assessments than in previous years. NASA appears to have received most, if not all, of them. The finished intelligence was frequently uncertain as to what extent the Soviets were really competing in the race to the Moon, and NASA's leadership occasionally complained about this ambiguity. However, there is no evidence that the CIA or any other intelligence agency ever changed any estimates or assessments.

NASA, of course, used the intelligence to evaluate the status of its space programs versus that of the USSR. In contrast to their predecessors, NASA's leaders during the Apollo era at times also utilized this information in budget negotiations with the White House and Congress or to update the latter on Soviet developments. The available evidence does not indicate that NASA provided any intelligence that these officials were not already receiving from the CIA and others, but in all likelihood NASA probably advanced some different interpretations. The only known instance of intelligence affecting any of NASA's missions was the decision to change Apollo 8 from an Earth-orbital to a lunar-orbital mission. Even in this case, though, it was just one of several factors.

Another area in which the existing interaction greatly increased was NASA's contributions to analyzing intelligence. It detailed two of its employees to the CIA's Foreign Missiles and Space Analysis Center, and several of its personnel sat on CIA panels which met once or twice a year to evaluate foreign space programs. The assistance, however, extended beyond the space field. At the CIA's request, a limited number of Langley Research Center personnel began a decades-long project examining the aerodynamic characteristics of Soviet aircraft, rockets, and missiles.

NASA Receives Increasing Amounts of Intelligence

Within a few days of Webb taking office, the CIA gave him a security briefing covering IDEALIST (the new codename for CIA's U-2 operations), CORONA, and ARGON. He was also informed about and signed the necessary documents to obtain TALENT, TALENT-KEYHOLE, and Special Intelligence clearances. Webb indicated that he wanted Seamans to receive the same information and clearances, and undoubtedly he did. One of the CIA personnel present commented very favorably on the new administrator in the memorandum to the record of the meeting:

> On the whole, I believe this was a very useful discussion. Mr. Webb strikes us as a completely receptive and energetic individual whose questions about our Projects and whose desires to know what is going on in the intelligence field are considerable. Mr. Webb indicated that he would shortly seek to have some substantive discussions with Mr. Dulles [the DCI] and Mr. Amory [the deputy director for intelligence] on subjects such as the status of the Soviet missile and space programs and the like.[8]

NASA continued receiving *National Intelligence Estimates* and *Special National Intelligence Estimates*. In May 1961, the CIA forwarded *NIE 11-5-61: Soviet Technical Capabilities in Guided Missiles and Space Vehicles* and its *Annexes A* through *F*, published the previous month. The *NIE* and *Annexes A* through *D* were Top Secret. Both the Top Secret/Restricted Data *Annex E* and Top Secret/Codeword *Annex F* remain classified today. NASA still did not have the requisite storage facility for *Annex F*, which was read by an assistant to Webb at NASA and returned to the CIA courier.[9]

NIE 11-5-61 was published shortly after Yuri Gagarin's historic flight and before Kennedy's announcement that the United States would land men on the Moon by the end of the decade. It noted at the outset that "The Soviet leaders clearly believe that achievements in space enable them to persuade the world that in the realms of science, technology, and military strength, the USSR stands in the very front ranks of world powers. In seizing an early lead and following it with a series of dramatic successes, they have sought to bolster, both at home and abroad, their claims of the superiority of the Soviet system." The estimate contained an extended discussion on the different components of the space program. With respect to future missions, the earliest possible dates for different space operations were the same as in *NIE 11-5-60*, except that a manned circumlunar flight was pushed back a year to 1966.[10]

Beginning later in 1961, NASA received a number of estimates, most of which concerned Soviet and Chinese military capabilities. During this period, the CIA's Board of Estimates disapproved transferring two estimates to NASA: *NIE 13-2-60: Chinese Communist Atomic Energy Program (Summary and Conclusions)* and another whose title remains classified.[11]

John McCone, who succeeded Allen Dulles as DCI in early 1962, sent Dryden the draft of *NIE 11-1-62: The Soviet Space Program* in November and requested that he review it before the U.S. Intelligence Board approved it.[12] Dryden, and perhaps some other NASA officials, met with the CIA on 19 November.[13] However, there is no information available on what their input was and how it possibly affected the final *NIE*. This is the only known instance in which NASA's input was solicited on a draft *NIE*.

The Board approved *NIE 11-1-62* in early December 1962 and NASA obtained a copy within days. Classified Secret, it was the first *NIE* devoted entirely to the Soviet space program and was over 30 pages in length. The *NIE* acknowledged the difficulty in predicting future Soviet space activities, stating, "Our evidence as to the future course of the Soviet space program is very limited. Soviet propaganda dealing with future space activities has canvassed the whole range of possibilities. Our estimates are therefore based largely on extrapolation from past Soviet space activities and on judgments as to likely advances in Soviet technology."[14]

NIE 11-1-62 noted that there was no confirmation that the USSR had a program for a manned lunar landing, but that it could be under way without U.S. intelligence collection systems being able to detect it. Soviet leaders had not stated publicly that they were competing with the United States in achieving a manned lunar landing, and it was highly unlikely that they would do so. However, it was believed the chances were better than even that the Soviets had a competitive program. The Soviets would probably "dispatch the lunar vehicle from an orbiting earth or lunar satellite than to attempt a direct flight from the earth." Either method would require major new vehicle development, facility construction, and supporting activities in many other fields.[15]

The progress of any such program would be evidenced in the coming years by many events: long duration manned missions in Earth orbit to develop improved life support systems, radiation shielding, and biomedical information; flight tests of multimillion pound thrust boosters, advanced upper stages, and advanced guidance equipment; development of new reentry techniques for the higher speeds in returning from the Moon; extensive unmanned lunar exploration. If the estimated dates of the availability of more powerful boosters and improved upper stages were correct, a manned circumlunar flight could occur in 1965–66, a manned lunar orbital flight in about 1966–67, and a manned lunar landing in 1967–69.[16]

The intelligence agencies at this point adopted a two-year schedule for producing NIEs on the USSR's space program and thus did not publish any during 1963 or 1964. Nevertheless, the CIA continued generating reports on the possible Soviet objective to place cosmonauts on the Moon. In April 1963, the Office of National Estimates sent the Top Secret/Codeword "Soviet Intentions Concerning a Manned Lunar Landing" to the DCI. The report's final assessment was virtually unchanged from that of NIE 11-1-62. Although it is very likely that NASA either received the report or was briefed on it, there is no evidence confirming this. The CIA sent Webb an analysis of the USSR's space program in October 1963 that was classified Secret.[17]

In February 1964, the Office of Research and Reports in the Directorate of Intelligence produced the Top Secret/Codeword "ORR Position on Soviet Manned Lunar Landing Program," which has been released in heavily redacted form. The report analyzed evidence in three areas: expansion of launch facilities at Tyuratam (the complex at which all human spaceflight and many unmanned launches took place), static engine test stand sites, public statements by Soviet officials. Its overall evaluation was as follows:

> Current evidence on the Soviet space effort does not permit firm conclusions to be drawn concerning the status of a manned lunar landing program and is not an adequate basis for judging whether it is competitive with the US program, or indeed whether such a program even exists in the USSR. . . . Accordingly, in the absence of firmer evidence than is now available, we believe it is premature to make a *confident judgment* regarding Soviet intention to achieve a manned lunar landing in this decade. If [redacted] indicates that a booster capable of accomplishing this mission is being developed, we should be able to judge with a fair degree of confidence by late 1965 or early 1966 that the Soviets are competing. On the other

hand, if [redacted] it could be safely concluded that the USSR would not be capable of accomplishing the lunar objective by 1970.[18]

Once again, it is likely that NASA received this report or was at least briefed on it.

Between Gagarin's April 1961 mission and April 1965, the USSR conducted seven additional human space flights. Notable accomplishments included the first woman in space, the first three-man crew, and the first spacewalk. After the March 1965 mission, there were no further human space flights for another two years. Five lunar probes were launched in 1963–64 and four in 1965, but all were unsuccessful.[19]

NASA's Use of Intelligence at the White House and Congress

Webb, in contrast to his predecessor, definitely employed intelligence in budget negotiations at the White House on at least one occasion and probably used it on others. An internal December 1965 CIA memo indicates Webb asked for and received the following four sets of graphic materials for meetings with President Johnson and the Bureau of the Budget (BoB) on NASA's Fiscal Year (FY) 1967 request for $5.6 billion:

1. U.S. and Estimated Soviet Civil Space Budget Plans, FY 1959–65 (Secret)
2. Distribution of Cumulative U.S. and Estimated Soviet Space Funding through FY 1965 (Secret)
3. Unmanned Lunar/Planetary Programs: Launches and Funding through FY 1965 (Top Secret/Codeword)
4. U.S.-Soviet Manned Lunar Landing Programming (Top Secret/ Codeword)[20]

In the end, the administration's budget requested only $5.26 billion for NASA and Congress appropriated $5.175 billion.[21]

As described in the previous chapter, there is no data that NASA could have furnished the president and other White House officials that the CIA was not already providing. The same held true for the BoB, which routinely received intelligence from the CIA on foreign space programs for its review of NASA and the DoD's budgets. As an example, in October 1964 personnel from the CIA's Foreign Missile and Space Analysis Center (FMSAC) briefed the BoB at the latter's request precisely for this reason. They briefed the director of the BoB in February 1969 on Soviet space developments for use in the review of NASA's budget. Additionally, from 1962 to 1965 the CIA regularly provided the BoB with a number of finished intelligence products that contained data on the Soviet space program, including the *Central Intelligence Bulletin* and all *NIEs*. Thereafter, it only sent those publications that the BoB specifically requested.[22]

NASA also began utilizing intelligence with Congress in connection with its budget reviews and to keep the members updated on the latest developments in the Soviet space program. Of course, Webb, Dryden, Seamans and others regularly testified at both the authorization and appropriations hearings. During the 1960s, members of Congress often asked them in open hearings to compare U.S. and Soviet space programs. They were careful in their responses so as not to disclose any classified information, often attributing the source of sensitive information to press reports or other open sources. Both members of Congress and NASA officials occasionally suggested holding executive sessions to discuss details of the USSR's programs that could not be disclosed in open sessions. Whether any such closed hearings were held, however, is not known.[23]

Separate and apart from hearings, NASA also briefed individual legislators and staffers on the USSR's space program. Since the CIA was the source of all this data, NASA maintained close liaison with it to ensure that there were no objections or problems. For example, Rep. Joseph Karth (D-MN), chair of the Subcommittee on Space Science and Applications of the House Committee on Science and Astronautics, requested a NASA briefing on Soviet unmanned spacecraft in June 1964. NASA informed the CIA of the request, and the latter decided to conduct the presentation itself. Col. Floyd Sweet of NASA's recently established Office of Defense Affairs contacted the CIA's FMSAC later that year after a classified NASA briefing of the staff director of the Senate Committee on Aeronautical and Space Sciences. Sweet stated that some of the questions the individual asked indicated some "possible familiarity with SI [Special Intelligence] information," and he wanted to know if the CIA had briefed the committee at that classification level. In June 1965, NASA's briefing of the same individual had to await the CIA granting him an unknown clearance.[24]

Just as with the White House and the BoB, NASA very likely did not furnish any information that the CIA was not already providing Congress. The CIA continued briefings on the Soviet space program to the CIA Subcommittees of the House and Senate Armed Services Committees and Appropriations Committees, including one to the CIA Subcommittee of the House Armed Services Committee in August 1964 which covered recent Soviet planetary probes. The July 1968 briefing to the combined CIA Subcommittees of the Senate Armed Services and Appropriations Committees discussed the USSR's large booster program, circumlunar missions, tracking networks, and a possible manned reconnaissance vehicle.[25]

CIA also continued giving briefings to NASA's House and Senate authorization and appropriations committees, although their nature changed. The CIA appearances at formal hearings apparently provoked too many questions on exactly what the United State knew and how it acquired the information. Along these lines, in 1962 Webb pointed out to the chairman of the House Committee on Science and Astronautics "the danger of continuing the practice of calling CIA officials" with the inevitable very sensitive questions asked of them. Webb suggested that just he and Deputy Secretary of Defense Roswell Gilpatric appear before the committee to avoid that sort of inquiry. It appears that except for a February 1968 briefing of the full Subcommittee on Manned Space Flight of the Committee on Science and Astronautics, the CIA stopped making formal appearances before the two committees in subsequent years and instead briefed members and staff individually or in small groups. No transcripts were made of these informal meetings.[26]

The following are just some of the briefings of members and staff set forth in the CIA's Office of Legislative Counsel records. CIA personnel gave several on the Soviet space program at the Top Secret/Codeword level to the chair and staff director of the Senate Committee on Aeronautical and Space Sciences during 1966 and 1967. The ranking Democrat and Republican members were invited but did not always attend. In March 1967, Carl Duckett, the deputy director for science and technology (DDS&T), and David Brandwein, the director of FMSAC, gave a presentation on the Soviet and Chinese space programs to the chair and several other members and staffers of the House Committee on Science and Astronautics. Two staffers from this committee came to FMSAC in January 1969 for a "thorough briefing on the status of the Soviet space program to date and their plans for the development of a new booster." In September 1969, Brandwein briefed the committee's chair on Soviet space events. Duckett and Brandwein gave a presentation in October 1970 on the Soviet space program to the senior Republican and several staff members of the Senate Committee on Aeronautical and Space Sciences.[27]

Conflicts with the CIA Regarding Intelligence

At the 10 May 1965 DCI morning meeting, DDS&T Albert Wheelon distributed a paper on the Soviet lunar probe shot the previous day. (This was *Luna V*, the first Soviet soft landing attempt, which crashed into the Moon two days later.) The summary minutes of the meeting indicate that there was a "lengthy discussion" on whether such papers should go to NASA and the United States Information Agency. It was decided that the DCI would send them to the agency heads, cautioning them that "sensitive sources and methods" were used in their preparation.[28]

NIE 11-1-65: The Soviet Space Program was published at the Secret level in January 1965. However, the new DCI, Adm. William Raborn, USN (Ret.), only sent it to Webb on 25 June. He wrote: "Should you consider it necessary to discuss the contents of this document with anyone who is not normally a recipient, I would appreciate your consulting me." Webb replied that he would comply.[29] This is the first known instance of such a request being made. Although the exact reasons for Raborn making it are unknown, earlier in June the deputy director of central intelligence had asked that a memo be prepared on the CIA's security problems with NASA and a letter be prepared for the DCI to send to Webb.[30] There is no information available on the nature of the problems or what was done to resolve them, and it is not known whether Raborn ever wrote Webb. However, this may explain at least in part the long delay in forwarding the *NIE* to NASA.

The *NIE* and its *Annexes A, B,* and *C* addressed all aspects of the USSR's civilian and military space programs. It concluded that it was almost certain that the Soviets had a manned lunar landing program. Primarily from CORONA and GAMBIT 1 photography of construction of the new Complex J at the Tyuratam launch complex, it was estimated that "a very large booster (about five million pounds of thrust) could become available for manned space flight in 1968" and could be used for placing humans on the Moon. However, based on the "appearance and non-appearance of various technical developments, economic considerations, leadership statements, and continued commitments to other major space missions all lead us to the conclusion that a manned lunar landing ahead of the present Apollo schedule probably is not a Soviet objective." The most likely date for an attempted lunar landing was in the early 1970s.[31]

In regards to a manned circumlunar mission, the *NIE* concluded that there was no specific evidence of one, but that "its relative simplicity compared to the manned landing mission, as well as its propaganda value as a major 'first' lead us to consider it (along with earth-orbiting space stations) as a prime Soviet goal." The earliest possible date the Soviets could attempt such a mission was 1967, but substantial technological problems would have to be solved in the next two years, including perfecting rendezvous and docking techniques.[32]

Webb wrote the DCI in late August 1965 requesting an updated estimate on the possibility and effects of a Soviet program in competition with Apollo. Although the letter has not been located yet, an internal CIA memorandum from the following month described it as requesting "a clear estimative focus on the likelihood and consequences of a Soviet program to land a man on the moon in competition with Apollo." The CIA memorandum also mentioned that former deputy director of central intelligence and now NASA consultant Gen. Cabell had drafted the letter at Webb's request.[33]

Exactly why Webb made his request is unclear. The most likely explanation is that he was not satisfied with the conclusions in the *NIEs* and other publications that the Soviets were probably not engaged in a lunar landing project competitive with the Apollo program and that this hurt NASA's attempts to increase its budget every year. There is at least one CIA document that lends supports to this—an internal memorandum from December 1967 that summarized a meeting between an assistant to NASA's Gen. Jacob Smart and CIA Directorate of Intelligence personnel as follows:

> Actually the discussion included a general rundown of Mr. Webb's and Gen. Smart's concern that NASA wasn't receiving the kind of support from the intelligence community that it needed for its own planning purposes. This is not a new story and in the past we have tended to discount this feeling as one flowing from Mr. Webb's disagreement with Agency estimates.[34]

Webb's request for an updated estimate was discussed at the DCI's morning meeting on 27 August, and some participants suggested that a *SNIE* be prepared on the subject. In the end, though, it was agreed that the

Board of National Estimates would take the lead in drafting a response but that it would not be formal *SNIE,* which would have to be submitted to and approved by the U.S. Intelligence Board.[35]

Raborn directed DDS&T Wheelon in early September to expedite the formation of a Space Intelligence Panel to contribute to the analysis. Composed of experts from outside the Intelligence Community, this was one of several DCI advisory bodies being established at the time. The Space Intelligence Panel would meet at least twice a year, with one of the meetings focusing on *NIEs* concerning the Soviet space program. The Panel was soon established with Dr. Simon Ramo of the Bunker-Ramo Corporation as its chair. Its other members were H. Guyford Stever, president of the Massachusetts Institute of Technology; Joseph Shea, NASA's Apollo spacecraft manager; Raymond Bisplinghoff, director of NASA's Office of Advanced Research and Technology; William Pickering, director of the Jet Propulsion Laboratory; Werner Kirchner of the Aerojet General Corporation; Brig. Gen. Harry Evans, vice director of the Air Force Manned Orbital Laboratory program; Benjamin Blasingame of General Motors' A. C. Electronics Division; Allen Donovan, senior vice president of Aerospace Corporation; and Harold Finger, manager of the Atomic Energy Commission's Space Nuclear Propulsion Office. All the members were cleared for access to TALENT-KEYHOLE and Special Intelligence material.[36]

Before the Space Intelligence Panel reported, the head of the Board of National Estimates gave the DCI the draft paper on the status of the USSR's lunar landing program. Not surprisingly, it contained the same conclusion as *NIE 11-1-65.* However, Raborn and Wheelon decided that this document would not be sent to NASA and that the official response would incorporate the findings of the upcoming meeting of the Space Intelligence Panel.[37] It appears that the CIA was ensuring that whatever estimate it produced would be written in part by experts (including two NASA employees) not involved in preparing *NIEs,* thereby helping to insulate it from any criticism from NASA.

On 14 October, the DCI met with Webb and others at NASA headquarters.[38] Although no details are available on the meeting, in all likelihood at least one of the subjects was the new estimate. The Space Intelligence Panel met for the first time on 28 October. Carl Duckett, director of FM-SAC at the time, quickly forwarded to the DCI its preliminary one-page draft finding on a possible competitive Soviet program. It read in part:

[Soviet] actions visible to us to date are not inconsistent with a manned lunar landing program competitive with Apollo. Neither are they inconsistent with a major manned space station program nor with other projects such as an early manned circumlunar flight aimed at reducing the world impact of a successful Apollo landing. Since this focus of the program is not as yet evident, the possibility that it may be competitive manned lunar landing cannot be ruled out. If this is in fact the case, we are quite certain they are not ahead of the United States in this program but are rather from 0 to 18 months behind.[39]

Raborn sent the finding to Webb on 17 November. Webb wrote back that the "information which you have furnished is, as you know, important to NASA top management in considering the many facets of our FY '67 budget preparation."[40] Sherman Kent, head of the Board of National Estimates, met with Webb and others at NASA on 29 November. One of the topics discussed undoubtedly concerned the finding and Soviet space program estimates in general.[41]

In early January 1966, the DCI asked Wheelon to consider making available to Webb the special FMSAC weekly report on Soviet space exploration programs. He informed Raborn the following month that Webb had one senior officer responsible for intelligence matters and the authority to store both TALENT-KEYHOLE and communications intelligence material (i.e., NASA headquarters finally had a Sensitive Compartmented Information Facility in which it could use and store codeword material). Wheelon added that NASA regularly received the *Central Intelligence Bulletin, FMSAC Daily Surveyor, FMSAC Weekly Surveyor, Scientific Intelligence Digest, Annual Soviet Space Handbook*, and all special space intelligence reports published by the Office of Scientific Intelligence and FMSAC on such subjects as Soviet communications satellites and the French space program. Furthermore, FMSAC personnel briefed Webb and the other leadership of NASA every two or three months on all space events and after "each major event such as a manned space operation."[42]

By this time it certainly appears that the CIA was providing NASA with virtually all the intelligence it produced on the Soviet and other foreign space programs. However, it is not known how widely it was disseminated below the top leadership at NASA headquarters or whether any officials at the NASA centers around the country had access to it. By August 1966 the agency had over 120 personnel with TALENT-KEY-HOLE clearances who, with a demonstrated need-to-know, had access to the photography from imagery intelligence satellites and the intelligence derived therefrom.[43] Of course, much of the information the CIA provided NASA was based on electronic and communications intelligence, which TALENT-KEYHOLE clearances did not give access to. Beyond the top leadership, it is not known how many NASA personnel had the other clearances which permitted access to these types of information.

Richard Helms, who replaced Raborn as DCI in 1966, placed tighter controls on what intelligence was passed to NASA. At NASA's request, the assistant deputy director for science and technology gave a briefing on 4 January 1967 of the possible Soviet Mars probe launch that day (no launch took place). The director of FMSAC gave a presentation to Webb in February 1967 on the USSR's space program, apparently in connection with his upcoming testimony on NASA's budget. In July 1967, unknown Directorate of Science and Technology personnel briefed NASA's leadership on the same topic. When the DCI learned of this, he asked that in the future he be informed beforehand. Later that month, the DCI approved NASA receiving data on the Soviet meteorological program previously requested.[44]

Beginning in 1968, the Office of Basic Geographical Intelligence began furnishing intelligence to NASA at the latter's request. According to an internal CIA memorandum from 1971, the relationship involved frequent intelligence assessments of Soviet developments in satellite geodesy and earth satellite resource surveying. These are used in developing policy for the expanding U.S.-USSR space research cooperation. OBGI support is provided in the form of written and oral comments on various NASA actions and proposals, informal consultations with NASA officials at their request, and papers prepared for other purposes but believed to be pertinent to NASA interests.[45]

Intelligence and Apollo 8

NASA began intensive planning in August 1968 to change the planned Apollo 8 crewed Earth-orbital mission scheduled for early December to a lunar-orbital mission later that month. It launched the spacecraft on 21

December and, after 10 orbits around the Moon, the crew of Frank Borman, James A. Lovell Jr., and William A. Anders returned to Earth on 27 December. The United States gained considerable international acclaim and prestige for the remarkable achievement of flying the first humans in the vicinity of the Moon.[46]

There were numerous technical justifications for Apollo 8. At the same time, there was also a competition with the Soviets to be the first to accomplish this feat. For several years, the intelligence agencies had been predicting a possible manned circumlunar flight in late 1967. They began monitoring the first test flights in the program in late 1967 and continued doing so with the subsequent flights the following year. Because of problems in the program, the intelligence agencies changed the probable date of the first crewed flight to late 1968 and in late November 1968 moved it back to early 1969. NASA's leadership regularly received intelligence on the program from the CIA. However, whether it affected the Apollo 8 decision and, if so, to what extent has been the subject of debate in recent years.

Two NASA histories written long before the release of any classified documents—*Chariots for Apollo: A History of Manned Lunar Spacecraft* and *Where No Man Has Gone Before: A History of Apollo Lunar Exploration Missions*—do not even discuss the Soviet circumlunar program in 1967–68 and the resulting competition to be the first to fly humans near the Moon, much less the possible impact of intelligence on the Apollo 8 decision.

In contrast to the NASA histories, several works by astronauts have discussed the possible effect of intelligence. Frank Borman recalled that in early August 1968 he was told in Houston by Deke Slayton, the director of flight crew operations, that the CIA had just informed NASA that by the end of the year the Soviets were planning a manned lunar flyby. Upon being asked if his crew could change from an Earth-orbital mission to a lunar-orbital mission, Borman assured Slayton they could.[47] Slayton and Alan Shepard stated in their joint work that "In mid-1968, intelligence agencies made NASA aware of the Zond program, indicated that, if all went well with it, the Soviets could dispatch a single cosmonaut on a circumlunar journey in December or January." They then assert that the lunar-orbital Apollo 8 mission was approved so the United States could beat the Soviets.[48] Buzz Aldrin wrote that the NASA personnel involved in the August decision to possibly fly a circumlunar or lunar-orbital mission well understood that the USSR might beat the United States in this regard, and they did not want that to happen. However, Aldrin concluded, "Just how large a role the Soviet's competitive pressure played in the Apollo 8 decision will probably never be fully known" because written records were not kept of the discussions involving intelligence on the USSR's efforts.[49]

Other writers have addressed the issue as well. Based on some of the recently declassified CIA documents discussed below, Dwayne Day concluded, "To date, the evidence supports the conclusion that although the Zond program was a factor in the Apollo decision, it was a *supporting factor, not the decisive one.*"[50]

NIEs on the Soviet space program from the early and mid-1960s had addressed a possible circumlunar project, as discussed above. The March 1967 Top Secret/Codeword *NIE 11-1-67: The Soviet Space Program* was the first since January 1965, and NASA probably received it soon after publication. It initially acknowledged that in the two years since *NIE 11-1-65* the USSR had accomplished some notable "firsts," but that these were "much less spectacular" than previous "firsts." It noted that the Soviets had not engaged in any manned space flights since March 1965 and had not conducted many activities believed essential to the manned space program—rendezvous and docking, long duration manned flight, high speed reentry tests. The *NIE* concluded that a manned circumlunar flight could be undertaken with existing hardware and might be "one of the few within their capabilities that could offset some of the propaganda value of a successful U.S. lunar landing." At least one unmanned flight using the same mission profile and hardware would precede a crewed flight. Test flights over six months or longer would be required to test the equipment and recovery techniques. Because the optimum launch window for such missions was the first five or six months of every year, the *NIE* believed that the earliest the Soviets would attempt such a flight was the first half of 1968. However, it acknowledged that it was possible that it might be attempted in late 1967 as an "anniversary spectacular" (commemorating the fiftieth anniversary of the Bolshevik revolution or the tenth anniversary of *Sputnik 1*).[51]

NIE 11-1-67 was not far off the mark. Unbeknownst to the intelligence agencies, the primary goal of the USSR's space program during 1966 and 1967 was to fly a crewed circumlunar mission in November 1967 to celebrate the fiftieth anniversary of the Bolshevik revolution. Early in 1967, the Soviets adopted plans to launch four unmanned 7K-L1 spacecraft followed by a manned vehicle in time to meet the deadline. Because of delays in testing the 7K-L1, though, the schedule could not be met.[52]

After the tragic Apollo 1 fire in January 1967, however, the possibility of competing with the Soviets to achieve this "first" was not a factor in NASA's planning that year. NASA quickly began drafting a new mission schedule to accomplish a manned lunar landing before 1970. Several Manned Spacecraft Center officials advocated insertion of a crewed lunar-orbital mission to learn more about such issues as communications, navigation, and thermal control in deep space. Such a flight was included in the schedule finalized later in 1967 and was to immediately precede the first manned lunar landing flight. Of course, the schedule was subject to change based on the results of each individual flight, spacecraft availability, and other factors.[53]

The intelligence agencies concluded that an upcoming April 1967 Soviet launch might be part of a circumlunar program. The CIA reported the following in this regard in *The President's Daily Brief* of 20 April:

> All our usual indicators point to a major space launch—probably another test in the Soviet manned space program—within the next few days. It could be a third test of the new capsule, possibly with cosmonauts on board, or a further test of a new upper stage propulsion system for boosting a spacecraft into a much higher orbit. An unmanned flight around the moon (with return to earth) is also a possibility. The best opportunity for such a flight during April will come up next week.[54]

The CIA wrote in *The President's Daily Brief* four days later that a Soyuz capsule, the new Soviet manned spacecraft which it incorrectly assessed had been developed for a circumlunar flight, had been launched on 23 April. It added that the capsule carried one cosmonaut and was "having serious difficulties" deorbiting.[55] (The cosmonaut, Vladimir Komarov, was killed when the parachutes malfunctioned during reentry about 27 hours after launch.)

At the 14 September 1967 DCI morning meeting, DDS&T Carl Duckett reported that "Soviet space-related activities suggest either a new manned event or an unmanned circumlunar flight within the next fortnight." On 28 September, the Soviets finally launched the first 7K-L1 on a circumlunar flight. It was to employ either a direct ballistic reentry into a recovery zone in the Indian Ocean or a guided reentry in Kazakhstan.

Problems with the Proton booster approximately one minute into the flight caused the capsule to separate, and it landed about 40 miles downrange in one piece. Another unmanned 7K-L1 spacecraft was launched on 22 November in the second attempted circumlunar flight, but problems with the Proton booster resulted in both the 7K-L1 and rocket crashing a little less than 200 miles downrange. Two declassified documents from this period describe the November shot as an unmanned circumlunar test.[56] Because they do not mention the failed September mission, it is likely that the intelligence agencies did not detect it.

NASA resumed flight operations in the Apollo program with three unmanned Earth-orbital missions from November 1967 to April 1968. Apollo 4 tested the first complete Saturn V, Apollo 5 the first lunar module, and Apollo 6 the second complete Saturn V. Although the flights were successful in many respects, NASA had to resolve major problems with the Saturn V before using it to launch astronauts. Shortly after Apollo 6 ended in April 1968, Webb issued a new mission schedule. Apollo 7 was to be the first crewed Apollo flight and was scheduled for October 1968. Using a Saturn IB, it would be an Earth-orbital mission and had as one of its primary objectives the testing of the command and service module that had been redesigned after the Apollo 1 fire. Employing a Saturn V, Apollo 8 was to be launched in December. It would be either an unmanned developmental Earth-orbital flight using boilerplate hardware if the rocket's problems had not been resolved or a crewed Earth-orbital flight with the new command and service module and a lunar module if they had been.[57]

Another unmanned 7K-L1 spacecraft, launched on 2 March 1968, several weeks outside a launch window for a circumlunar flight, flew as planned in the opposite direction of the Moon to a distance of some 186,000 miles. On that date, the USSR's press reported that the vehicle (which it designated *Zond 4)* was being sent outward "to study outlying regions of near-earth space." However, that was the last mention of the mission in the Soviet media. The DDS&T noted at the DCI's morning meeting of 11 March that it had been recovered, but there was "not sufficient evidence to judge whether the shot was a success." In fact, the reentry capsule failed to separate from the service module, and the angle of reentry was too steep. Ground controllers set off the self-destruct mechanism over the Gulf of Guinea. Undoubtedly based on further analysis of the data, Duckett mentioned at the morning meeting a week later that there were "additional indications that *Zond-4*'s reentry was a failure."[58]

The Office of National Estimates received the DCI's permission to send Webb in May 1968 a copy of the brief Top Secret *Memorandum to Holders National Intelligence Estimate 11-1-67* published the previous month. It stated that the Soviets would probably attempt a manned circumlunar flight both as "a preliminary to a manned lunar landing and as an attempt to lessen the psychological impact of the Apollo program." Based on the failed November 1967 mission (*Zond 4* flew too recently to be included in the update), the document estimated that "a manned attempt is unlikely before the last half of 1968, with 1969 being more likely."[59]

The intelligence agencies detected the next circumlunar test a few weeks later. A one-page National Security Agency report of 23 April stated, "A Soviet spacecraft, launched from the Tyuratam Missile Test Range (TTMTR) at approximately 2301Z on 22 April 1968 on a probable circumlunar mission apparently failed early in flight prior to orbital insertion. . . . The failure was caused by a premature shutdown of the second stage of the SL-12 launch system."[60] This was, in fact, the fourth unmanned 7K-L1.

The Soviets quickly set a new goal of flying a crewed circumlunar mission by October 1968, which was to be preceded by three unmanned flights in July, August, and September. Four days before the July launch, the booster and spacecraft were undergoing testing on the launch pad at Tyuratam when a second stage fuel tank exploded. It took several weeks to remove the wreckage and, as a result, the July and August shots were scrubbed. Only two more unmanned flights were now planned, and the target for the manned mission under the best circumstances was moved back to December. In *The President's Daily Brief* of 20 July, the CIA reported the preparations for the launch of an unmanned circumlunar mission and that "we think the launch will occur in about five days." It stated in this document four days later that "the Soviets apparently are postponing their latest attempt to get off an unmanned circumlunar flight" and that "we do not know just what caused the delay."[61] Thus the intelligence agencies knew at this point of the attempted test but did not know about the launch pad explosion that ended it.

LM-3, the Apollo 8 lunar module, arrived at Cape Canaveral in June. However, NASA concluded by August 1968 that it would not be ready for flight that year because of numerous deficiencies. Maj. Gen. Samuel Phillips, NASA's Apollo program director, and others discussed in early August the possibility of employing the Apollo 8 Saturn V and command and service module in a circumlunar or lunar-orbital mission after learning that a mock lunar module could be substituted for the LM-3, the Saturn V problems were almost completely resolved, and the allocated Saturn V could be launched by 1 December. Before meeting with Deputy Administrator Thomas Paine in Washington, NASA selected a crew for the possible mission and set the target date for 20 December when potential landing sites on the Moon for later Apollo flights would be in the best light to be photographed and DoD assets were available for astronaut recovery.[62]

At the meeting with Paine on 14 August, the participants unanimously agreed to proceed with a lunar-orbital mission if Apollo 7 were successful. They were informed that the allocated Saturn V could now be launched by 6 December. Webb and George Mueller, associate administrator for manned space flight, were quickly contacted at a United Nations conference in Vienna. After some initial skepticism, both agreed to the proposal. On 16 August, Webb instructed Paine to begin planning for a lunar-orbital Apollo 8 that would be launched on 6 December. At the same time, he ordered that public statements be limited to describing it as an Earth-orbital mission. Paine issued a directive to that effect three days later, and NASA announced that the flight was an expansion of Apollo 7, but that the particulars of the mission had not been decided.[63]

Zond 5, the first of the two remaining unmanned circumlunar flights originally planned to be flown before the crewed mission, was launched on 14 September, and it orbited the Moon taking photographs. Twelve days later, the reentry capsule flew through the Earth's atmosphere and, due to guidance system failures, did not carry out a guided reentry to land in the USSR. Instead, it flew a ballistic reentry and, with its payload of film and biological specimens, was recovered by Soviet ships in the Indian Ocean. U.S. Navy ships in the recovery zone observed the placement of the capsule aboard a Soviet vessel. *Zond 5* was by far the most successful flight to date in the circumlunar program. The intelligence agencies closely monitored the entire flight, and they issued a number of reports during it. The press correctly assessed the mission as an important milestone on the road to a manned flight to the Moon or a manned lunar landing. Soviet officials now hoped to fly two more unmanned vehicles in November and December and the crewed mission in January.[64]

NASA resolved the remaining Saturn V problems, and in September the vehicle allocated to Apollo 8 was cleared for manned flight. Apollo 7, launched by a Saturn IB on 11 October, met all of its primary objectives, and its crew returned after nearly 11 days orbiting the Earth.[65]

Duckett mentioned at the DCI morning meeting on 10 October that "13–15 constitute the most appropriate dates this month for a Soviet circumlunar shot although there are no signs of the kind of activity generally associated with a launch in the Soviet Union. It would appear at least that there will be no manned shot at the moon this month." Although these dates were within the known launch window, the intelligence agencies obviously had no information that the USSR was not even planning a flight during that month. At the 23 October morning meeting, he confidently stated that "it is plain that NASA's plan for a manned circumlunar launch in December is a direct product of an earlier intelligence briefing on Soviet space intentions." On 25 October the USSR launched the unmanned Soyuz 2, and on the following day it placed Soyuz 3 carrying cosmonaut Col. Georgy Beregovoi in orbit. The two vehicles rendezvoused twice, and both were eventually recovered. It was the first successful manned flight in the new Soyuz spacecraft and the first Soviet manned mission since the Soyuz 1 accident in April 1967.[66]

Zond 6 was launched on a lunar flyby mission on 10 November. Carrying a variety of sensors, a biological payload, and film cameras, it successfully landed in the USSR seven days later. However, depressurization within the return capsule, first identified by ground controllers when the spacecraft was on its way to the Moon, caused the single parachute to deploy early. As a result, it crashed and broke into many pieces. However, the Soviet press noted the successful recovery of the vehicle and stated that it was launched "to perfect the automatic functioning of a manned spacecraft that will be sent to the Moon." In the United States, the media covered the mission extensively and quoted unnamed officials that there was a good chance that the USSR would fly a manned spacecraft around the Moon in early December that would beat Apollo 8.[67]

Two days into the flight, DDS&T Duckett noted at the DCI morning meeting that "*Zond 6* contains no new developments and that the most likely issue for the Soviets is whether they will be ready to go for a manned circumlunar shot." At the 15 November morning meeting, he stated that there were probably living organisms aboard the spacecraft and that this "tends to confirm the notion that the Soviets are indeed racing to be first with a manned circumlunar flight." The day after *Zond 6* landed, Duckett mentioned at this meeting that "the chances of a Soviet manned circumlunar flight in December are high." The Space Intelligence Panel (as noted previously, a body of experts from other government agencies and contractors that the DCI had established in late 1965 to personally advise him) met on 19 November and reached a consensus that the USSR would try a manned circumlunar mission in December.[68]

A high-level NASA panel meanwhile had approved in early November all the components of the Apollo 8 mission—spacecraft, launch vehicle, launch complex, mission control network, and spacesuits. The next review was by the chief executives of 16 major Apollo contractors, who met on 10 November with NASA's top officials and supported a lunar-orbital mission for the new target date of 21 December. George Mueller wrote Acting Administrator Paine, who had assumed this position after Webb's sudden resignation in October, the following day. He strongly recommended approval of the lunar-orbital mission on several technical grounds critical to the eventual lunar landing mission, including providing valuable operational experience for flight, ground, and recovery personnel; permitting the evaluation of onboard navigation in lunar orbit and command and service module communications and navigation systems at lunar distances; completing the final verification of ground support elements and onboard computer programs; increasing the knowledge of thermal conditions in deep space and in the vicinity of the Moon; and providing additional photography of the Moon for scientific and operational purposes. Overall, Mueller stated that a successful lunar-orbital flight would "represent a significant new international achievement in space," advance the progress of the Apollo program, and greatly increase the morale of the program's workforce. Paine approved the lunar-orbital mission that same day to take place on 21 December at the earliest.[69]

Apparently additional intelligence collected and analyzed in later November caused the CIA to change their estimate of a high probability of a December manned flight. In *The President's Daily Brief* of 26 November 1968, the CIA wrote:

> Rumors out of Moscow notwithstanding, it now looks as if the Soviets will not attempt a flight around the moon in December. Movements and activities of key space support and recovery ships [one line redacted] make early 1969 a more likely period. Soviet announcements on the achievements of the *Zond-5* and *Zond-6* flights suggest that the next circumlunar capsule may well be manned.[70]

The available evidence demonstrates that the intelligence agencies acquired extensive data on the Soviet circumlunar program and detected and closely monitored all the unmanned flight tests, except for the first one in September 1967. NASA certainly received much if not all of this data. It still had two personnel detailed to the CIA's Foreign Missile and Space Analysis Center (FMSAC), obtained *NIEs* and FMSAC reports such as the daily *Missile and Space Summary*, and received periodic briefings from the CIA on significant events in the Soviet space program.[71]

The only conclusion at this point is that intelligence on the USSR's program was only one factor in the Apollo 8 decision. Although NASA received this intelligence beginning in late 1967, there is no indication that this caused any concern at NASA. It was not ready at the time to enter the competition for this "first" and would not be for some time because of the delays resulting from the Apollo 1 fire. Among many other things, NASA had to conduct further tests of an unmanned command module at lunar return velocities, ensure the redesigned command and service module was safe for human spaceflight, and qualify the Saturn V to launch astronauts.

The extensive planning beginning in August 1968 for changing Apollo 8 to a lunar-orbital mission took place in large part for sound technical reasons that would make it more likely that a manned lunar landing would occur before 1970. By this time, NASA had received considerable intelligence that the USSR was aggressively pursuing its circumlunar program and was probably preparing for a manned circumlunar flight later that year. It recognized that because of several factors, including the need for a successful Apollo 7 mission to qualify the redesigned command and service module, the earliest it could attempt a lunar-orbital mission was December. It also knew that between August and the initial Apollo 8 launch date of 6 December there were three launch windows for a Soviet circumlunar flight.

Even after the successful Apollo 7 flight there was still no guarantee that Apollo 8 would fly to the Moon in December. Pre-flight testing and crew training were continuing. Additionally, there were still several levels of review to take place, including the final one by NASA's acting administrator. On 11 November, shortly after *Zond 5* and in the middle of *Zond 6*, Paine gave approval for the flight to take place on 21 December. This date still afforded one more chance for the USSR to beat the United States because the next launch window for a Soviet circumlunar flight was 8 to 14 December, which NASA was well aware of.

In short, there were sound technical reasons for the Apollo 8 lunar-orbital mission. There is no question that the flight was important in advancing the Apollo program toward a manned lunar landing before

1970. The fact that it just might beat the Soviets in achieving the first manned flight to the vicinity of the Moon was a factor but by no means the sole or determinative one.

Continuing Intelligence on the Soviet Manned Lunar Landing Program

The March 1967 Top Secret/Codeword *NIE 11-1-67: The Soviet Space Program*, the first one on the subject since January 1965, stated that construction at Complex J at Tyuratam was clearly for the purpose of launching the new large booster needed for placing cosmonauts on the Moon and would be completed during the first half of 1968 at the earliest. The conclusions with respect to a Soviet manned lunar landing program were the same as in *NIE 11-1-65*—it was probably not competitive with the Apollo program. Although the earliest the USSR could attempt a manned lunar landing was mid- to late 1969, the more likely date was sometime during 1970 or 1971.[72]

U.S. imagery intelligence satellites first detected a launch vehicle on a Complex J pad in December 1967 that the intelligence agencies immediately recognized as the rocket the USSR was developing to land cosmonauts on the Moon. They designated it the "J-vehicle" (the Soviet designation was N-1). Remaining on the pad for about three weeks, the vehicle was in fact a nonflight version.[73] NASA was undoubtedly quickly informed of this discovery.

The Top Secret *Memorandum to Holders National Intelligence Estimate 11-1-67* published in April 1968 stated that the manned lunar landing program was not competitive with Apollo and that the earliest possible attempt would be in 1970, but that it was more likely that it would be in the latter half of 1971 or 1972.[74]

U.S. imagery intelligence satellites detected another J-vehicle on the launch pad at Complex J in August 1968. Although unknown to the intelligence agencies at the time, this was an actual flight test vehicle. FM-SAC prepared a *Scientific and Technical Intelligence Report* entitled "Complex J at Tyuratam and Its Role in the Soviet Manned Lunar Landing Program" in September 1968. Presumably, NASA quickly learned of this development.[75]

At Paine's request, the CIA sent projections on future Soviet programs in aeronautics and space in January 1969. Imagery intelligence satellites detected another J-vehicle on the pad in December 1968. The CIA reported this in *The President's Daily Brief* of 3 January 1969 and stated that "the first flight test could occur within the next few months." Once again, NASA also must have been informed of this almost immediately. What the intelligence agencies missed, however, was the first actual flight test on 21 February, which ended in failure when the first stage engines shut down at an altitude of about 17 miles and the rocket fell back to Earth.[76]

The Top Secret/Codeword *NIE 11-1-69: The Soviet Space Program* was published in June 1969, which was soon sent to NASA. Since the previous year's *NIE*, four successful manned Apollo missions had taken place. The first successful flight to land a human on the Moon, Apollo 11, was less than one month away. With respect to the USSR's efforts to land cosmonauts on the Moon, *NIE 11-1-69* concluded that 1971 was now the earliest probable date because of problems with developing the J-vehicle and the lack of successful lunar return and recovery tests. A manned circumlunar flight was unlikely because the successful Apollo 8 lunar-orbital mission "has removed the primary incentive for such an attempt as a spectacular."[77]

Shortly before the launch of the historic Apollo 11 mission in July 1969, the USSR once again tested the J-vehicle. This time the rocket blew up on the pad, causing massive damage. Based on the destruction photographed by their satellites, the intelligence agencies concluded there had been a failed test flight.[78] Undoubtedly, the CIA quickly informed NASA.

With the successful Apollo landings on the Moon, the United States won the space race, and the competition effectively ended. Nevertheless, the intelligence agencies continued to monitor the USSR's space program. They published *NIEs* on it in 1970, 1971, 1973, 1983, and 1985 and provided them to NASA. Beyond this, however, there are very few accessible records documenting additional intelligence received by NASA and virtually none on NASA's use of it.[79]

NASA as an Analyst of Intelligence

The earliest known contribution NASA made to intelligence analysis and production is briefly described in the recently declassified *Soviet Land-Based Ballistic Missile Program, 1945–1972: An Historical Overview*. In 1961 and 1962, the director of the National Security Agency asked NASA, the Defense Atomic Support Agency, and the three services for assistance "in resolving certain technical problems—in particular those pertaining to Soviet missile- and space-related telemetry." The history gives no details on the support provided or who furnished it, except to say that it was very helpful to the National Security Agency in the telemetry analysis field.[80]

The next known assistance was to a report for the U.S. Intelligence Board's Guided Missiles and Astronautics Intelligence Committee on reconstructing and evaluating the Soviet's August 1962 dual launch of the manned Vostok 3 and 4 capsules.[81] DCI McCone approved a proposal by John Rubel, deputy director of the Office of the Director of Defense Research and Engineering, and NASA's Seamans that an ad hoc subcommittee do the report. NASA's Joseph Shea, deputy director of the Office of Manned Space Flight (and later a member of the DCI's Space Intelligence Panel), and representatives from the Air Force, CIA, Defense Intelligence Agency, National Security Agency, and undisclosed contractors sat on the subcommittee. This was apparently the first instance in which an interagency group had examined data from a Soviet space shot on a timely basis. The subcommittee quickly prepared its report, which expressed unanimous agreement on various aspects of the missions and what they meant for the future. Because the ad hoc subcommittee could not continue indefinitely, Webb, Dryden, and Seamans met with the DCI and offered to detail NASA employees to a permanent interagency body to be formed that would analyze all-source intelligence on Soviet space missions on a continuing basis.[82]

No such interagency group was created for reasons that are unclear, but in February 1963 the DCI proposed to the U.S. Intelligence Board that a NASA representative become a voting member of its Guided Missiles and Astronautics Intelligence Committee (a step it had considered shortly after NASA's creation but had dropped for unknown reasons). At its 13 March, 28 May, and 10 June meetings, the Board discussed the matter. The CIA, Department of State, Navy, National Security Agency, Atomic Energy Commission, and Federal Bureau of Investigation members approved. For unknown reasons, the Defense Intelligence Agency, Army, Air Force, and Joint Staff members believed that NASA should participate only on an ad hoc basis. The two sides did not resolve their differences, and the CIA decided to proceed on its own in employing NASA's expertise.[83]

NASA's assistance in analyzing intelligence was not restricted to its headquarters personnel. The Jet Propulsion Laboratory (JPL) had a contract with the CIA to help analyze video transmissions from Soviet Cosmos satellites during 1962 and 1963. The transmissions from *Cosmos 9* in

September 1962 were the first from which photos could be reproduced. CIA specialists consulted with the U.S. Weather Bureau and concluded that the spacecraft was an experimental weather satellite that electronically scanned the film. It presented its findings to the other intelligence agencies and the JPL in December. Utilizing the most advanced equipment then available, JPL personnel then examined all the Cosmos video intercepts to date and were able to develop three overlapping images in total. From the total land area in each, they determined that an 85 degrees wide-angle lens was used, which would be expected in a meteorological satellite. The CIA used the information JPL developed to brief NASA before the August 1963 implementation of an agreement with the Soviets to begin a regular exchange of data from weather satellites.[84]

After the failure to gain membership for NASA on the Guided Missiles and Astronautics Intelligence Committee in early 1963, the CIA and NASA soon agreed that NASA personnel would work full-time at the Ballistic Missiles and Space Division within the Office of Scientific Intelligence. The program was evidently given the code name GALAXY. William Taylor, assistant director for engineering studies in the Office of Manned Space Flight, and Tom Hagler were the first NASA detailees in early 1964.[85]

Carl Duckett, then head of FMSAC within the Directorate for Science and Technology, was critical of NASA's contributions at the time. He set forth his concerns in a November 1964 memo following a recent meeting with the deputy director of central intelligence and NASA's Gen. William McKee, USAF (Ret.), assistant administrator for management development. Duckett believed the present detail of two NASA personnel to his office was inadequate and that "they are almost like members of our staff and have very little more contact with the real brains at NASA than some of our regular staff employees." He favored NASA being given some sort of limited membership on the U.S. Intelligence Board and its Guided Missiles and Astronautics Intelligence Committee and concluded as follows:

> I am convinced that we need more help from NASA than we currently are receiving. No small part of the problem is the general reluctance to have people cleared for sensitive intelligence data. Of course, here I am referring specifically to clearing some truly competent scientists and engineers who have their fingers directly in the technical programs and are not spending their time shuffling

papers. If you could get to the bottom of the NASA attitude on this point, it might be the easiest way to predict the likelihood that there is a serious intent on their part to perform a real intelligence job. A number of people have told me that Bob Seamans is the individual who argues against NASA having any more people cleared.[86]

James Shaw from the Office of Manned Space Flight replaced William Taylor at the Ballistic Missiles and Space Division in 1965. Shaw and Hagler moved to FMSAC in late 1965 when it took over the organization. Jerome Terplitz from the Office of Manned Space Flight apparently replaced Hagler in 1966. NASA continued detailing two personnel to FMSAC at least through 1968 and perhaps in later years as well.[87]

Seamans wrote DDS&T Albert Wheelon in May 1966 that he understood that while the NASA engineers at FMSAC were not formally detailed to CIA that "they work on a daily basis in the FMSAC and receive direction from a designated staff member of that office. Their function is to assist CIA in preparing analyses and evaluation of Soviet space intelligence by providing technical support in the area of manned space flight and by furnishing useful data to FMSAC from OMSF [Office of Manned Space Flight] and its Centers." Seamans had no problems with the NASA engineers assisting the Guided Missiles and Astronautics Intelligence Committee at the pleasure of its chairman, but warned, "Our policy continues to be that any direct association of NASA with the activities of the USIB is inappropriate."[88]

Seamans and Gen. McKee proposed to the DDS&T in late January 1965 that NASA's contributions to the CIA be expanded beyond the two personnel detailed to FMSAC. McKee argued that NASA had important expertise to share with the CIA on foreign space events, NASA could profit immensely from intelligence on Soviet space technology, and the CIA needed NASA's input before advising the president on the USSR's space programs. All three agreed that CIA-sponsored panels with senior NASA personnel as their members and meeting as needed would be ideal, but they left the details to be worked out.[89]

The following month, Wheelon proposed a joint seminar on the USSR's human space flight program to George Mueller, associate administrator for manned space flight. In a follow-on memo to Mueller (with copies to Seamans and McKee), he wrote:

All of this has been somewhat overtaken by a broader approach to the space intelligence problem, suggested by General McKee. However, his proposal has as one of its logical parts a CIA/NASA group focused on the Soviet Manned Space Program. Assuming that the two are congruent, it seemed important to keep our previous ball in play. If we wish to do something this summer, it is not too early to begin identifying people so that appropriate clearances can be arranged. Let me know your reaction when you have examined the concept further.[90]

The proposal for a series of panels soon eclipsed that of having a single one on human space flight. Wheelon wrote Seamans in April that five separate panels would be best—Manned Space Flight, Launch Vehicles, Launch and Test Facilities, Scientific and Technical Satellites, and Lunar and Planetary Probes. Each would have a CIA chair and up to 10 NASA technical experts. It was expected that the panels would meet at least twice a year at the relevant NASA centers and prepare written reports. The CIA would consider the NASA personnel involved as consultants and provide them with the necessary clearances.[91] Seamans replied a few weeks later, expressing concerns about some of the details:

> First, I believe it is very important to identify carefully the specific and limited role that the panels are to play. The underlying problem in this area is the possibility that our technical personnel might become *de facto* working elements of a community for which they have inadequate training; as you know, our enabling legislation is specific on the subject of full disclosure of data. I therefore wonder whether it is appropriate for the panels to prepare written reports of their findings themselves, it perhaps being preferable to have your senior officer use the panel as a resource but retain the final responsibility for the documented output. We would prefer that the groups not be formally designated as NASA panels.[92]

In July 1965, Seamans and Wheelon signed an agreement concerning the panels. It set up the following eight panels: Manned Space Flight, Launch Vehicles, Launch and Test Facilities, Scientific and Technical Satellites, Lunar and Planetary Probes, Aeronautics, Advanced Research and Technology, Tracking, Data Acquisition, and Reduction. The agreement provided generally that each panel would meet no more than twice a year and that each set of meetings would last no more than two days. NASA members would act solely as consultants and advisors and would not be expected "to furnish or participate in formal analyses or evaluations of intelligence." CIA officials would prepare all reports and other written documentation.[93]

There is very little information on the subsequent history of the panels. The first meeting of the Advanced Research and Technology Panel was scheduled to be held at CIA headquarters in March 1966, and the tentative agenda items were electronics, chemical propulsion, space power, electronic propulsion, and materials.[94] A heavily redacted three-page document is the only record on the first meeting of the Lunar and Planetary Probe Panel in June 1966. The agenda included a presentation of the purpose of the panel, a CIA review of the USSR's programs, a panel discussion, and two still-classified topics. Most of the CIA briefing and panel discussion sections also remain classified.[95]

The DCI established the Space Intelligence Panel in the fall of 1965 to personally advise him. All of its nine members were from outside the Intelligence Community. Joseph Shea and Raymond Bisplinghoff, the initial NASA members, each had four-year terms. The group's charter covered the following aspects of the USSR's space programs: human space flight; lunar and planetary probes; scientific payloads; meteorological, communication, reconnaissance, and navigation satellites; orbital bombardment; launch vehicles and facilities; tracking and communication networks; and reentry and recovery developments. It also was to look at the status and capabilities of other foreign space programs. The panel was charged with reviewing and making recommendations on Intelligence Community assessments, evaluating and making recommendations on collection systems and analytical techniques, and conducting a detailed review of a specific problem or issue as requested by the DCI. It was to meet twice a year, with one of the sessions devoted to the *NIEs* on the Soviet space program.[96]

The Panel's first product was the aforementioned one-page report on the Soviet manned lunar landing program issued late in 1965. There are only two brief documents available shedding light on the subsequent history of the group. The first is an undated one-page summary by an unknown author that contains handwritten notes in the margins indicating that at a February 1966 meeting it concluded that the USSR was "even" with the United States in the Moon race. These notes also state that at its July 1966 meeting the Panel was asked by the White House to evaluate the Soviet space effort and with respect to a manned lunar landing concluded that the earliest the Soviets could achieve this was after Apollo did in 1969. Their estimate was based in part on the fact that the USSR had not flown any manned capsules since the previous year and had not tested any larger spacecraft necessary for a lunar landing. The second document lists the agenda items for the 19 November 1968 meeting as strategic weapons systems, space activity in 1968, Soyuz rendezvous and docking, the new large Soviet booster, and the circumlunar program.[97]

Shea left NASA in 1967, and the following year he was still serving on the Panel. Bisplinghoff left NASA in 1969, and the following year his membership on the Panel ended.[98] There is no information available on how long the Panel continued in operation or whether any other NASA employees ever became members.

A small group of specially cleared personnel at NASA's Langley Research Center began a decades-long project in 1963 assisting the CIA and subsequently other intelligence agencies in evaluating the aerodynamic properties of Soviet aircraft, missiles, and spacecraft. Led by M. Leroy Spearman and Mark Nichols, they used open photography or classified imagery to build models of the vehicles, which were then tested under secure conditions in one or more of the facility's wind tunnels. The classified photography included at times TALENT-KEYHOLE imagery from overhead platforms. In these cases, the NASA researchers had to review it at a Sensitive Compartmented Information Facility at the nearby Langley Air Force Base or a CIA office in the Washington, D.C., area since the Langley Research Center did not have one. They set forth their findings and conclusions in reports entitled Langley Working Papers. These documents did not mention the foreign vehicle being studied and, if classified, it was done so only at the Confidential level. Occasionally, the researchers just forwarded the raw data from the tests and did not prepare any formal reports.[99]

Langley provided its assistance under an informal working arrangement during the first few years. In early 1965, NASA and the CIA signed a formal agreement for aerodynamic consultation. Requests for assistance from the Defense Intelligence Agency, the Air Force's Foreign Technology Division, and other DoD intelligence organizations soon followed.

Working through NASA's Office of Defense Affairs, the CIA first requested advice in 1963 on what their analysts believed was an antiballistic missile. Because it was quite large and was very different from other known surface-to-air systems, the United States and its NATO allies had not assigned it a standard surface-to-air missile designation but instead codenamed it Griffon. CIA personnel initially brought photographs of the missile taken at a November 1963 Moscow parade and asked for an assessment based solely on them, which was provided. They then agreed to Langley's proposal to build and test a model to obtain more data. The results indicated very good stability and control characteristics and good maneuverability, which NASA reported to the CIA in January 1964. As it turned out, the missile was never deployed.

In the next few years, the CIA asked Langley to analyze numerous Soviet aerospace vehicles. With respect to the SA-2 surface-to-air missile, which became a major threat to U.S. air forces over North Vietnam, the researchers initially reviewed wind tunnel data from tests of a small-scale model done by a contractor in the late 1950s. They noted some errors and, using current imagery, constructed a larger model and tested it in several wind tunnels beginning in August 1965. Soviet modifications to the missile—including relocating the fuel tank well forward, external fuel lines, and asymmetric wings—led to the building and testing of new models. Some analysts at first concluded that these changes would result in decreased performance. However, Langley's investigations determined that the aerodynamic center moved forward with higher speeds, and when the fuel was burned, the center of gravity also moved forward so that a low stability level could be maintained and maneuverability increased. They also demonstrated that the asymmetry introduced by the altered wings exactly offset the roll caused by the external fuel lines.

Langley studied the MiG-21 extensively beginning in 1966. This was primarily done because of its use by North Vietnam against U.S. air forces. The initial wind tunnel tests helped improve the tactics used against the fighter. Subsequent tests analyzed the effects of battle damage on the plane and its use of Atoll air-to-air missiles. During this period, Langley also analyzed the Vostok capsule, which carried Soviet cosmonauts. The CIA apparently provided Langley with photographs of one after its landing and gave a briefing on the Vostok. In its human spaceflight programs, the United States employed capsules with a conical shape and a complex control system to ensure that the base with the heat shield was forward during reentry. The Vostok, in contrast, was a slightly blunted spherical shape with the heat shield on the blunted surface. This surface had to face forward during reentry. Langley's wind tunnel tests demonstrated that the capsule was stable with the blunted heat shield surface forward. As a result, the Vostok automatically reentered with it forward, and the spacecraft did not need any elaborate control system as U.S. capsules did.

Langley's analyses of the aerodynamic properties of Soviet aerospace vehicles were so important that Spearman and Nichols soon began giving classified briefings to high-level officials throughout the national security community. After a rehearsal before Webb, Paine, and several others at NASA headquarters in May 1968, they gave their first formal presentation the following month to DDS&T Carl Duckett and others from this office.[100] Later in June, Spearman and Nichols briefed DCI Richard Helms and other high-level CIA officials. Helms quickly wrote the following to Secretary of Defense Clark Clifford on the matter:

> I have just heard a briefing by technical personnel of NASA Langley. Their conclusions, if borne out by further analysis, could have critical implications for U.S. defense planning. The subject matter concerns the capabilities of Soviet aircraft, missiles, and spacecraft. . . . The results to date are disturbing. For example, aircraft such as FISHBED and FOXBAT are found to be superior in most important performance characteristics to U.S. combat aircraft now in service. Design features on several Soviet missile systems show similar advances. The general conclusion is that Soviet aerodynamic design capabilities are far advanced, and at the present rate of effort the USSR may well enjoy an ever increasing advantage over our aerospace vehicles.[101]

Spearman and Nichols made the presentation to more than 30 other audiences before the end of the year, including to DDR&E John Foster and the Air Force, Army, and Navy's assistant secretaries for R&D in July. After the briefing, Foster wrote to the DCI stating that it was an "excellent analysis of Soviet aircraft developments." He added that they considered it "of sufficient import to cause us to re-examine the purposes and procedures which are currently in use in U.S. aircraft technology and airframe developments. I am grateful to the CIA for having initiated and supported this analytic work at NASA."[102]

NASA also assisted the intelligence community outside the channels described above. In September 1965, the DCI requested a NASA presentation on nuclear propulsion in space. Later that month, three NASA personnel working on the joint NASA-Atomic Energy Commission Nuclear Engine for Rocket Vehicle Application program conducted the briefing. It covered the advantages of nuclear power, the development effort to date, and "the nuclear rocket test and operating signatures that might serve to identify similar activities elsewhere."[103]

NASA continued contributing to the analysis of Soviet programs after the 1960s. However, all that is known in this regard is that the Langley Research Center's assistance to the CIA and others in analyzing the aerodynamic properties of Soviet aircraft and missiles continued into the 1980s and that beginning in 1970 NASA finally sat on the U.S. Intelligence Board's Guided Missiles and Astronautics Intelligence Committee and its successor.[104]

Summary

The USSR's space program became an even more important intelligence target in the 1960s with the race to the Moon, and as a result NASA's interaction with the national security agencies in this area grew dramatically. Its leadership regularly received the CIA's highest level reports and estimates as well as oral briefings. They occasionally took issue with some of the conclusions, but there is no evidence that their objections caused the CIA to change them. NASA made much wider use of the intelligence than it had done earlier, using the information in budget negotiations at the White House and Congress. However, the individuals with whom NASA was dealing also had access to the intelligence, and it is unlikely that these efforts had any effect. Apollo 8 was the only known mission influenced by intelligence on the Soviet space program, but only in part as there were also valid programmatic reasons for changing it from an Earth-orbital to a lunar-orbital flight. NASA also greatly expanded its contributions to the analysis of data on the Soviet program. These ranged from establishing a formal project at a NASA center to evaluate Soviet hardware to detailing personnel to work at CIA headquarters.

NASA increased its ties with the defense and intelligence agencies in other areas as well during the 1960s. The next chapter examines several of these and the successes and problems it experienced in trying to meet their requirements.

Expanding Interaction in Old and New Areas

NASA and the national security agencies maintained their existing relationships during the 1960s in many fields besides foreign intelligence
and developed close ties in entirely new areas. It satisfied more defense
and intelligence requirements but also faced the first of many conflicts
over the classification of scientific data and the challenges faced in trying to preserve an image of openness while protecting national security.

NASA continued to be involved in cover stories for intelligence projects after the May 1960 U-2 shoot down. Although it refused to take a
public role as it had with the U-2, the CIA and others solicited its advice
on developing a cover story for the A-12 reconnaissance aircraft and perhaps other vehicles. Beginning in the early 1960s, NASA participated
with the national security agencies in recovering and analyzing U.S. and
foreign space fragments that had returned to Earth, an activity that had
major intelligence and foreign policy implications. There were major
interagency disagreements over which agencies should be responsible
for obtaining U.S. and foreign fragments and under what circumstances
the latter should be returned. NASA and the CIA were allies in these
struggles, which ended when all the interested agencies finally agreed to
a government-wide policy after ratification of the Outer Space Treaty of
1967.

The close ties regarding the acquisition and dissemination of space surveillance information continued. NASA had the same access as before to all unclassified data and classified data with a demonstrated need-to-know and remained the conduit for the release of unclassified data to the public. This occasionally created problems when the press and others demanded more information on Soviet space shots than it could disseminate because of classification. What classified space surveillance data NASA received and what it was used for is unknown. Its Space-flight Tracking and Data Acquisition Network and Manned Spaceflight Network supported a growing number of DoD missile tests and orbiting satellites. The majority of these networks' overseas stations could only provide unclassified assistance due to technical limitations or political restrictions imposed by the host government. However, these were by-passed at times, and they furnished classified support. Domestic sta-tions generally could and did provide both unclassified and classified assistance when needed. NASA successfully opposed DoD proposals to consolidate in whole or part the civilian and military command and con-trol networks on the basis that its networks were an integral part of its spaceflight programs and that the promised economies would not materialize. There was extensive collaboration on developing their sepa-rate data relay satellite systems, designed to greatly increase the time satellites in low-Earth orbit were in contact with ground stations and the amount of information that could be transmitted between them. The NRO launched its first two data relay satellites in 1976, while NASA placed its first one in orbit in 1983.

The joint National Operational Meteorological Satellite System was created in 1961 to meet all civilian and national security requirements, but it suffered numerous delays and had little chance of achieving its objectives. As a result, the DoD quickly withdrew and developed its own classified weather satellites. NASA continued the effort as a research and development program and maintained close liaison with the DoD on these and its other polar-orbiting weather satellites, which provided data to an expanding network of military ground stations to support tac-tical operations. NASA also agreed in principle during 1960 to what was ultimately designated the National Geodetic Satellite Program, which planned to meet all civilian and military requirements for geodetic infor-mation from space and in which there was another clash between civilian and national security interests. It initially refused to participate until the DoD under pressure from the White House and civilian scientists agreed that all the data collected from the first-generation satellites would be unclassified. There was close collaboration on the follow-on satellites, which NASA built, launched, and operated. They frequently carried DoD geomeasuring devices and flew specific orbits the DoD requested to provide the best data for improving the accuracy of long-range missiles. Despite NASA's objections, the DoD classified some of the raw data its ground stations acquired from these spacecraft and the finished product.

NASA's Participation in Cover Stories

NASA's embarrassment from its involvement in the attempted cover-up of the loss of Gary Powers's U-2 in 1960 did not stop it from continuing to work with the national security agencies on cover stories for other projects. In late 1961, Dryden nominated John Stack, director of aeronautical research at NASA headquarters, to serve as the NASA member on the proposed interagency SUMAR Committee for supporting the cover story for the CIA's A-12 reconnaissance aircraft (the planned successor to the U-2). CIA personnel soon briefed Dryden and Stack. The latter suggested publicizing various aspects of the A-12 program that would be useful in developing civilian aircraft such as the supersonic transport.[1]

It is unclear whether the SUMAR Committee was ever formally established. CIA personnel met with Dryden again in June 1962 on several issues, including the technical progress of the A-12 project and its proposed cover story. (Stack's involvement ended when he left NASA earlier that year.) Dryden received the new version of the cover story and believed it was an improvement. He recommended one change, suggesting the fallback position not identify the aircraft as a "satellite launching system" because the public would demand to know what types of satellites it could launch. Dryden proposed that the fallback position be that the A-12 was for launching ballistic missiles, but the CIA attendees thought that this was too close an association with offensive weaponry.[2] Whether the meeting resulted in any changes to the proposed cover story is unknown.

The DoD recommended in late 1962 that the Air Force interceptor version of the A-12 be publicly revealed to provide a cover for any crashes or sightings of the A-12 and help explain the rise in Air Force spending. DCI McCone, as well as the leaders of the President's Foreign Intelligence Advisory Board, opposed the move. They met with Kennedy and Secretary of Defense Robert McNamara in January 1963 and successfully argued against any public disclosure for the time being.[3]

During the summer of 1963, the Interdepartmental Contingency Planning Committee was created with the mandate to review and approve contingency and cover plans for all manned and unmanned reconnaissance vehicles operating over the Sino-Soviet Bloc. The NRO director was the chair, while the other members came from the White House, State Department, Joint Chiefs of Staff, CIA, and Air Force. NASA's Seamans was one of four consultants to the group.[4]

Supporters of disclosure in late 1963 advanced the argument that revealing the A-12 program was necessary to disseminate the supersonic technology for other Air Force projects and the civilian supersonic transport then on the drawing boards. As a result, Kennedy directed McCone in early November 1963 to develop a plan for public disclosure but to not implement it until further discussions.[5]

McCone, McNamara, and several others briefed President Lyndon Johnson on the CIA's A-12 program and the associated Air Force projects later that month. While McNamara strongly advocated revealing some version, McCone persuaded Johnson that further study was needed. By early the next year, McCone was advocating disclosure, and the National Security Council unanimously approved this action in late February. Johnson quickly held a news conference at which he announced the development of the "A-11" and that it had flown at more than 2,000 mph and at altitudes over 70,000 feet. However, the actual aircraft described was the Air Force's YF-12A interceptor, which had already been canceled. Following the news conference, the Air Force moved its versions of the A-12 to Edwards Air Force Base in California, and from that point on they were a matter of public record. In July 1964, President Johnson revealed the existence of the Air Force SR-71 reconnaissance aircraft. At no point was the CIA's involvement in the program or the existence of its A-12 aircraft ever made public.[6] The role, if any, of the Interdepartmental Contingency Planning Committee in the events leading to this disclosure and creating cover stories for other vehicles is unknown.

Collection and Analysis of Space Fragments: Conflict and Cooperation

NASA began a long involvement with the national security agencies in the early 1960s concerning the retrieval and analysis of U.S. and foreign space debris that had survived reentry and returned to Earth. The debris included pieces of rockets, missiles, and their payloads. Acquiring foreign fragments was essentially a continuation of the long-standing U.S. practice of obtaining foreign military equipment for exploitation. Equally important to the United States was retrieving its own debris for analysis and keeping it out of the hands of the Soviets. For many years, the division of responsibilities and the accompanying cover story was a contentious issue between NASA and the CIA on one hand and the DoD on the other.[7]

It was several years after *Sputnik 1* before the United States had an opportunity to gather any Soviet space debris. The USSR only conducted 13 orbital launches from October 1957 through 1959, and the only known object to return to Earth from them was *Sputnik 1*'s booster in December 1957, which landed in Mongolia. Soviet suborbital tests during this period did not offer any opportunities for retrieval either, as all the landing zones were within the USSR.[8]

Prospects for obtaining Soviet debris improved in the early 1960s. It performed 10 orbital launches in 1960, 11 in 1961, 21 in 1962, and 22 in 1963. Additionally, it began in January 1960 to conduct some suborbital tests in designated landing zones in the Pacific (one northwest of Midway Island and the other southeast of Johnston Island). *Sputnik 4* fragments that landed in Wisconsin in September 1962 were apparently the first Soviet ones the United States collected. The Air Force, CIA, and other government agencies analyzed them, and the debris was returned to the Soviets in January 1963.[9]

The first U.S. debris from an orbital launch that is known to have survived reentry and fallen on land was from the upper stage of the Atlas-Able rocket that launched the *Pioneer P-30* lunar probe in September 1960. Due the failure of its engines to fire, it reentered, broke up, and fell in South Africa. The world's press gave the story little attention, and with the assistance of the CIA the debris was quietly collected and returned to NASA. Two months later, the first stage of a Navy Thor Able-Star fell in Cuba after the Cape Canaveral range safety officer had to destroy the rocket. In February 1962, pieces of the Atlas booster that launched John Glenn on the first orbital Mercury mission fell in Brazil and South Africa. Debris from the Atlas that launched the third orbital Mercury mission eight months later landed in Upper Volta and the Ivory Coast.[10]

The increasing number of American fragments that were surviving reentry and landing in other countries, as well as the willingness of the

USSR to negotiate on a wide range of issues concerning outer space, led the U.S. government to try to establish a uniform policy on acquiring, analyzing, and returning debris. In November 1962, representatives of the National Security Council, NASA, DoD, Atomic Energy Commission, State Department, and the Department of Justice met under the auspices of the National Aeronautics and Space Council. They unanimously agreed to drafts of two policy documents, the first of which was a public statement that read as follows:

> The National Aeronautics and Space Administration is designated as the Government agency responsible for receiving, analyzing, and identifying any fragments or suspected fragments of space vehicles of all types, including the contents thereof, which fall upon the territory of the United States or which otherwise come into its possession. Fragments identified as coming from vehicles launched by foreign countries will be retained by NASA pending final arrangements for their return to the launching country or for other appropriate disposition. In performing these functions, NASA will fully coordinate with other Government agencies.[11]

The second policy draft was a classified guideline with the same first sentence, but which in addition directed NASA to turn over debris from DoD launches to the DoD and to "give the DoD and other responsible agencies access, as expeditiously as possible, to fragments which are readily identifiable as coming from space vehicles launched from foreign countries." After completion of this analysis, NASA was to get the debris back "pending final arrangements for their return to the launching country or for other appropriate disposition." DoD would notify NASA when debris was expected to land and whether collection teams were successful in obtaining it. The State Department was to provide policy guidance on the disposition of foreign fragments, and NASA was charged with coordinating and releasing all public information on them.[12]

The National Aeronautics and Space Council sent the two drafts to the agency heads for approval, and all gave it except for McNamara. He maintained that the issue involved intelligence responsibilities and procedures and therefore had to be considered by the U.S. Intelligence Board. The matter was placed on the agenda for a Board meeting in December 1962, but was withdrawn at the last minute. For reasons that are not entirely clear, the dispute continued and became bitter. An internal Council memo from May 1963 stated that Deputy Secretary of Defense

Roswell Gilpatric had stopped trying to talk with DCI McCone on the issue and the DoD "will not buy the proposition that 'McCone and NASA' had agreed to." Arnold Frutkin, NASA's associate administrator for international affairs, told Council staff the same month that the DoD and Air Force want to use NASA as a "cover" and only return pieces, while the "CIA and NASA get along fine and can do so in the future on this matter."[13]

NASA and the CIA rejected a DoD proposal during this period to take away NASA's responsibility to analyze fragments. Although there were subsequent plans to finally bring the matter before the U.S. Intelligence Board, it is not known if and when this was done.[14]

The serious disagreements between the CIA-NASA and the DoD that prevented adoption of a uniform policy in 1962–63 persisted for several years. Nevertheless, the United States continued to enjoy success in retrieving both its own and foreign fragments. From 1963 to 1967 there were more than 20 known instances of U.S. debris impacting in various nations. In most instances, the recoveries went smoothly with the State Department or defense attachés obtaining the debris from the host governments and forwarding them to NASA or the DoD. Increasing numbers of Soviet fragments were also returning to Earth, and the United States was recovering more of them. For example, it obtained three metal spheres from the booster for the *Luna 8* lunar probe, which fell in Spain in December 1965. The Air Force reported that the analysis enabled a detailed comparison of U.S. and Soviet welding techniques. A Soviet titanium spherical vessel landed in Wisconsin in October 1966 and, without the knowledge of the State Department, the Defense Intelligence Agency conducted destructive testing. When the DoD subsequently contacted the State Department about possibly returning the vessel to the Soviets, the State Department decided against it, since the item had been "mutilated."[15]

President Johnson in May 1966 called for adoption of a treaty incorporating the principles of the General Assembly's Resolution 1962, the Declaration of Legal Principles Governing the Activities of States in the Exploration and Use of Outer Space, passed three years earlier. On the matter of space debris, the nonbinding Resolution 1962 had provided in part that "ownership of objects launched into outer space, and of their component parts, is not affected by their passage through outer space or by their return to the Earth" and that "such objects or component parts found beyond the limits of the State of Registry shall be returned to that State."[16]

The United Nations' Outer Space Committee soon considered the proposed treaty, and then the General Assembly approved what was formally called the Treaty on Principles Governing the Exploration and Use of Outer Space, including the Moon and Other Celestial Bodies (Outer Space Treaty). It was opened for signature in Washington, London, and Moscow in January 1967. After the United States, Soviet Union, and more than 60 other nations signed, the treaty came into force in October. Article VIII concerns the ownership and right of return of space debris:

> Ownership of objects launched into outer space, including objects landed or constructed on a celestial body, and of their component parts, is not affected by their presence in outer space or on a celestial body or by their return to Earth. Such objects or component parts found beyond the limits of the State Party to the Treaty on whose registry they are carried shall be returned to that State Party, which shall, upon request, furnish identifying data prior to their return.[17]

The State Department used the treaty and its binding obligations as leverage to finally forge a policy on at least some aspects of the debris issue. Clearly, it did not want to see a repeat of the recent experience involving the Soviet fragment that underwent destructive testing without a prior determination by all the interested agencies that the intelligence value outweighed the possible negative foreign policy repercussions. Foy Kohler, deputy undersecretary of state, contacted the DCI, the secretary of defense, and NASA's administrator in June 1967 proposing that the State Department be immediately notified when an object of foreign or questionable origin was offered to or came into the possession of the United States. He stated that this would ensure that the international obligations of the United States would be taken into consideration in deciding initially whether even to obtain the debris and, if acquired, what type of testing to subject it to. Kohler also stressed that what the United States did would seriously impact how successful it would be in recovering its own objects.[18]

The State Department, DoD, and NASA soon came to an agreement setting forth the procedures to be followed when space fragments landed in a foreign country. (Although there is no available record of it, the CIA also must have given its approval.) The State Department issued a classified "Joint State-Defense-NASA Message" to all of its posts in February 1968 setting forth these steps. Diplomatic personnel were to report to Washington the presence of any debris and try to determine its origin. If necessary, and if the host government permitted it, U.S. experts were to be brought in for that purpose. The Outer Space Treaty was to be referenced if the object was American and the host government refused to return it for any reason. If the fragment was clearly foreign or its origin could not be ascertained, the interested agencies would decide whether to try to obtain it, depending on its potential intelligence value and "attendant foreign policy considerations." The directive did not address who was to perform the testing or whether destructive testing would be permitted. The available evidence indicates that after the various agencies reached this agreement they were finally cooperating.[19]

The Outer Space Treaty had little impact on the United States retrieving its own debris. In the years immediately following it, pieces of various NASA boosters and spacecraft landed in Australia, Angola, Columbia, and the Caribbean. No major problems were encountered in getting any back.[20]

The treaty did affect the recovery and return of foreign objects in some cases, however. A piece of a Soviet spacecraft landed in Finland in December 1967 and was readily identified as such. It was returned to the Soviets, but unbeknownst to them the Finnish government retained a portion and offered it to the United States. The offer was refused on the basis of the Outer Space Treaty and the questionable intelligence value. In September 1968, a spherical vessel fell in Alaska that was determined to be of Soviet origin. It was returned to the USSR eight months later, but undoubtedly only after the intelligence agencies had examined it. In the fall of 1969, a spherical vessel landed in Sweden, and the government soon determined that it was of Soviet origin. The U.S. embassy requested permission to examine the object, but ultimately Sweden refused on the basis that it was Soviet property and under the Outer Space Treaty a nation had no right to inspect a fragment when it was clear that another nation had launched it. The vessel was eventually returned to the USSR.[21]

The number of domestic and foreign fragments in space and those that returned to Earth increased dramatically in the following decades as more orbital launches were conducted. However, except for NASA's participation in an interagency task force under the National Security Council to deal with the decaying Soviet nuclear-powered *Cosmos 954* and its eventual crash in Canada in January 1978, there is virtually no information available on the interaction between NASA and the defense and intelligence agencies on the issue.

Acquiring and Distributing Space Surveillance Information

Implementation of the January 1961 NASA-DoD agreement dividing responsibility for conducting space surveillance and establishing NASA as the conduit to the public for all unclassified information proceeded smoothly for the most part. There were some problems in a few cases, though, such as Gherman Titov's August 1961 Vostok 2 flight. The Soviets announced it about two hours after launch, and the press immediately contacted NASA for details. However, NASA did not receive any unclassified data from the North American Air Defense Command (NORAD) for many hours, and it finally took a call from Dryden to obtain it.[22] Additionally, there were some who criticized the decrease in information being released to the public. Congressman John Moss (D-CA), chair of the Government Information Subcommittee of the House Committee on Government Operations, wrote to Webb in February 1961 complaining of the new restrictions. Webb replied quickly and emphasized that NASA would disseminate all information it received from NORAD. This quieted the critics for the moment.[23]

A November 1961 agreement between the Goddard Space Flight Center and NORAD clarified procedures to eliminate the few delays in disseminating information that had occurred to date. It required NORAD to furnish Goddard all information on U.S. scientific satellites as soon as possible and all information on foreign scientific satellites as soon as positive identification was made. Goddard was to provide NORAD with prior notification of all NASA civilian scientific satellite launches, instant notification of countdown and immediate post-launch information, and timely notification when its spacecraft became inactive. It was responsible for publicly releasing all unclassified data on U.S. and foreign spacecraft. Information on U.S. military satellite launches classified by the launching agency and information on foreign military launches would be given to Goddard on a need-to-know basis. NASA and NORAD were to meet periodically to review existing procedures and modify them if needed.[24] NORAD also regularly distributed classified raw data and classified finished intelligence reports to the Air Force, NRO, and others, but it is not known whether it sent any to NASA.

Congressman Moss again raised the matter of the restricted space surveillance data being released by NASA as part of hearings in 1963 on the broader issue of the increasing government secrecy surrounding national security space programs. He specifically criticized DoD and NASA for withholding information on Soviet space failures the previous year and NASA's delays in publishing the Satellite Situation Report, which was supposed to contain details of them. Moss noted that the Soviets certainly knew their missions had failed and that NASA's networks independently of NORAD's had collected sufficient data to establish the failures. NASA's associate administrator for public affairs insisted that the classification decision was up to NORAD and that there was nothing NASA could do about it. Although the hearings opened up the secrecy issue to public scrutiny, in the end no policies or practices were changed.[25]

NORAD sent a proposed new agreement to Seamans in October 1964 to replace the existing one. It would bring the recently established Deep Space Network and the Baker-Nunn cameras operated by the Smithsonian Astrophysical Observatory for NASA under its jurisdiction, as well as expand the technical data NASA was to provide on its launches and active and inactive spacecraft in orbit. Seamans replied in December that NASA had several major objections to the proposed agreement. These included that the proposed space surveillance priorities in times of national emergency were not covered by the January 1961 NASA-DoD agreement, much of the new technical data NASA was to furnish was not obtainable without extraordinary cost, and some of the work levied on NASA stations overseas could not be performed under the terms of the agreements with the host governments. NASA, DDR&E, and NORAD officials apparently held meetings in early 1965 to resolve the differences, but there is no evidence that a new agreement was ever reached.[26]

Details of subsequent collaboration between NASA and NORAD in the acquisition and dissemination of space surveillance information are not available. However, the liaison must have remained close as the number of objects in orbit (particularly low-Earth orbit where many NASA spacecraft operated) and the resulting threats they posed grew.

NASA Network Support to DoD Spacecraft and Missile Tests: Development of Data Relay Satellites

NASA was building three separate tracking, data acquisition, and command and control networks in the early 1960s. Its Minitrack network grew to 15 foreign and 5 domestic stations and serviced robotic satellites in any inclination. The Manned Space Flight Network (MSFN) supported the human spaceflight program and had 10 foreign and 7 domestic stations. The Deep Space Network serviced lunar and deep-space probes and had one domestic and two foreign stations. A number of Baker-Nunn cameras operated around the world by the Smithsonian Institution under contract to NASA assisted in tracking.[27]

The DoD's main network was the Air Force Satellite Control Facility with seven domestic and foreign stations designed primarily to support reconnaissance satellites in high-inclination orbits. It also provided assistance to other vehicles such as the Vela nuclear test detection satellites and Midas early warning satellites in similar orbits. Much smaller networks were built to provide support to individual programs, such as the one for the Navy's Transit navigation satellites. Not only did the DoD's ranges monitor boosters and their payloads in the early phases of flight but they also serviced spacecraft in orbit when needed. The Atlantic Missile Range (renamed the Eastern Test Range in 1964) included a network of more than 10 tracking, data acquisition, and command and control stations stretching from Florida through Puerto Rico and other islands and ending in South Africa. Ships and aircraft augmented them. The Pacific Missile Range (part of which was designated the Western Test Range in 1964) included a network of over 10 stations stretching from California to Wake, Midway, and Eniwetok Islands. It too employed ships and aircraft to assist them.[28]

Many of the DoD's tracking, data acquisition, and command and control assets began furnishing extensive help to NASA in the early 1960s. Western Test Range assets provided support during the initial phases of every NASA launch from California, as did those of the Eastern Test Range for every NASA launch from Florida and Wallops Island in Virginia. These assets also helped track and provide telemetry and communications support to the Mercury, Gemini, and Apollo capsules in orbit. Stations from both ranges occasionally commanded and received data from other NASA vehicles in orbit, such as Tiros weather satellites.[29]

The first known NASA network support to a DoD satellite started in June 1960 with selected Minitrack stations tracking and receiving telemetry from GRAB 1. Minitrack and MSFN assistance to DoD space programs grew dramatically in the early 1960s, even though they operated under restrictions that the DoD's did not. Most NASA domestic and foreign stations could not transmit classified data over the leased commercial communication circuits they utilized. Additionally, several of the agreements between NASA and the host governments establishing foreign stations expressly prohibited any assistance to military programs, and almost every such agreement required that indigenous personnel be employed, access be given to host government personnel, and all data collected by the station be made available to the host government. These factors obviously limited the classified and in some instances the unclassified support that could be given to DoD satellites.[30]

NASA support generally involved one or more of the following: tracking, telemetry reception, and command transmission. The length of assistance ran from one day up to one year. By 1964, there were nine foreign Minitrack and MSFN facilities (Australia, Bermuda, Canada, Canton Island, Chile, Ecuador, Peru, South Africa, and United Kingdom) that furnished unclassified support to DoD spacecraft, two (Spain and the Canary Islands) that could likely provide unclassified assistance if needed, and four (Malagasy, Mexico, Nigeria, and Zanzibar) that in all probability could not furnish any help because of political restraints. All 12 Baker-Nunn cameras in foreign nations operated by the Smithsonian Institution for NASA, with the exception of the one in Argentina, provided unclassified tracking assistance to DoD missions.[31]

To better coordinate the growing DoD demand for NASA support, the director of defense research and engineering (DDR&E) ordered in late 1961 that all requests go through that office to NASA's Office of Tracking and Data Acquisition. Once approved, DoD personnel communicated directly with NASA's Goddard Space Flight Center (which managed Minitrack and MSFN) as they had been doing previously. At the time, DDR&E decided that one Air Force program, one Army program, and five Navy programs qualified for NASA support.[32]

DDR&E formally contacted NASA in early 1962 asking that Minitrack receive telemetry from and send commands to the Navy's *Solar Radiation 4B* and *5A* satellites to be launched later that year on NASA Scouts from Vandenberg. (These spacecraft carried an open solar radiation experiment and the highly classified GRAB electronic intelligence package. It is extremely doubtful that any NASA personnel in the Office of Tracking and Data Acquisition or Goddard had the necessary clearances and knew of GRAB's existence.) As was typical with these requests, DDR&E furnished the orbital inclination, nominal altitude, apogee, perigee, and telemetry and command frequencies. It estimated that the on-orbit life for each was expected to be one year and sought maximum coverage for that period on a non-interference basis. As it turned out, *4B* failed to reach orbit and *5A* apparently was never launched.[33]

In subsequent months DDR&E forwarded other requests for NASA's assistance. Among others, the Navy wanted selected Minitrack stations to acquire telemetry from *Injun 3* (which measured solar radiation), several Navy Transit navigation satellites, and the ANNA geodetic satellite. The Air Force desired similar assistance with respect to a series of Tetrahedral Research Satellites that carried experimental nuclear test detection sensors. Although the Air Force wanted the data from these to remain classified, it recognized that the agreements establishing NASA's foreign stations effectively prohibited this. Accordingly, the Air Force authorized NASA to release certain information as unclassified but urged NASA that it be released "minimally" and that "every effort should be made to protect the information administratively, and to persuade the particular government involved to keep the information in a need-to-know, official-use-only status, if at all possible." From the available evidence, it appears that NASA provided the requested support for all of these spacecraft that reached orbit.[34]

NASA reported in early 1965 that between 15 and 20 percent of the workload of the former Minitrack network (now designated Space Tracking and Data Acquisition Network, or STADAN) was in support of DoD. It did not object to this, but believed DoD had to submit requests for assistance earlier and had to develop priorities for NASA's help when they sought it for multiple payload launches.[35]

STADAN's support increased during the next few years as it tracked, received telemetry from, and/or commanded additional Air Force Tetrahedral Research Satellites, Air Force Environmental Research Satellites, an Air Force Octahedral Research Satellite conducting a cold welding test in space, Navy satellites carrying gravity gradient and TEMPSAT experiments, and Army Sequential Collation of Ranges navigation/geodetic satellites. The available documents indicate that all the assistance was unclassified. At times DDR&E submitted a request to NASA without identifying the satellite, which possibly meant that the program was classified. For example, in August 1970 it asked that the STADAN station in Alaska acquire telemetry from an unspecified spacecraft during six passes early the following month. NASA also occasionally provided support for just the DoD launch vehicle and not the payload. Along these lines, in 1970 DDR&E asked that the MSFN stations in Australia track and receive telemetry from the Titan IIIC boosters carrying Defense Satellite Communications System satellites beginning later that year. All the data was unclassified.[36]

DDR&E asked in 1967 that the MSFN's station on Bermuda (a British possession) give tracking support to testing both Poseidon submarine-launched ballistic missiles and Minuteman III intercontinental ballistic missiles at the Eastern Test Range. Unspecified data would be classified, but DDR&E was confident that the Navy could work with NASA to devise secure handling procedures and that the British would not object to the use of the Bermuda station for military purposes. In an exchange of notes with the United States, the British government approved. There were 19 Minuteman III tests from 1968 to 1970 and well over 100 Poseidon tests from 1968 into the 1980s. The MSFN station in Bermuda provided assistance to most, if not all. It also furnished assistance to an unknown number of Trident I submarine–launched ballistic missile tests beginning in 1977. Tracking support for selected missile tests at the Eastern Test Range was also given by the Wallops Island launch complex. The MSFN Kokee Park station in Kauai provided a wide range of assistance to DoD missile tests, including tracking intercontinental ballistic missiles launched at the Western Test Range from Vandenberg Air Force Base to Kwajalein and tracking and receiving telemetry from surface-to-air and air-to-air missiles at the Pacific Missile Range.[37]

Relations between the DoD and NASA were not always cordial and cooperative regarding the issue of management and operation of their networks. In 1963 and 1964, the Aeronautics and Astronautics Coordinating Board's Space Flight Ground Environment Panel recommended several actions to try to achieve economies, including consolidating several NASA and DoD ground stations that were co-located or in close proximity and possibly standardizing NASA and DoD ground station tracking and telemetry systems. The DoD favored such steps, believing that substantial economies could be achieved. However, NASA's leadership strongly believed that their networks were an integral part of their spaceflight programs and should remain totally under NASA control. As a result, none of the recommendations was implemented.[38]

The issue of consolidation and duplication resurfaced in 1965 when Congress reviewed NASA's request to build a MSFN station in Antigua, where there was already an Eastern Test Range station. The chair of the House Subcommittee on Military Operations wrote McNamara and Webb asking about the possibilities of expanding the existing Air Force station to serve NASA and whether studies had been made of single management of other co-located facilities at Ascension Island and Grand Bahama Island. Months of negotiations followed between NASA, the Air Force, and DDR&E. In May 1965, Seamans and Brown signed the "Agreement between the Department of Defense and the National Aeronautics and Space Administration Regarding Their Respective Land-Based Tracking, Data Acquisition, and Communication Facilities," which largely settled the matter in NASA's favor. Among other things, the agreement provided that separate maintenance and operation of co-located facilities would continue when an organization sought this but that logistical support would be the responsibility of one or the other. The agreement also transferred the Navy's Kokee Park station in Kauai to NASA, with the provision that NASA would satisfy DoD requirements for support from it. Kokee Park had previously operated about 90 percent of the time in support of NASA and the remainder for the Navy and Air Force in support of Western Test Range and Pacific Missile Range activities. Similarly, pursuant to the agreement, NASA assumed a minor Eastern Test Range station in Bermuda and a Western Test Range station on Canton Island.[39]

Webb and Clark Clifford, the new secretary of defense, discussed in the summer of 1968 various means of jointly cutting costs, one of which involved possibly combining the civilian and DoD networks. John Foster, DDR&E, and Homer Newell, associate administrator of NASA, soon signed an agreement directing the Aeronautics and Astronautics Coordinating Board's Space Flight Ground Environment Panel to conduct a study looking at what consolidations could be made within each organization's network and between the two organizations' networks.[40]

The study, completed later that year, recommended only minimal consolidations based primarily on the different mission objectives, orbital characteristics, technical configurations, and concepts of operations. The most significant result of the study was reducing NASA assistance from 14 to 7 DoD satellites on the basis that sufficient data had been collected to accomplish mission objectives.[41]

Details of NASA network support to DoD satellites after the early 1970s are extremely sketchy. All that is known in this regard is that NASA informed Congress in 1974 that it had reduced the assistance from the 1960s and that all of it was unclassified.[42]

NASA and DoD independently began feasibility studies on data relay satellites in the mid-1960s. These spacecraft were intended to perform the tracking, data acquisition, and command and control functions of ground stations and eliminate the need for many. The promised benefits over ground stations were many: increasing the rate of data transmission to and from spacecraft, decreasing the vulnerability of transmissions to interception and jamming, providing continuous contact between the network control centers and satellites (which was impossible at the time with low-altitude satellites due to the brief periods of contact by the ground stations), and eliminating the possibility of military and political action against overseas ground stations.[43]

NASA employed RCA and Lockheed as contractors, while DoD used TRW and General Electric. The Space Flight Environment Ground Panel acted to ensure coordination and full exchange of information between NASA and DoD in the area, although it was not anticipated that a single data relay satellite system would be built to service both NASA and DoD. Beginning in 1966, personnel from both organizations attended all the contractor presentations and had full access to all the technical reports generated by the other. The Space Flight Environment Ground Panel's Development Sub-Panel sponsored a lengthy review of NASA and DoD developments at the Jet Propulsion Laboratory in early 1969.[44]

While there is little information available on subsequent cooperation regarding data relay satellites, it must have been substantial. NASA launched *Applications Technology Satellite-6* into geosynchronous orbit in 1974. It was the first known spacecraft to track and relay data from another (NASA's *Nimbus 5* and *Geodynamics Experimental Ocean Satellite 3*). In 1976, the NRO launched the first two Satellite Data Relay System satellites into highly elliptical orbits at an inclination of 63 degrees to relay data to and from the KH-11 digital return imagery intelligence satellites. It placed additional ones in orbit during the following years. NASA deployed its first dedicated data relay satellite, *Tracking and Data*

Relay Satellite 1, in 1983. It successfully launched eight more from 1988 to 2002.[45]

Although it is widely believed that the Tracking and Data Relay Satellite System was built and operated for civilian purposes, there is some evidence that the national security agencies were key users. This includes an Office of Management and Budget document from 1979 that stated, "A draft MOU was developed in early 1979 by DOD and sent to NASA regarding the DOD use of the TDRSS. Discussions have subsequently taken place on the TDRSS MOU. It is our understanding that the major point-of-difference involves resource allocation in an emergency situation." A 1982 NASA "TDRSS Program Security Classification Guide" mandated that "association of TDRSS and associated network control systems with specific military satellite programs" be classified at the Secret level. Various imagery intelligence satellites launched beginning in the late 1980s reportedly utilized the Tracking and Data Relay Satellite System.[46]

NASA's Weather Satellites Attempt to Meet Defense Requirements

NASA's weather satellites became much more capable during the 1960s. It launched *Tiros 3* in July 1961 and *Tiros 4* in February 1962 into the same 48 degree orbits as the first two. Their operational lifetimes were a little over 100 days. From June 1962 to June 1963, it placed the next three into 58 degree orbits, and *Tiros 7* was the first to operate for over a year. Even at the higher inclination, however, the satellites still could not sense all of the polar regions. Each carried one or two television cameras and occasionally infrared radiometers that could be used to determine temperatures of clouds, their height, and moisture content. They did not have a direct readout capability, and the daytime cloud cover imagery and other data acquired continued to be recorded on tape and transmitted to the ground stations in New Jersey or Alaska when within range. These stations in turn forwarded the data to the U.S. Weather Bureau's analytical center in Suitland, Maryland. *Tiros 8*, placed in a 65 degree orbit in late 1963, had one television camera and the first automatic picture transmission camera. The latter continuously broadcast its daytime cloud cover photographs. More than 45 inexpensive ground stations (including several U.S. military ones) around the world read them out when the satellite was within range. Because the original satellite design was still used, however, the cameras were still pointed toward the Earth only 25 percent of the time. NASA launched the last two satellites in the program, *Tiros 9* and *10*, in 1965. They were the first to be boosted into near-polar orbits and also the first to utilize the cartwheel configuration (developed in the Defense Meteorological Satellite Program discussed below), which permitted nearly complete coverage of the Earth on a daily basis. Each carried two television cameras, whose imagery continued to be recorded on tape and downlinked to one of the two U.S. ground stations.[47]

International participation in Tiros and the follow-on programs grew tremendously during the early 1960s. Many foreign nations acquired meteorological analyses directly from the U.S. Weather Bureau via a special facsimile network. A number of countries installed equipment to obtain the imagery directly from the automatic picture transmission camera on *Tiros 8* when it was within range. The United States and the Soviet Union signed an agreement in June 1962 to begin the exchange of data from their weather satellites, to be followed in 1964–65 by the co-ordinated launching of a system of operational spacecraft. Progress was slow, however. A Washington-Moscow Meteorological Circuit was built in 1964, but the Soviets did not launch their first successful weather satellite until June 1966, and the United States withheld sending any data until this time. Thereafter, it sent cloud photography and analyses from the follow-on Environmental Science Services Administration (ESSA) satellites. The Soviets sent only limited amounts of data, which was generally of poor quality due to technical problems with their satellites and the circuit. In 1968, the two countries held talks that led to improvements in the circuit. This fact, along with more capable Soviet spacecraft, resulted in a greater exchange of data.[48]

The Tiros program did not satisfy many national security requirements, of which the most vital was acquiring weather data over Sino-Soviet Bloc targets shortly before the imagery intelligence satellites overflew them to program their cameras and increase the amount of cloud-free photography collected. However, civilian requirements dictated that the Tiros satellites cross over the Eurasian landmass at times that were not optimal to achieve this. Additionally, they had limited coverage until adopting the cartwheel configuration. Meteorological data for tactical applications was an important but secondary national security requirement. In this regard, only *Tiros 8* carried an automatic picture transmission camera that could be read out by operating forces around the world. However, the image quality was poor and the coverage restricted because the satellite did not utilize the cartwheel configuration.

The defense and intelligence agencies closely collaborated with NASA and the U.S. Weather Bureau on a proposed National Operational Meteorological Satellite System designed to eliminate these limitations and meet all civilian and national security requirements. The interagency Panel on Operational Meteorological Satellites in early 1961 issued an unclassified plan for the system. Under it, NASA was responsible for the development, launch, and operation of the planned Nimbus satellites, which would be placed in near-polar orbits with an automatic picture transmission camera, advanced vidicon camera, and various infrared sensors. The U.S. Weather Bureau was to set the meteorological objectives and process and disseminate the data. The proposal quickly won White House and congressional approval, and General Electric was selected as the contractor.[49]

During the first phase from mid-1962 through the end of 1963, the plans were to have one satellite in orbit most of the time providing global cloud cover imagery at about noon and midnight local time. Three ground stations (Alaska, the eastern United States, and Western Europe) would acquire the stored daytime cloud cover imagery from the advanced vidicon camera system and transmit it for analysis to the U.S. Weather Bureau. U.S. government organizations and foreign nations could establish small fixed or mobile stations with inexpensive equipment anywhere in the world that could directly read out daytime cloud cover photographs of their region from the automatic picture transmission camera. High-resolution infrared radiometers would map the nighttime cloud cover, and this data would both be recorded for downlinking to the three main ground stations and broadcast continuously for direct readout by the fixed and mobile automatic picture transmission camera stations. Two Nimbus would be in orbit most of the time during the second phase in 1964–65, which together would provide global cloud cover photography four times daily. These spacecraft would also begin carrying additional sensors to measure surface pressures, winds, sea conditions, atmospheric temperatures, and other conditions. Two satellites would be in orbit at all times during the third phase beginning in 1966, and it was planned that the system would be declared operational at this point.[50]

The Joint Chiefs of Staff set forth the military requirements for the National Operational Meteorological Satellite System in a separate, classified annex. To satisfy both strategic and tactical needs, these dictated that the program at a minimum must provide global cloud cover photographs twice a day, direct readout of cloud imagery over local areas of tactical interest four times a day, high-resolution infrared measurements of nighttime cloud cover, and surface and atmospheric temperatures. They also strongly recommended that the system incorporate security equipment to prevent enemies from taking command and control of the satellites, acquiring the data, or jamming them.[51]

The NRO was skeptical of the claims made for Nimbus, both with respect to its capabilities and expected operational date of 1966. Because of this and the inability of Tiros to meet the critical strategic requirement of timely and accurate data over the Sino-Soviet Bloc targets, in the summer of 1961 the NRO started an interim program (ultimately called the Defense Meteorological Satellite Program, or DMSP) to build and launch four meteorological spacecraft to begin providing such information the following year. The program would continue only until the National Operational Meteorological Satellite System proved it could meet this strategic requirement.[52]

The NRO selected RCA, the manufacturer of Tiros, to build the satellites and directed that technology from NASA's program be used whenever possible. Launched into near-polar orbits on Scout boosters obtained from NASA, DMSP was the first to employ the cartwheel configuration that enabled almost continuous monitoring of the Earth. The television cameras would provide 100 percent daily coverage of the Northern Hemisphere at latitudes above 60 degrees and 55 percent coverage at the equator. In contrast to NASA's satellites, DMSP would be a secure system through encryption of downlinks and possibly uplinks. Imagery would be recorded on tape and downlinked to ground stations at Loring Air Force Base in Maine or Fairchild Air Force Base in Washington when the satellite was within range. The two stations would relay the photography to the Air Weather Service's Global Weather Center at Offutt Air Force Base in Nebraska for analysis. Personnel there would combine the DMSP data with that from NASA satellites and other sources and prepare weather forecasts that would be forwarded to the Air Force Satellite Control Facility (which controlled imagery intelligence satellites) and other military commands around the world.[53]

The first launch in April 1962 failed, but the second in August of that year succeeded. It provided photographs of the Soviet Union each day at noon until it stopped operating in March 1963. Two of the three launches in 1963 failed, and the third put the spacecraft in the wrong orbit. All utilized NASA's Scout booster. As a result, the NRO never employed them again and began using several versions of modified Thor intermediate-range ballistic missiles as the launch vehicle. The new boosters successfully placed both DMSP payloads in orbit in 1964.[54]

DMSP was a Special Access Program that apparently required its own clearance and a demonstrated need-to-know to learn of the existence of the program or to access the product. This was evidently done, since authorities believed the United States was obligated under various international agreements to freely share its weather satellite data, and there was absolutely no intention of doing so with DMSP data. A NRO history discloses that only three senior administrators of NASA (presumably Webb, Dryden, and Seamans) had the necessary clearance at the beginning of the program. It strongly implies that these officials opposed any additional NASA personnel receiving them but does not explain why. This caused major difficulties in working with the agency on the procurement of Scout boosters and other issues because NASA personnel were ignorant about the ultimate use.[55] More NASA personnel undoubtedly received the necessary clearance to gain access to the DMSP program before it was declassified in 1973.

The Nimbus program was meanwhile experiencing major problems in developing the camera systems and some other components. Webb wrote DDR&E Harold Brown in early 1963 asking for a DoD assessment on its possibility of meeting the requirements previously established by the Joint Chiefs of Staff. John Rubel, Brown's deputy, replied that although the program would meet the requirement for direct readout of local weather data at noon by military forces around the world, it would not satisfy the growing need for direct readout of local weather data during the early morning and late afternoon. He noted that modifying the satellites to meet these additional requirements might prove prohibitively expensive and interfere with civilian missions. As a result, it might be necessary to supplement Nimbus with a new, simple, and less costly NASA or DoD weather satellite program incorporating the automatic picture transmission camera system. The DoD was in the process of purchasing 27 readout stations that could be used with either Nimbus or a newly proposed satellite system. With respect to the security concerns, Rubel stated that the satellites should have a classified disabling or destruct device until a more advanced protection system could be built. The National Security Agency was developing such a system and would continue working with the three services and NASA on the matter.[56]

Extensive discussions among DoD, NASA, and U.S. Weather Bureau officials in 1963 focused on modifying Nimbus to provide the data needed for tactical operations. The Joint Meteorological Satellite Advisory Committee examined the issue and concluded that the program could not furnish it. Because of this and the severe delays, the U.S. Weather Bureau and the DoD withdrew and NASA decided to continue it as an R&D effort. It launched *Nimbus 1* in August 1964 into a near-polar orbit with three advanced vidicon cameras, one automatic picture transmission camera, and a high-resolution infrared radiometer. Employing the cartwheel design developed for DMSP, it was able to provide total photographic coverage of the Earth. There is no evidence that the spacecraft had a classified emergency disabling or destruct device. Although it was the most capable meteorological satellite NASA had launched to date, it ceased functioning after about two months due to solar panel problems. NASA did not launch *Nimbus 2* until May 1968.[57]

With Tiros and Nimbus able to meet few of the DoD's tactical weather requirements, the Joint Chiefs of Staff essentially recommended to the secretary of defense in early 1964 that the DoD build and operate its own meteorological satellite program to do so. No approval was obtained; however, the newly established Joint Meteorological Satellite Program Office within the Air Staff persuaded the NRO to conduct limited tests of DMSP support for tactical forces. These were successful, but improvements in photographic resolution, ground terminals, and other areas were necessary before an operational system could be achieved.[58]

With extensive input from the Joint Meteorological Satellite Program Office, NASA and the U.S. Weather Bureau were developing plans for the second-generation Tiros Operational System (TOS) because it was clear that Nimbus was not going to meet many of its objectives. NASA would procure and launch the satellites and then turn over their management to the U.S. Weather Bureau once they were checked out on orbit, a practice that would continue through the decades with all polar-orbiting civilian weather satellites. DDR&E concluded in April 1964 that TOS would provide an "acceptable minimum interim capability for the DoD" if three stored data and three automatic picture transmission satellites were launched between October 1965 and October 1966, it employed infrared sensors for nighttime weather observations as soon as possible, it carried classified equipment for "a measure of security in command and control for use during periods of tension or hostilities and to limit unauthorized access to the spacecraft or data," and its power level was increased to provide protection against jamming and make it easier for military automatic picture transmission stations to receive signals. Even with TOS, however, DDR&E emphasized that DoD might need its own weather satellite program for use in emergencies.[59]

Brockway McMillan, NRO director, forwarded to the Joint Meteorological Satellite Program Office in early 1965 the weather satellite requirements for both strategic and tactical applications. Among other things, the spacecraft had to provide global and continuous coverage, one nautical mile ground resolution, coverage twice a day, prevent "enemy intrusion or control of satellite," prevent if feasible "passive acquisition of data by enemy," and provide if possible security against "enemy jamming." Strategic applications did not require a direct readout capability, only a stored data capability for use with the command and control ground stations in the United States. In contrast, tactical applications required both.[60]

NASA launched the first TOS satellite in February 1966. As with all of the spacecraft in the series, it employed the cartwheel design originally developed in the DMSP program and flew in a near-polar orbit that enabled complete coverage of the Earth. Like Nimbus, the program fell far short of meeting the national security requirements established by DDR&E in 1964 and the NRO the following year. Among other things, it had fewer satellites in orbit than needed, their crossing times over the Eurasian landmass were not optimal for programming imagery intelligence satellite cameras, their resolution was poorer, and they did not carry any security devices.

The first TOS was designated *ESSA-1* for the new Environmental Science Services Administration, which had replaced the U.S. Weather Bureau. ESSA assumed command and control once it was declared operational shortly after launch. The satellite carried two advanced vidicon cameras designed for Nimbus that produced daytime cloud cover imagery at a higher nadir resolution than any Tiros. It transmitted the photographs stored on tape to one of the two ground stations in the United States, which in turn relayed them to ESSA and the Air Weather Service's Global Weather Center for analysis.[61]

NASA placed *ESSA-2* with an automatic picture transmission system in orbit later in February, and it continued operating until June 1968.

It continually transmitted daytime cloud cover pictures covering an area of about 2,000 square miles each at a nadir resolution of about two miles, which could be directly read out by small ground terminals when within range. U.S. civilian and military organizations had 50 such terminals worldwide, and foreign governments operated another 30. In the critical Southeast Asia theater, *ESSA-2* provided valuable morning cloud pictures directly to Air Force, Army, and Navy ground terminals, but it could not furnish any afternoon pictures, which were also needed. A few aircraft carriers also installed terminals to give them access to the imagery from the automatic picture transmission camera. DDR&E asked NASA in March 1966 if *Nimbus 2* could provide the afternoon photos, citing NASA's prior assurances that this would not affect the program. However, the launch of *Nimbus 2* was delayed until May 1968 and failed. NASA launched eight more ESSA spacecraft from 1966 to 1969. The odd-numbered ones continued carrying only advanced vidicon cameras, while the even-numbered ones continued carrying only automatic picture transmission cameras. *ESSA-8*, launched in late 1968 and deactivated in March 1976, had the longest operational life.[62]

With the failure of NASA's satellites to satisfy any strategic and many tactical weather requirements, DMSP became a permanent program. The NRO successfully launched four of six Block 1 and Block 2 satellites during 1965 and 1966. Their vidicon camera produced a resolution along the flight path as high as 3.5 miles. With two satellites finally in orbit at the same time, one passed over the USSR at about 7:00 a.m. local time and the second about 11:00 a.m. local time. These crossing times, which best satisfied the primary mission of programming the cameras on imagery intelligence satellites, were continued in all the subsequent flights. DMSP first began supporting tactical forces when the one that reached orbit in March 1965 transmitted Southeast Asia weather data to a new ground station in Saigon for the three months it was in operation. The NRO launched a special Block 3 with direct readout capability in May 1965 to exclusively serve U.S. forces in that region, although naval vessels could not directly access the data because they could not accommodate the trailer-sized ground terminal. This satellite, the only one ever launched which did not acquire data for strategic applications, operated until February 1967. The NRO successfully launched all seven Block 4 spacecraft from 1966 through 1969. Each carried two vidicon cameras, which provided a greatly improved nadir resolution of 0.9 miles and a resolution of 3.5 miles at the edges of the 1,725-mile swath. They also flew various infrared sensors, which began furnishing some nighttime data, and continued to transmit stored pictures of the entire northern hemisphere to one of the two ground stations in the United States and local imagery directly to tactical forces equipped with ground terminals. It is important to note that during this period, DMSP ground terminals could not receive data from ESSA satellites and vice versa.[63]

NASA and ESSA were developing the more capable third-generation Improved Tiros Operational Satellite (ITOS) and scheduled the first launch for late 1968. In part to meet the new, classified military requirements for meteorological satellite data established by the Joint Chiefs of Staff in late 1966, the spacecraft was to carry both an automatic picture transmission camera and the advanced vidicon camera system, provide even higher resolution photographs, and carry a scanning radiometer for nighttime coverage. Adm. Walter F. Boone, NASA's assistant administrator for defense affairs, wrote DDR&E in early 1967 stating that the launch planned for late 1968 would be delayed by six months unless DoD upgraded the project's priority rating from DO to DX (the latter was reserved for programs of the highest national interest). He stated that the contractor, RCA, was having difficulty acquiring parts such as quartz crystals and relays with the DO rating. DDR&E quickly replied that although DoD eagerly awaited the increased capabilities of ITOS, it could not give it the higher rating. It believed that if NASA and RCA utilized the expedited procurement procedures under the DO rating, a launch delay could probably be avoided.[64] As it turned out, NASA did not launch *ITOS 1* until January 1970.

Improved ground terminals had to be developed to receive the ITOS automatic picture camera signals because those used with *ESSA-2, ESSA-4, ESSA-6*, and *ESSA-8* could not handle the increased volume of data. In late 1967, DDR&E directed the Joint Meteorological Satellite Program Office to work with the individual services and NASA in this effort to ensure that military units would be able to receive photography from the ITOS automatic picture cameras.[65]

NASA and ESSA also flew some meteorological experiments onboard the former's Applications Technology Satellites (ATS). These experiments provided the first meteorological data from geosynchronous orbit and tested sensor systems for possible incorporation in the planned Geostationary Operational Environmental Satellites. *ATS-1*, placed in orbit in December 1966, carried a spin scan cloud camera that provided continuous photographic coverage of much of the Pacific Ocean and the western United States between the latitudes of 52 degrees north and south. The resolution was a little less than two miles. It operated until 1978. NASA launched *ATS-2* with the same camera in April 1967, but it failed to reach the correct orbit and was quickly deactivated. *ATS-3*, put in orbit in November 1967, carried a spin scan cloud camera that produced color imagery for the first time. It too operated until 1978.[66]

Following the success of the ATS experiments, NASA planned to acquire two Geostationary Operational Environmental Satellites and to place the first in orbit above the Western Hemisphere in 1972 (its launch was delayed until 1975). In March 1970, the Joint Chiefs of Staff assessed the potential military value of the system as minimal. Among other things, it stated that a single satellite in the planned orbit would be limited in coverage to North and South America and the adjoining oceans and that few DoD overseas military weather stations could receive its data. Moreover, the photographic resolution of most areas in North and South America would be inadequate for tactical requirements. A three- or four-satellite constellation in orbit would greatly lessen the communication limitations with overseas stations but would not alter the poor photographic resolution. Additionally, the Joint Chiefs of Staff believed that the planned international participation in the program introduced "security and operational control considerations which, for national security purposes, militate against reliance on such a system."[67]

NASA launched the third-generation polar-orbiting *ITOS 1* in January 1970. When declared operational six months later, it turned over the operation to the new National Oceanic and Atmospheric Administration (NOAA). It placed the next five, designated *NOAA 1* through *NOAA 5*, in the same orbit from late 1970 to 1974. Although *ITOS-1* and *NOAA-1* carried two automatic picture cameras, two advanced vidicon camera systems, and two radiometers, the subsequent ones only flew three improved radiometers that measured such conditions as atmospheric temperatures and moisture and for the first time produced both daytime and nighttime cloud imagery. The highest resolution along the flight path was a little over one mile. The sensors enabled global cloud cover observations every 12 hours, compared with the 24-hour period of the predecessor *ESSA-1* through *ESSA-9* satellites. Data from the automatic picture cameras and selected radiometers was directly read out by numerous U.S. civilian and military ground terminals worldwide. Data from the advanced vidicon camera systems and other radiometers was transmitted to one of the two ground stations in the United States and then relayed to NOAA and the Air Weather Service's Global Weather Center for analysis.[68] There is no evidence that the satellites carried any security devices that the DoD had recommended.

By the early 1970s, DMSP satellites were apparently satisfying nearly all strategic weather requirements and many tactical ones as well. Their resolution (due to better cameras and lower orbits), pointing accuracy, nighttime capability in the visible light range, flexibility, and data delivery times were greater than the civilian satellites. Nevertheless, the latter were helping to meet at least some tactical weather requirements as evidenced by the large number of DoD direct readout terminals for ITOS-1 and its successors.[69]

The Joint Chiefs of Staff issued a new list of military requirements for data from both DMSP and NASA satellites in the summer of 1970. These were forwarded to the secretary of commerce and NASA's administrator for use in the design of the fourth-generation TIROS-N satellites, of which the first was to be launched in 1974 or 1975.[70]

The Joint NASA-DoD National Geodetic Satellite Program

The science of geodesy determines the Earth's size, shape, and mass, variations in its gravity, and the distances between and locations of points on it. It has both civilian and national security applications. During the Cold War, geodesy was critical to the DoD because of its importance in the guidance and targeting of long-range missiles by determining the location of the targets in relation to the launch points and determining gravitational anomalies that affected trajectories. This was especially true in the case of hardened targets, which required the warheads to detonate very close to them to ensure destruction.[71]

The science had always relied on surface and airborne measurement techniques, but they could only produce datums such as the North American, European, and Tokyo datums, which were continentally based and nationally centered. Neither a unified world reference system nor accurate mapping of the Earth's gravity field was possible using these two methods. In the late 1950s, the Joint Chiefs of Staff established a Geodetic Objective Plan to produce increasingly accurate world reference systems (designated World Geodetic Systems) and gravity maps over the next 20 years. The DoD generated the first in 1960, almost exclusively from surface and airborne measurements.[72] It is unclear whether all or part was classified.

Satellites were essential to achieving these objectives because they enabled more accurate geodetic measurements by acting as reference points in space for triangulation calculations. Additionally, the orbital perturbations of satellites could be measured to detect variations in the Earth's gravitational field, since these cause dips and rises in their flight path. Scientists began obtaining geodetic data from spacecraft during the International Geophysical Year in 1957–58 when they imaged *Sputnik 2* and *Vanguard 1* against the star background. NASA launched the *Beacon 1* and *Beacon 2* inflatable balloon satellites in 1958 to be photographed by NASA and DoD cameras, but neither reached orbit. Both organizations imaged NASA's *Echo 1* passive communications satellite after its successful launch in 1960.[73]

This limited photography clearly demonstrated the value of satellites to geodesy, and both civilian scientists and the DoD desired to greatly expand this effort. DoD components soon developed various geomeasuring devices for placement onboard spacecraft—flashing beacons by the Air Force, Doppler transmitters by the Navy, and Sequential Correlation of Range (SECOR) transponders by the Army. Each of the services began building its own worldwide network of mobile ground-based instruments to gather data from their particular device.[74]

The NASA-DoD Geodetic Satellite Working Group prepared a report in early August 1960 on a joint geodetic satellite project designated ANNA (Army, Navy, NASA, Air Force) and presented it at a meeting of the Aeronautics and Astronautics Coordinating Board's Unmanned Spacecraft Panel later that month. The spacecraft would carry a Navy Doppler transmitter, Air Force flashing beacons, and an Army SECOR transponder. Among other things, the report concluded that a successful program would likely reduce the total geodetic error on a global basis to 500 feet in three to five years. It recognized the DoD's strong desire to keep most raw and finished geodetic data classified and NASA's opposition to participating in the program under these circumstances. The report recommended that this dilemma might be resolved in part by reserving 50 percent of the lifetime of ANNA's flashing beacons for NASA's use in unclassified scientific projects and the possible release to NASA of selected data acquired by the DoD.[75]

The Unmanned Spacecraft Panel continued to consider ANNA at subsequent meetings, and the DoD representatives maintained that some geodetic data must be classified because the United States was ahead of the USSR in the field, it was in a better position to obtain worldwide data because it had an international network of geodetic stations while the Soviet Union did not, geodetic data was critical to ballistic missile accuracy, and the USSR would not reciprocate by releasing its data. Since NASA would not drop its opposition to the classification of data, however, the panel ultimately recommended that the DoD undertake a geodetic satellite program. It also recommended the following: (1) data from the DoD program should be made available on a classified basis to NASA and any unclassified scientific results that could be published openly but without the supporting data, (2) declassification of data from the DoD program would be made on a case-by-case basis, and (3) NASA should participate in the planning of the DoD program and the analysis of the data it acquired.[76]

The DoD proceeded with ANNA and initially planned to launch four spacecraft into orbits at different inclinations and altitudes. It complemented a separate DoD effort involving placing Doppler transmitters on the Navy's Transit navigation satellites (the first of which was launched in April 1960) and the launching of the Army's SECOR satellites (of which 13 were launched from 1962 to 1969).[77]

NASA participated minimally in ANNA due to the DoD's policy on classifying the data. Civilian scientists from outside the government voiced strong criticism of this policy to Jerome Wiesner, special assistant to the president for science and technology. DDR&E Harold Brown wrote to Wiesner in late 1961 that with Soviet advances in geodesy the ANNA data would assist them very little and that a classified geodetic satellite program was in direct conflict with ongoing international efforts to share information in the field. However, he maintained that there were a few narrow categories of information that required continued classification.[78]

Under pressure from these various parties, the DoD reluctantly announced in April 1962 that it would no longer classify the data, and NASA quickly agreed to participate in the program. As best as can be determined, the decision covered all the ANNA data, including the narrow categories that the DDR&E had argued a few months earlier must remain classified. The following month the DoD launched *ANNA 1A*, but it failed to reach orbit. In November, it successfully placed *ANNA 1B* in orbit. It carried the Navy Doppler, Air Force flashing beacons, and Army SECOR transponders. Around the world, 16 Air Force stations with cameras, 14 Navy sites with Doppler equipment, 4 Army stations with SE-COR equipment, and 26 NASA sites with cameras acquired data from *ANNA 1B*. Although the SECOR transponder soon malfunctioned, the Doppler and flashing beacon systems continued to furnish data for several years. The DoD quickly turned over the spacecraft's management to NASA.[79]

The two organizations soon agreed that NASA would assume responsibility for a national geodetic satellite program and created the joint Geodetic Satellite Policy Board to coordinate efforts in this area. NASA's Goddard Space Flight Center would receive all the raw data acquired under the national program by NASA's cameras and some of the data under it collected by DoD's cameras and other ground instruments. However, the DoD was not going to rely solely on this joint effort to meet its geodetic requirements. It continued the Transit and SECOR satellite programs; ground, ocean, and airborne surveys; and acquisition of data from foreign governments. The DoD would unilaterally decide on the public release of the raw data from its worldwide network of stations acquired from NASA and DoD satellites, the techniques for analyzing the data, and the finished product.[80]

The DoD submitted its requirements in 1963 for future NASA geodetic satellites under the national program. NASA's plans to launch several active and passive geodetic satellites in the next four years met these and civilian scientific requirements, but the DoD urged NASA to accelerate the program because it required the data for weapons targeting and other purposes as soon as possible.[81]

Seamans and DDR&E signed an agreement in early 1964 creating the National Geodetic Satellite Program, whose primary goals were developing a unified, worldwide geodetic reference network with an accuracy of 10 meters, creating a more precise model of the Earth's gravity field, comparing and correlating results from the different geodetic instruments on satellites, and improving positional accuracies of satellite tracking sites and calibrating tracking equipment. In March 1964, NASA attempted to launch the first satellite under the program, *Beacon Explorer I*, but the booster's third stage failed. The satellite was to be placed in an orbital inclination of 80 degrees at the Navy's request. NASA successfully launched *Beacon Explorer II* seven months later into that orbit to satisfy the Navy's requirement. Along with equipment for ionospheric research, the spacecraft carried a Doppler transmitter and was the first to be equipped with optical reflectors, which enabled ground-based lasers to track it for geodetic purposes. NASA placed *Beacon Explorer III* into an orbit with an inclination of 40 degrees in April 1965. It was the last in the series and carried the same instrumentation for both ionospheric and geodetic research.[82]

NASA launched the *Geodetic Earth Orbiting Satellite I* in November 1965 at an inclination of 59 degrees. It carried the most complete set of geomeasuring devices flown on one spacecraft up to that time. More than 100 DoD, NASA, and foreign ground stations received data from them. A December 1966 command system failure rendered all the devices inoperable, except for the optical reflectors observed by ground-based lasers. NASA placed the *Passive Geodetic Earth Orbiting Satellite,* a 30-meter plastic balloon coated with aluminum, into a near-polar orbit during June 1966. It carried no instruments and was simply a target for the cameras of 12 teams of Coast and Geodetic Survey and Army Map Service personnel moving among 40 sites around the world over the next five years. Due to the orbit, it allowed for the first time simultaneously viewing across the spans of oceans and contributed greatly to the establishment of a single worldwide geodetic reference system. NASA launched *Geodetic Earth Orbiting Satellite II* into a near-polar orbit in January 1968. It carried the same instruments as the first in the series did, along with an experimental transponder to test the effectiveness of C-band radar tracking to satellite geodesy. A large number of NASA, DoD, and foreign stations received data from the various instruments, although the flow was reduced after satellite power supply problems surfaced in 1969.[83]

This was the last vehicle under the National Geodetic Satellite Program. NASA planned to launch *Geodetic Earth Orbiting Satellite III* in 1970 or 1971 as the last satellite, but due to several factors this spacecraft did not reach orbit until 1975 under a successor NASA program.[84]

The DoD continued its own separate satellite geodesy program utilizing Transit and SECOR. In 1966, the U.S. Intelligence Board had established that by 1970 the positioning of Soviet targets should be accurate to within 450 feet horizontal and that elevations should be accurate to within 300 feet vertical, both with 90 percent assurance relative to the World Geodetic System. (The DoD had completed the second and largely classified World Geodetic System that same year, based in large part on optical and electronic tracking of satellites.) Whether the requirements for positioning of Soviet targets were achieved by 1970 and, if so, the contributions of the data DoD acquired and processed under the National Geodetic Satellite Program and its other efforts to this cannot be determined. By the end of the 1960s, the DoD had concluded that the Doppler system was best for satellite geodesy, and it continued placing the transmitters on Transit and several other satellites.[85]

The classification of certain geodetic information continued to be an issue between the DoD and NASA. A Defense Intelligence Agency manual from the late 1960s set forth the relevant guidelines. It mandated classification of DoD models of the Earth's gravity field more accurate than those of the National Geodetic Satellite Program, "selected parameters" of the DoD World Geodetic System, and geodetic data obtained from sources other than National Geodetic Satellite Program (which included DoD satellites, DoD airborne, ground, and ocean surveys, the oil exploration industry, and foreign nations). However, selected gravity data and geodetic data acquired from these other sources could be released on a case-by-case basis.[86]

For unknown reasons, Deputy Secretary of Defense Paul Nitze wrote President Johnson concerning the matter in November 1968, copying NASA's administrator, the special assistant to the president for science and technology, the deputy undersecretary of state for political affairs, and an unknown fourth official. Nitze noted that the core of the problem was that the DoD primarily required geodetic data for improving the accuracy of strategic weapons systems and accordingly classified the processes involved and the final product. NASA and the scientific community needed essentially the same data for such purposes as studying continental drift and determining the size and shape of the Earth and maintained that all the data and analytical techniques must be unclassified. Nitze pointed out that the USSR and its allies classified all of their geodetic and gravimetric data and did not publicly distribute any. He recommended that the National Security Council establish a policy giving security protection for gravitational models of the Earth and intercontinental positional relationships that were more accurate than provided by the National Geodetic Satellite Program, as well as advanced data reduction techniques and instrumentation technologies.[87] There is no evidence that either President Johnson or the National Security Council took any action.

The dispute between NASA and the DoD over classification and other aspects of the National Geodetic Satellite Program continued into 1969 and centered on three areas. First, NASA believed the number of control points (ground stations taking measurements) in the program had to be nearly doubled to achieve the goal of establishing a unified, worldwide geodetic reference network with an accuracy of 10 meters. The DoD considered additional stations to be valuable but unnecessary. Second, NASA wanted more extensive and accurate gravity models openly published and an additional satellite (*Geodetic Earth Orbiting Satellite III*) to provide needed data so that the National Geodetic Satellite Program's objectives in this area could be met. The DoD strongly believed its gravity models based on data from a number of sources were more accurate than any derived solely from National Geodetic Satellite Program data and firmly refused to release theirs because of the assistance they could give foreign nations in improving their long-range missile accuracy. It did not feel that NASA's proposed satellite would provide much additional data. Third, NASA wanted DoD to provide more of its observational data of the National Geodetic Satellite Program spacecraft and the products it derived. The DoD maintained that almost all of its raw data from tracking these satellites was being furnished to NASA. However, limited amounts from certain Air Force cameras and Navy Doppler sites could not be because of the potential for improving unclassified gravity models that would compromise DoD's classified models.[88] There was apparently no resolution of these issues at the time.

Summary

NASA expanded its existing relationships with the defense and intelligence agencies and developed new ones as well during the 1960s. Several did not involve NASA's spaceflight programs, including contributing to cover stories, supporting DoD missile tests and satellites, and collaborating on the recovery of U.S. and foreign space debris. Much of this assistance and cooperation was hidden or classified. In its applications satellite programs, NASA tried to build a weather satellite under a joint program to satisfy both national security and civilian requirements, but this effort failed. It continued developing its other polar-orbiting satellites in close coordination with the defense and intelligence agencies, and they furnished increasing amounts of data for tactical applications. NASA also launched geodetic satellites under a joint program to serve both national security and civilian interests, but they were only partially successful in satisfying the former, and conflicts arose over the classification of selected data.

The clash between national security and civilian interests was much more complex and intense in NASA's spaceflight programs designed to perform remote sensing of the Earth. The following chapter examines this interaction through the early 1970s and the severe restrictions under which NASA operated.

Restrictions on Remote Sensing from Space

NASA's programs to observe the Earth from space had both tremendous potential to benefit a wide range of civilian disciplines and massive national security implications. These concerns first arose with the Tiros weather satellites. However, the limited measures adopted to address them ended quickly after the national security agencies determined that the imagery did not reveal anything of intelligence value and did not provoke any international reaction.

The interaction concerning NASA's systematic land remote sensing programs designed to produce much higher quality photography was much more complex and, at times, strained. The national security community would not tolerate any threat to its growing fleet of reconnaissance satellites and had two concerns about NASA's plans: political and technological. With respect to the former, the Soviet's very public but unsuccessful propaganda campaign against reconnaissance from space ended in 1963 and it tacitly accepted the activity thereafter. Nevertheless, the defense and intelligence agencies were extremely worried about any actions of NASA—ranging from inadvertent public statements on the capabilities of the classified programs to releasing imagery of sensitive foreign sites—that would result in a renewed international effort to limit observing the Earth from space. Their second concern revolved around the possibility of NASA openly using classified technologies that would reveal U.S. capabilities, assist other nations in improving theirs, or help them in developing countermeasures. As a result, all of NASA's space-based Earth-imaging activities were closely monitored and numerous restrictions were imposed on them. No other NASA program operated under such severe constraints.

The U.S. government advanced the principle of freedom of space to protect the unrestricted operation of reconnaissance spacecraft whose capabilities and importance were growing. At the same time it was also building a wall of secrecy around these programs. NASA planned to conduct systematic land remote sensing with very sophisticated sensors in the Apollo Applications Program, the planned successor to the Apollo program in the human spaceflight field, and restrictions were placed on them. The first was a 1965 NASA-NRO agreement that limited the resolution of any image-forming sensors NASA could fly. A high-level interagency group on which NASA's deputy administrator sat reviewed the entire issue of NASA's remote sensing plans the following year. Although the NRO and others strongly argued that aircraft could better conduct Earth observation for civilian purposes, the group gave a cautious endorsement to a space-based program. It also reaffirmed the technical limitations previously imposed but recommended that these thresholds be relaxed as the state of commonly used technologies and worldwide acceptance of photography from space advanced. As it turned out, the thresholds remained in place for over a decade, although exceptions were granted in several cases. Two joint committees established under a NASA-DoD agreement reached later in 1966 monitored all of NASA's space-based imaging activities into the early 1970s. Their reach was very broad and extended from examining articles NASA proposed for publication to research and development on image-forming sensors.

NASA switched to a robotic land remote sensing satellite program as the Apollo Applications Program was severely cut back beginning in the late 1960s. The new program, subject to the same sensor restrictions, experienced delays in getting approved, and NASA did not launch the first satellite until 1972. To compensate in part for the fact that NASA was prohibited from acquiring any high-quality photography of the Earth, Project Argo was created in 1968 at the behest of President Johnson's science advisor and with the full backing of the national security agencies. It gave cleared researchers in certain federal civilian agencies access to selected classified overhead imagery, most of which was limited to the

United States. The project's record was mixed, but when it was disestablished in 1973, the effort continued under new mechanisms.

The Importance of Freedom of Space to U.S. Reconnaissance Satellites and Early U.S. Actions to Protect Their Unrestricted Operation

A key goal of the United States at the beginning of the space age was to ensure freedom of space to permit the highly classified satellites under development to operate without any international political restrictions. These included the Air Force's SAMOS imagery and electronic intelligence satellites, the Air Force–CIA's CORONA imagery intelligence satellite, the Air Force's MIDAS early warning satellite, and the Naval Research Laboratory's GRAB electronic intelligence satellite. These platforms would be absolutely essential in acquiring data on the growing Soviet military threat. Even before the United States launched its first successful reconnaissance satellites in 1960 (GRAB 1 and two CORONA), however, the USSR had initiated a vigorous propaganda campaign against these platforms. Not surprisingly, every attempt to reach agreement with the Soviet Union to guarantee freedom of space failed.[1]

Although all these satellite programs were classified, unofficial and official publicity revealed details of some. Several newspapers and periodicals in the late 1950s described SAMOS as a space-based platform to photograph the USSR and MIDAS as a means of giving early warning of a surprise attack. Senior Air Force officers also made statements along the same lines before Congress and in other forums. James Killian, special assistant to the president for science and technology from 1957 to 1959, warned that the publicity was forcing the Soviets to take action against them and strongly urged that the official publicity stop. However, a total blackout would not occur for several years.[2]

There were a few State Department officials during this time who called for greater openness concerning at least the imaging portion of the SAMOS program. They argued that the fact of photoreconnaissance from space should be admitted and perhaps some imagery be released as well, believing that these steps would lead to increased international acceptance of this activity and rebut Soviet criticism of it. Releasing photography that could be used for civilian purposes such as mapping and geology would be particularly useful in this regard. However, the CIA and DoD strongly disagreed, based on their long-standing opposition to revealing intelligence sources and methods and the belief that the release of any information would only cause the Soviets to intensify their campaign. Reflecting the sensitivity of this activity, President Eisenhower created the TALENT-KEYHOLE special security control system shortly after the first successful CORONA flight in August 1960. This system, managed by the DCI, gave additional protection to both the fact of photoreconnaissance from space and its product.[3]

During the same year that the classified satellites were enjoying their initial successes, NASA began conducting several open space-based photography programs. The first systematic effort started when it placed the *Tiros 1* weather satellite in orbit in April 1960. NASA launched nine more Tiros in the next four years. Earth and weather photography began in the human spaceflight program in 1960 with fixed cameras in several Project Mercury test flights. Astronauts in both the subsequent suborbital and orbital missions took more than 800 terrain and weather color photographs using 35-mm and 70-mm handheld cameras. The Earth images were particularly useful for mapping and geological studies. NASA's leadership was fully aware of the classified reconnaissance programs and, beginning in the early 1960s, a number of additional NASA personnel acquired the requisite clearances and learned of them.[4]

When John Kennedy became president there was still no comprehensive, uniform policy on what, if any, information should be disclosed concerning the classified satellite programs. SAMOS, MIDAS, and DISCOVERER (the official cover story for CORONA) continued receiving substantial official and unofficial publicity. To better coordinate the nation's strategic reconnaissance activities, the CIA and DoD agreed in September 1961 to create the National Reconnaissance Office. It was to manage the newly designated National Reconnaissance Program that included all "satellite and overflight reconnaissance projects whether overt or covert." The initial directors were the undersecretary of the Air Force and the CIA's deputy director (plans). Everything concerning the office was classified, even its existence and mission. The Kennedy administration soon moved to reduce the publicity about the classified space-based programs. When its initial actions did not significantly reduce the information released, the DoD issued a directive in March 1962 that ended the disclosure of virtually all information concerning its classified and even unclassified programs.[5]

The United Nations' General Assembly in December 1961 approved a resolution proposed by the United States that called on nations to register all objects launched into Earth orbit or beyond. The resolution did not specify the data to be supplied. Because the State Department had not coordinated the action with the CIA and other intelligence agencies, the measure took them by surprise and caused them great concern. A series of high-level meetings took place in the spring of 1962 trying to formulate a policy to comply with it. While the State Department advocated that such information as payload and mission be provided, the intelligence agencies strongly opposed this and favored furnishing only minimal data. Their position was adopted.[6]

The confusion and lack of coordination on the launch registration question illustrated the need for a permanent high-level committee to establish policy on such matters. President Kennedy signed National Security Action Memorandum 156 in May 1962 to establish one. It directed that the State Department form an interagency group to examine the pending space and disarmament negotiations with the Soviets "with a view to formulating a position which avoids the dangers of restricting ourselves, compromising highly classified programs, or providing assistance of significant military value to the USSR and which at the same time permits us to continue to work for disarmament and international cooperation in space."[7]

Formally known as the Committee to Establish National Policy on Satellite Reconnaissance, it was commonly referred to as the NSAM 156 Committee. The first chair was U. Alexis Johnson, deputy undersecretary of state for political affairs. Its other initial members were NASA's Robert Seamans; Joseph Charyk, undersecretary of the Air Force and (covertly) NRO director; Herbert Scoville, deputy director (research) at the CIA; Roger Hilsman, director of the State Department's Bureau of Intelligence and Research; Paul Nitze, assistant secretary of defense (international security affairs); Adrian Fisher, deputy director of the Arms Control and Disarmament Agency; and Carl Kaysen, deputy special assistant to the president for national security affairs.[8]

The group's July 1962 report contained 18 policy recommendations, of which all but one were approved by the National Security Council as NSC Action 2454 that same month. They included that the United States must maintain that outer space is free and that reconnaissance activities are legitimate and are "peaceful uses" of outer space. The report endorsed the present policy of not identifying individual launches by mission or purpose, but also stated there should be a "more open (but not more detailed) public reference to the general over-all military program" and its objectives should be described "in broad and general terms." However, it cautioned that there should be no disclosure of the "status, extent, effectiveness, or operational characteristics" of the reconnaissance programs at the current time. The report strongly advocated that studies be made on whether there is any releasable data, such as mapping information, which could help gain wider acceptance of photography from space.[9]

That same year the DCI established the BYEMAN special security control system to give additional protection to certain aspects of the reconnaissance satellite program. It covered sensitive technologies (including image-forming sensors), the manufacturing processes for selected components of these technologies, and certain operational details.[10] By the end of 1962, the government had constructed a huge wall of secrecy around its military and intelligence space programs. It released virtually no information other than the numerical or alphabetical designators of an individual launch. All unclassified discussions of payloads or missions had ceased.

Carrying out its recommendation from the previous year, the NSAM 156 Committee in June 1963 examined the possibility of releasing some degraded ARGON mapping camera photography to foster wider acceptance of imaging from space. (CORONA satellites carried 12 ARGON cameras from 1961 to 1964.) Based on the increasing international acceptance of the U.S. position that observation from space was lawful and on the fear of provoking the Soviets, though, the group unanimously opposed any disclosure.[11]

The strident Soviet campaign against reconnaissance from space ended in 1963 for reasons that were not completely clear to U.S. officials. Although the USSR was now tacitly accepting space-based reconnaissance, this did not alter the position of most U.S. officials against even admitting the activity. They now believed that to do so would probably force the Soviets to renew their campaign.[12]

Based on Premier Nikita Khrushchev's indications to several U.S. visitors in May 1964 that the Soviets now accepted reconnaissance from space, the NSAM 156 Committee met several times over the summer of 1964 to examine proposals to have an international organization receive U.S. and Soviet reconnaissance photography, privately share with the USSR some information concerning U.S. reconnaissance programs, or discuss reconnaissance from space with the Soviets at the United Nations. However, based on the almost unanimous opposition to even admitting the existence of reconnaissance from space, it ended up rejecting all of them.[13]

NASA's Initial Plans for Remote Sensing of the Earth and Restrictions on Them

The University of Michigan's Institute of Science and Technology began holding annual symposia on remote sensing from space in 1962, stemming in part from the scientific knowledge gained from Project Mercury photography. During the next several years, the National Academy of Sciences and the National Research Council formed three separate remote sensing committees, with their members coming from universities, industry, NASA, the Naval Oceanographic Office, and the Departments of Agriculture, Commerce, and the Interior. Each of the committees called for a greatly expanded program. Except for a small number of NASA personnel, however, the individuals involved did not have TALENT-KEY-HOLE or BYEMAN clearances and thus did not know of the existence or details of the covert imaging programs or have any access to their products or technologies.

NASA took steps in 1964 toward developing a formal remote sensing project. It began funding studies at universities and several federal civilian agencies, including the Agriculture Research Service, the U.S. Geological Survey, and the Naval Oceanographic Office. Because aircraft were excellent test platforms for Earth-observation instruments being designed for spacecraft, in late 1964 NASA acquired a Convair 240A for its newly designated Earth Resources Survey Aircraft Program. It acquired a Lockheed P-3A and Lockheed C-130 in subsequent years. The aircraft carried a number of sensors, including mapping and multispectral cameras and radar and infrared imaging systems. Working with other federal civilian agencies and university researchers, the planes flew over test sites in the United States to evaluate sensors and develop techniques for processing data.[14]

NASA planned at this time to conduct systematic Earth resources surveys as part of the Apollo Extension Systems human spaceflight program (soon renamed the Apollo Applications Program, or AAP). Designed as the follow-on to Apollo, AAP was to use Apollo launch vehicles, capsules, and other hardware to drastically reduce the cost. The program envisioned missions beginning in 1968 of up to 45 days in Earth orbit and subsequent missions as long as 28 days in lunar orbit and up to 14 days on the lunar surface. Both low-inclination and polar orbits would be utilized in the Earth-orbital flights, which would include scientific experiments involving astronomy, biomedicine, meteorology, and materials testing.[15]

Although NASA also had a modest effort studying remote sensing with robotic satellites, it greatly preferred performing this activity in the AAP because human spaceflight programs were larger and much more visible and prestigious. Furthermore, the Earth images in Project Mercury certainly had proved the value of conducting Earth resources surveys in a human spaceflight program. The approximately 1,400 Earth photographs taken in the crewed Project Gemini missions in 1965–66 would confirm this.[16]

Researchers from NASA and the other civilian agencies, as well as from outside the government, advocated utilizing some very sophisticated image-forming devices for the AAP Earth resources surveys, and this greatly concerned the national security agencies. Secretary of Defense Robert McNamara wrote James Webb in May 1965 expressing concern for the security of the National Reconnaissance Program technologies and proposed that the Air Force acquire and test all sensitive hardware in Earth orbit for NASA. In an interim response, Webb stated that he shared the security concerns but that he wanted to review the August 1963 Top Secret/BYEMAN "DOD/CIA-NASA Agreement on NASA Reconnaissance Programs" on security for NASA's lunar photography program (which is discussed at length in the following chapter) and certain proposed NASA transactions with the Departments of Agriculture and the Interior on remote sensing before making any final decisions. Webb sent his final reply in June. He described the interest of various federal civilian agencies in Earth resources surveys and gave assurances that prior to beginning any study or project with them involving sensors approaching reconnaissance quality that NASA would consult the NRO director. While assuring McNamara that any contracts or agreements with the federal civilian agencies would prohibit them from developing any hardware, Webb rejected his proposal for the Air Force to acquire and test hardware for NASA.[17]

Seamans and Brockway McMillan, the NRO director, reached a Top Secret/BYEMAN agreement in August 1965 limiting the sensors NASA could employ in its space-based imaging programs and setting forth the procedures to review NASA's activities in this area. There is no information available on exactly what led to the agreement. It can only be surmised that since Webb had rejected McNamara's proposal for the Air

Force to acquire and test all of NASA's planned high-quality sensors, the NRO and others wanted specific guidelines on what NASA could and could not do in this area and establish a mechanism to review them.

Under the agreement, three NASA employees would receive BYEMAN clearances, identify all NASA reconnaissance-related activities, and report them to Seamans (who would discuss them with the NRO). NASA had to inform the NRO of all expenditures of "NASA research and development money with a university or industry, or the transfer of NASA money to another government agency" for "the study, design, development, fabrication, test of reconnaissance-like sensors, or significant components thereof, for use in orbital systems, or studies of the use of such sensors in orbital systems." "Reconnaissance-like sensors" were defined as "image-forming sensors having an angular resolution of 0.1 milliradian or finer, or an optical or infrared image forming system with a physical aperture greater than 30 cm and an optical figure controlled to better than ¼ wave length." (This is the equivalent to a ground resolution of approximately 20 meters from low-Earth orbit.) The NRO also wanted to be informed of the "development or test of pointing, tracking, and stabilization techniques or systems to be used with satellite-bearing sensors, in which the pointing accuracy is better than 20 microradians, or the unstabilized rate is less than 20 microradians per second" and "the development or test of new recording media for use with reconnaissance-like sensors."[18]

Seamans quickly selected Robert Garbarini (Office of Space Science and Applications), Edward Gray (Advanced Manned Missions Office), and Frank Sullivan (Office of Advanced Research & Technology) to serve on the committee to review NASA's activities and report to him. During the fall of 1965, Seamans sent McMillan NASA's proposed agreements with the U.S. Geological Survey, Navy, and Department of Agriculture under which NASA would fund the study of instrument requirements for AAP experiments. (The NRO's final decision on them is not known.) Alexander Flax, the new NRO director as of October 1965, soon concluded that NASA was not always following the procedures established under the August agreement and that NASA, the other interested federal civilian agencies, and contractors were continuing to engage in public discussion and activities that were damaging to the National Reconnaissance Program.[19]

Adm. Walter F. Boone, head of NASA's Office of Defense Affairs, delivered a proposed U.S. Geological Survey–NASA Geography-Cartography Program Definition Study and Work Statement to Flax in January 1966. However, the NRO did not take any action on it for the time being. Boone and several other NASA personnel soon gave Flax and other NRO staff an extensive briefing on NASA's remote sensing plans.[20]

The Manned Space Flight Policy Committee (MSFPC), established under an agreement signed by McNamara and Webb in early January 1966, immediately began examining the possible civilian use of National Reconnaissance Program data and technologies. Replacing the long-inactive Manned Space Flight Panel of the Aeronautics and Astronautics Coordinating Board, the MSFPC's assigned responsibilities included resolving issues concerning the participation in and support of the other's human spaceflight programs and reaching agreements on top policy issues. Seamans and Director of Defense Research & Engineering (DDR&E) John Foster were cochairs. The other members were Flax, Daniel Fink (DDR&E's deputy director for strategic and space systems), George Mueller (NASA's associate administrator for manned space flight), and Homer Newell (NASA's associate administrator for space science and applications).[21]

NASA gave an extensive briefing on the AAP remote sensing experiments at the MSFPC's initial meeting in January 1966. Flax commented that the procedures established under the August 1965 NASA-NRO agreement were ineffective in restraining damaging public discussion and actions by NASA and its partners in the Earth observation field. He agreed to prepare a document setting forth the DoD's security concerns regarding NASA's plans and guidelines for the joint review of NASA's planned employment of reconnaissance-quality sensors. In the meantime, NASA and the NRO met on the U.S. Geological Survey Geography-Cartography Program Definition Study and Work Statement previously supplied. Although the NRO had no problems with certain aspects of the proposal, it strongly objected to the planned use of sensors "substantially equivalent" to those in classified programs and the goal to collect mapping data these programs were already acquiring. The NRO gave NASA permission to proceed with the U.S. Geological Survey after making unknown modifications to the Program Definition Study and Work Statement.[22]

Flax prepared and distributed his memorandum on security concerns in early April 1966. At the outset, he recommended that NASA inform the NRO of more activities than listed under the August 1965 NASA-NRO agreement, including all of NASA's Requests for Proposals, Requests for Program Recommendations, and plans for symposia or conferences. Flax cautioned that disclosure of the extent and success of overhead reconnaissance programs or the open use of classified technologies would seriously threaten the permissive international political environment that enabled U.S. reconnaissance satellites to operate without restrictions. The memorandum also expressed concern that there might be compromises of security or interference with the covert programs if NASA acquired high-quality sensors from the limited number of firms that manufactured them. All proposed NASA "reconnaissance programs" should be reviewed if they involved "development of systems, sensors, techniques or related equipment closely duplicating those already developed or being developed by the NRO" or "development of systems, sensors, techniques and related equipment to collect data which can be collected or have already been collected by the NRO." Flax recommended establishment of a NRO-DDR&E-NASA Committee on Reconnaissance Sensors to "review and analyze proposed NASA activities involving satellite-borne image-forming sensors with a view to identifying those reconnaissance activities having a potential impact on the NRP." The NRO would have the final authority to determine whether existing or planned reconnaissance-like sensors were needed to meet NASA's requirements and, if so, the security limitations on their use. If they were to be employed, the NRO and NASA would follow the procedures established under the August 1963 "DOD/CIA–NASA Agreement on NASA Reconnaissance Programs" applicable to NASA's lunar photography programs concerning their study, analysis, development, and/or acquisition.[23]

The MSFPC reviewed the NRO director's memorandum at its second meeting on 14 April. It decided that the three NASA personnel selected the previous year under the August 1965 NASA-NRO agreement (with Leonard Jaffe of the Office of Space Science and Applications having replaced Robert Garbarini) would examine NASA's programs in light of the memorandum and the procedures it recommended be followed.[24]

NSAM 156 Committee's Review of NASA's Plans and Subsequent Actions

The review by the three NASA personnel never took place as the matter was soon placed before the NSAM 156 Committee. (It is not known why this step was taken, but it was probably because the Committee had a much broader representation, including from the CIA and State

Department.) Charles Schultze, director of the Bureau of the Budget, and Donald Hornig, special assistant to the president for science and technology, wrote Secretary of State Dean Rusk within a few days of the 14 April MSFPC meeting asking him to convene the group to review the subject. They asked that it address several issues, including the possibility of declassifying some CORONA and ARGON photographs for NASA's use, permitting it to access other still-classified imagery, and whether NASA should be allowed to plan and openly conduct remote sensing of the Earth using high-resolution sensors and, if so, under what circumstances. Rusk quickly complied, and the NSAM 156 Committee first met the following month. Its membership had changed in part since its establishment in 1962. Flax had replaced Joseph Charyk as the NRO member, Huntington Sheldon had replaced Herbert Scoville as the CIA member, John McNaughton had replaced Paul Nitze as the Office of the Secretary of Defense member, and Spurgeon Keeney and Charles Johnson had replaced Carl Kaysen from the White House. There was no longer any representative from the State Department's Bureau of Intelligence and Research.[25]

NASA, the NRO, and the State Department submitted papers to the committee in May. NASA's focused on its collaboration with the Departments of Agriculture and the Interior and the Naval Hydrographic Office in the Earth resources program, the program's phases, and the sensor requirements. The first three phases involved analytical studies, laboratory experiments, airplane flight tests with sensors over calibrated target areas, and the planning of orbital systems. The fourth would begin in 1969 with AAP orbital missions at inclinations up to 48 degrees over test sites. Minimum requirements for sensors during these flights included multispectral synoptic cameras with a resolution of 30 meters and a wide-range spectral scanner capable of 200 meters. Missions beginning in 1971 would fly in polar orbits and carry approximately 20 sensors in the visible, infrared, ultraviolet, and microwave ranges selected by the user agencies. Their capabilities clearly exceeded the limits of the August 1965 NASA-NRO agreement. For example, from an altitude of 125 nautical miles the resolution requirements for metric cameras were greater than 12 meters, for panoramic cameras more than 4 meters, for high-resolution telescopes in excess of 2 meters, for synoptic multiband cameras greater than 30 meters, and for Synthetic Aperture Radars approximately 15 meters. In contrast to optical systems, Synthetic Aperture Radars could image through cloud cover and at night. The requirement for them was undoubtedly based in part on information NASA learned from the NRO on Project QUILL, the highly classified radar imaging satellite flown in December 1964. In the four days of operation, it achieved a maximum resolution of 7.5 feet while imaging parts of the northeastern and western United States.[26]

The State Department's paper addressed the political aspects of disclosing reconnaissance capabilities. It noted that in the four years since the establishment of the NSAM 156 Committee the emphasis had changed from "building world acceptance of space observations" to "actions which will preserve the present wide tacit acceptance of such activities." Although the Soviets had muted their criticism of reconnaissance from space in recent years, it was imperative that the United States not do anything to force them or other nations to renew the campaign against it. Civilian remote sensing "would stimulate wider and deeper awareness of the capabilities of reconnaissance," and these indirect disclosures must be evaluated to determine the risks to the covert programs and the best means of reducing them. The best justification for NASA's programs and minimizing their possible negative effect "will be valid scientific or economic payoff in which other countries can share."[27]

The NRO's paper initially compared the technologies NASA planned on using to those of the National Reconnaissance Program. With respect to photographic sensors, it stated that the proposed use beginning in 1971 of the best technology available implied "a virtually complete declassification of NRP sensors or a parallel development of sensors of equal quality." (The next two sections on other image-forming sensors—presumably radar and infrared—are redacted.) The NRO had no objections to NASA's intended utilization of non–image forming sensors such as magnetometers or radiometers. Although conceding the assessment of risk from NASA's planned program was subjective and depended in part on conditions unknown at the present time, the NRO vigorously argued that there were three major risks. First, it would likely stimulate "ill-timed discussion of space reconnaissance activities in the international arena which could produce unfavorable reactions from neutral, hostile or even friendly nations and could confront the Soviet Union with a condition in which it would be forced to take a hard position on observation satellites." This could result in a renewed propaganda campaign against reconnaissance from space, calls for international controls on space-based Earth observation programs, or proposals to consolidate all national reconnaissance activities into a single internationally controlled program. Second, the public disclosure of the scope or quality of current U.S. reconnaissance activities could lead to other nations adopting passive countermeasures such as camouflaging sensitive sites. Third, such revelations could lead the Soviet Union and other nations to develop anti-satellite weapons or other active countermeasures. The NRO believed that many of NASA's "high resolution photographic objectives" could be achieved sooner and cheaper through the use of aircraft. Nevertheless, it could support a NASA space-based remote sensing program if the orbital inclinations were below 30 degrees and the image-forming sensors had resolutions less than 30 meters.[28]

The committee discussed the three papers at its second meeting in late May. In a memo to the president's national security advisor, Spurgeon Keeney described the differences within the committee as "substantial" and expressed doubt that it would resolve them. While the CIA and NRO strongly opposed NASA's planned space-based remote sensing activities and maintained that most of the objectives could be achieved by using aircraft, the State Department believed that a program gradually introduced with proper preparation could be undertaken. The White House representatives took a neutral position in the deliberations.[29]

The committee issued its unanimous 11-page Top Secret/BYEMAN report entitled "Political and Security Aspects of Non-Military Applications of Satellite Earth-Sensing" on 11 July, which U. Alexis Johnson forwarded to Schultze and Hornig. It was a compromise between the positions of NASA and the national security agencies. The report noted, "Direct disclosure of satellite reconnaissance for the purpose of gaining world acceptance of the principle of space surveillance is both unnecessary and liable to provoke adverse reactions from the USSR and other states." Nevertheless, there were potentially great benefits from civilian remote sensing and that "the best justification for such programs and the best general basis for calming any alarm over their effects will be valid scientific or economic payoff in which other countries can expect to share." To protect the classified reconnaissance programs, NASA "should proceed gradually through current aerial experimentation, to unmanned and manned satellites, and in general moving from less to more precise ground resolution." The report stated that the sensor limits created under the August 1965 NASA-NRO agreement should be retained and that NASA should for the next five years restrain its discussion of future systems to those having a ground resolution of 10 to 15 feet. It recommended that a National Security Action Memorandum be issued requiring the other interested federal agencies to coordinate their Earth observation plans with NASA and that they analyze the advantages and disadvantages of aerial versus space-based remote sensing and the use of unmanned versus manned spacecraft. Personnel from these agencies should be given access to selected U-2 and CORONA photography in this process. To assist in this regard, the U.S. Intelligence Board should consider downgrading selected imagery for the use of these cleared personnel or, as a less desirable alternative, give more NASA personnel TALENT-KEYHOLE clearances so they could review it in the present codeword channels. Lastly, the report recommended that the DCI, in consultation with the NRO director, establish detailed guidelines on what cameras and other sensors NASA could utilize for observing the Earth from space and the applicable restrictions.[30]

In contrast to its 1962 report, the NSAM 156 Committee's July 1966 report was not incorporated into a National Security Council or other high-level directive. Nevertheless, it established policies for important aspects of civilian remote sensing that remained in place for many years.

The U.S. Intelligence Board quickly directed its Committee on Overhead Reconnaissance (which approved and set the targets for every National Reconnaissance Program imagery intelligence mission) to examine downgrading some photography or issuing more clearances to NASA. The Committee strongly opposed the former but supported the latter, adding that NASA should form a panel of cleared personnel to intensively study using TALENT-KEYHOLE photography for civilian purposes. In late August 1966, the U.S. Intelligence Board adopted this position. Although some additional NASA personnel received TALENT-KEYHOLE clearances, NASA never established the panel.[31] This was probably due to the fact that NASA's role in space-based remote sensing was building, launching, and operating spacecraft, and the other federal agencies that would actually use the imagery needed to have access to it as well.

The NSAM 156 Committee's report led to more restrictions on NASA's remote sensing activities and, for the time being, precluded any possibility of a systematic program. Moreover, NASA clearly could not employ any sophisticated sensors if and when a program was approved. Following the report, NASA initially sought to create a comprehensive review mechanism to replace the ineffective one set up under the August 1965 NASA-NRO agreement. Along these lines, Seamans and DDR&E John

Foster signed the Top Secret/BYEMAN agreement entitled "DoD-NASA Coordination of the Earth Resources Survey Program" in early September 1966.

This agreement had many provisions similar or identical to those that Flax had recommended to the Manned Space Flight Policy Committee (MSFPC) earlier in the year. It established a multilevel review process. At the bottom was the new Survey Applications Coordinating Committee (SACC). Its responsibilities included monitoring NASA's program "on the project and technical level to insure observance of guidelines and ground rules" and "reviewing all studies, experiment plans, work statements, project contracts, etc., for security considerations, overloading of available industrial capacity in the area of advanced state-of-the-art in remote sensors, and avoidance of unnecessary duplication in hardware development, and production and data acquisition." Above the SACC was the existing MSFPC. Its duties included ensuring that NASA's program was within the established guidelines, making recommendations on increasing the resolution of sensors NASA was permitted to employ, and referring any issues it could not resolve to NASA's administrator and the secretary of defense. The NRO was required to keep the NASA members of SACC and MSFPC informed of relevant National Reconnaissance Program activities.

NASA was required to inform SACC and MSFPC of all "activities of interest," which included the "expenditure of NASA R&D monies with a university or industry or the transfer of monies to another organization involving the study, design, development, fabrication, or test of reconnaissance-like sensors, or significant components thereof, for use in orbital systems, and studies of the use of such sensors in orbital systems." The new agreement incorporated the definition of "reconnaissance-like sensors" from the August 1965 NASA-NRO agreement. As in the earlier agreement, NASA was obligated to apprise SACC and MSFPC of its development or testing of "pointing, tracking, and stabilizing techniques or systems in which the pointing accuracy is better than 20 microradians or the unstabilized rate is less than 20 microradians per second" and "recording media for use with reconnaissance-like sensors." NASA was also required to bring before the SACC and MSFPC all "RFPs, requests for program recommendations, and plans for symposia or conferences where the subject matter is or could evolve into an activity of interest." It had to inform them in the conceptual phase of any missions using reconnaissance-like sensors flying polar or near-polar orbits. NASA's procurement of reconnaissance-like sensors was to follow the procedures established by the August 1963 "DOD/CIA–NASA Agreement on NASA Reconnaissance Programs" applicable to NASA's lunar photography programs.

Except in cases where there was an overriding national interest, the agreement prohibited NASA from developing equipment already existing or under development by the NRO, collecting data that could be acquired by existing or planned NRO systems and made available to NASA in a usable form, or collecting data that had already been acquired by the NRO and made available to NASA in a usable form.[32]

The review mechanism created by the "DoD-NASA Coordination of the Earth Resources Survey Program" remained in place into the early 1970s. During this period, it extensively scrutinized NASA's remote sensing programs and placed strict controls on them.[33]

The SACC first met in late September 1966. Its initial members were NASA's Edward Gray, Leonard Jaffe, and Frank Sullivan and the NRO's Col. Paul Worthman, Col. David Carter, and another individual whose identity is still classified. Personnel from NASA's Office of Defense Affairs served as executive secretary. The issues discussed at the initial meeting ranged from the need to get TALENT-KEYHOLE and BYEMAN clearances for a small number of employees at the Naval Oceanographic Office and the Departments of Agriculture, Commerce, and the Interior to the locations NASA personnel could use to examine TALENT-KEYHOLE imagery. At the second meeting in November, NASA submitted for review a draft prospectus for the joint NASA–National Academy of Sciences 1967 Woods Hole Summer Study on Space Applications (it was approved shortly).[34]

Secretary of the Interior Stewart Udall shocked NASA and the national security agencies on 21 September 1966 when, without any prior coordination with them, he publicly announced his department's plans to develop its own Earth Resources Observation Satellite carrying sophisticated sensors. (Udall did not have a TALENT-KEYHOLE or BYEMAN clearance at the time and could not have known much about the national security implications of civilian remote sensing. Similarly, he could not have been aware of the NSAM 156 Committee's report, which expressly recommended that a White House directive be promulgated requiring coordination with NASA.) Seamans quickly contacted the White House and issued a statement which, among other things, described the ongoing experiments with instrumented aircraft in conjunction with the

Department of the Interior and others and the need to complete these tests before an operational satellite system could be built. Within a few days, Seamans sent a draft National Security Action Memorandum to the White House which directed that NASA was the single point of co-ordination for all civilian activities in space and prohibited other agencies from initiating civilian space projects or activities without NASA's prior approval. White House staff revised it, but Seamans did not feel the changes were sufficiently strong.[35] In the end, however, no National Security Action Memorandum or other directive was issued. This was apparently due to the fact that Seamans privately communicated to Udall and others the sensitivities of the subject and the potential damage resulting from unilateral statements or actions.

Civilian Agency Access to Classified Imagery under Project Argo and Continued Review of NASA's Activities

Donald Hornig took action in November 1966 to implement the NSAM 156 Committee's recommendation that civilian agency personnel be granted access to selected TALENT-KEYHOLE imagery. He proposed to DCI Richard Helms and Deputy Secretary of Defense Cyrus Vance that the Army Corps of Engineers and a contractor work with cleared personnel from NASA, the Agency for International Development, the Department of Agriculture, the Environmental Science Services Administration (the predecessor to the National Oceanic and Atmospheric Administration within the Department of Commerce), and the U.S. Geological Survey to examine the use of this photography for civilian purposes. Helms wanted the program to remain directly under his authority for security reasons and thus recommended that the CIA's National Photographic Interpretation Center work with the civilian agencies. Supported by the Defense Intelligence Agency, Vance favored Hornig's proposal. The CIA, Defense Intelligence Agency, and Donald Steininger of the White House Office of Science and Technology finally agreed that the Army Corps of Engineers would manage the initial review by a small team of civilian agency personnel. They would have access to U-2 and CORONA photography of areas outside the Sino-Soviet Bloc, but not to any high-resolution GAMBIT 1 imagery. Within six months, the team was to produce a resource inventory of an area in Central or South America. To address the DCI's security concerns, the CIA would provide the photography and would classify all the team's products. NASA was informed of the arrangement and supported it.[36] This concept fell far short of the goal of NASA and the other civilian agencies to operate their own spacecraft and freely share the data with the world, but it would satisfy some of their needs.

The AAP received a huge boost in late 1966 when the White House approved a large funding request for it in the FY 1968 budget. President Johnson strongly supported the program in his January 1967 budget message to Congress. NASA's leadership testified before the House and Senate authorization committees in the coming months that there was now a firm plan for the first three AAP missions. NASA would launch the initial one to conduct science, meteorology, and Earth resources experiments in a flight lasting up to two weeks. A dual-launch mission lasting up to four weeks, the second was designed to demonstrate the feasibility of orbital workshop operations and conduct biomedical, science, and technology experiments. The third was also a dual launch mission that would stay in orbit up to eight weeks and perform astronomical, biomedical, science, and technology experiments.[37]

Both SACC and MSFPC focused extensively on the AAP in 1967. SACC's DoD members were very concerned about unclassified NASA and contractor documents which set forth the requirements for some AAP camera resolutions as high as two meters or greater and statements in a NASA publication on the Woods Hole Summer Study concerning the potential benefits of remote sensing and the cost-effectiveness of conducting it with satellites versus aircraft. This led to a letter from DDR&E John Foster to Seamans in July that initially recommended a joint review of cost-effectiveness studies and suggested that Hornig's proposed project to give cleared civilian agency personnel access to some TALENT-KEY-HOLE photography might lead to an expanded NASA aircraft program and render a space-based program unnecessary. Foster also severely criticized SACC's NASA members for not coordinating fully, resulting in the DoD often learning about actions in violation of the guidelines only after the fact. Jaffe wrote Seamans the same month that "there seems to be a growing impression on the part of DOD-types that it is NASA's job to discourage not only the use of space for ERS but to discourage discussion of the subject or somehow at least ignore it." He also complained that the issue of cost-effectiveness of satellites versus aircraft was not within the purview of SACC."[38]

In response to the concerns of SACC's DoD members and Foster, NASA quickly ordered that the sensitive AAP photographic experiments be deleted from all documents and issued a directive that mandated that all Earth resources survey documents be submitted to the Office of Space Science and Applications before publication or distribution. Seamans and Foster also discussed in the following months such steps as expanding the membership and role of the MSFPC, issuing a classified NASA directive explaining the sensitive aspects of remote sensing, and possibly conducting joint cost-effectiveness studies. Other than NASA drafting (but apparently not officially promulgating) a Top Secret directive on the national security implications of and the restraints on the Earth resources survey program for internal use, it is not known what measures, if any, were taken as a result of these discussions.[39]

The DoD members examined a wide range of other NASA materials during 1967, including documents related to projects NASA funded at other federal agencies and universities and a paper on space-based oceanography for the National Council on Marine Resources and Engineering Development. As best as can be determined, the DoD members did not make any changes to them.[40]

In September 1967, the DCI approved and forwarded to the NSAM 156 Committee the report it had requested the previous year on what cameras and other sensing equipment NASA could employ and the applicable restrictions. A small group of CIA, NRO, Defense Intelligence Agency, and National Security Agency personnel had prepared it. The report concluded that NRO camera technologies producing angular resolutions greater than 0.1 milliradian could be provided for NASA's unclassified use in Earth observation on a case-by-case basis, but special security arrangements to protect the hardware's origin and prior use would be required in each case. However, there should be no lowering of the 0.1 milliradian limit until the international reaction is determined from disseminating NASA imagery acquired in compliance with it, Soviet reconnaissance program technologies and countermeasures capability have been accurately assessed, and a firm requirement has been established for a major U.S. initiative in the Earth resources field. The report stated that if the threshold is lowered, it must be done gradually to protect the classified U.S. programs. Additionally, it directed that data revealing the optical characteristics of any system designed to produce resolutions greater than 0.1 milliradian and data on any system technology bearing on National Reconnaissance Program vulnerabilities must retain at least a Secret classification. BYEMAN security protection should continue for information relating to the origin and prior use. The sensitive resolution threshold of satellite radar-imaging systems should remain at 100 feet until their intelligence potential has been determined. (It is unclear when this threshold was established. As discussed above, NASA's May 1966 submission to the NSAM 156 Committee listed a requirement beginning in 1971 for a Synthetic Aperture Radar with a resolution of 15 feet. However, there is no further discussion of this sensor in the subsequent documents that have been released.) The report stated that command and control procedures used for NASA and National Reconnaissance Program Earth observation satellites must be sufficiently different "in order not to facilitate hostile system vulnerability analysis of the latter." Lastly, it concluded that NASA's experiments with passive microwave and radiometry equipment needed further definition and requested that the NRO director assess any efforts in this area and consult with the DCI to determine any impact on the National Reconnaissance Program.[41]

NASA Turns to a Robotic Satellite

The AAP's future at the beginning of 1968 was much bleaker than a year before. Primarily due to the overriding budget requirements of the Vietnam War and Great Society programs, the FY 1968 appropriations bill Congress sent to the White House in August 1967 only provided $300 million for AAP (out of an authorized $454 million). The FY 1969 budget requested $439 million for AAP, but Congress ended up authorizing only $253 million and appropriating a little less. These drastic reductions led NASA to postpone the first mission to 1970 (and then soon thereafter to 1971) and severely curtail the number of subsequent flights.[42]

The AAP was no longer a viable option for conducting systematic remote sensing from space because of the cutbacks. As a result, in 1968 NASA greatly increased its previously modest efforts to develop a robotic satellite to perform the mission. The Goddard Space Flight Center prepared a study on technologies for Earth resources satellites early that year, and in a written response SACC's DoD members strongly objected to its consideration of sensors having resolutions greater than the 0.1 milliradian limit. (Because neither document has been located, it is not known what sensors exceeding the threshold the Goddard study examined.) It quickly dropped all plans for flying such sensors and focused on those that were within the limit. Keeping the SACC, MSFPC, State Department, CIA, and Office of Science and Technology informed of its actions, NASA then asked the Bureau of the Budget to approve a little over $19 million in the FY 1970 budget to begin hardware development. It hoped to launch the first spacecraft in 1971 and the second one several years later. The request explained that the most likely sensor would be a Return Beam Vidicon television system being developed by RCA, capable of a maximum resolution of 300 feet from the planned orbital altitude of 500 miles. NASA had not yet determined what the second sensor would be. The request also argued that the present angular resolution limit was "arbitrary and may pose serious future problems for the development and exploitation of non-military earth sensing activities" and that "new sensitivity thresholds should be established."[43]

NASA also formed an Earth Resources Survey Program Review Committee in late 1968 to ensure greater coordination between NASA and the other federal civilian agencies in the field. Along with Homer Newell of NASA's Office of Space Science and Applications, its members were George Mehren, assistant secretary of agriculture; William Pecora, director of the U.S. Geological Survey; Robert White, administrator of the Environmental Science Services Administration; and Robert Frosch, assistant secretary of the Navy for research and development. Newell wrote the NRO director that each of the non-NASA members had TALENT-KEYHOLE clearances, but that only Frosch also had a BYEMAN clearance. He asked whether it was possible to give the other three BYEMAN clearances so that sensor technologies could be discussed in detail. If not, Newell stated that the group would conduct its business at the less-preferable Top Secret level.[44]

Other issues the SACC focused on during 1968 included the review and approval of the 1967 Woods Hole Summer Study reports. The DoD members determined that five volumes were releasable immediately and two could be released with their recommended changes. Both they and the NASA members agreed that the volume with the Geodetic Panel report had so many violations of the guidelines that they gave it an interim Secret classification. These sections were ultimately eliminated or modified and SACC approved the volume's release later in 1968. NASA submitted the plans for the 1968 Woods Hole Summer Study on Space Applications to SACC's DoD members, who evidently approved them with few, if any, changes.[45]

Meanwhile, the effort (now designated Project Argo) that Donald Hornig had initiated the previous year to allow cleared federal civilian agency personnel access to some classified overhead photography moved forward. Cleared scientists from several federal civilian agencies with expertise in agriculture, geology, hydrology, geography, marine sciences, mapping, and engineering completed the initial report in early 1968. The Army Corps of Engineers, Army Map Service, Defense Intelligence Agency, and National Photographic Interpretation Center provided advice and technical assistance. The scientists examined over 2,100 satellite photographs of selected areas in South America and areas in the United States for which good collateral data was available. They had come from the CORONA panoramic and index cameras, the ARGON mapping camera, the GAMBIT 1 strip and index cameras, and apparently the GAMBIT 3 strip and index cameras. The scientists believed the broad area coverage provided by CORONA would meet most of their needs, although the higher-resolution GAMBIT 1 and GAMBIT 3 could be utilized for some applications. Color photography was particularly useful. Some applications could be met by exploiting existing photography, but the majority would require new coverage at varying frequencies. The scientists had concerns about the effect of security restrictions and the potential costs of acquiring and exploiting the imagery.[46]

Hornig characterized the report as a strong endorsement of the value of classified imagery for civilian applications. With the DCI's approval, in March 1968 he sent copies to the heads of the participating civilian agencies and two new ones that had expressed interest in Project Argo, the Department of Transportation and the Office of Emergency Planning (all of these officials had recently been given TALENT-KEYHOLE clearances to enable them to see the report). Hornig proposed that the Argo Standing Committee be formed to collect and convey their requests for photography and handle security issues. He asked that each agency prepare a more detailed report setting forth further uses of the imagery, the geographic areas to be covered, and the frequency of coverage needed. The NRO soon received a copy of Hornig's memorandum and had no objections. An internal NRO memorandum to its director noted, "It appears to us that if this project showed promise it might eventually serve as a constraining force on the NASA Earth Resources Program."[47]

The civilian agencies quickly agreed to Hornig's proposal, and the Argo Standing Committee first met in June 1968. Along with Donald Steininger of the Office of Science and Technology at the White House and the DCI's personal representatives, individuals from the CIA, NRO, Defense Intelligence Agency, National Photographic Interpretation Center, NASA, Office of Emergency Planning, Agency for International

Development, the Department of Agriculture, U.S. Geological Survey, and the Environmental Sciences Services Administration attended. Steininger noted initially that President Johnson had been notified of Project Argo and supported it. After the user agencies reiterated their proposed employment of the classified imagery, discussion took place on the allocation of the costs of collecting photography and the process of forwarding their requirements to the U.S. Intelligence Board's Committee on Imagery Requirements and Exploitation. It was clearly understood that the civilian requirements would be on a non-interference basis with the primary intelligence mission. The meeting concluded with a call for detailed studies by each user agency on how they would specifically employ the product.[48]

The Bureau of the Budget approved NASA's request for $19 million in FY 1970 funding for hardware development for ERTS-A and ERTS-B (later known as Landsat 1 and 2) and included it in the president's budget transmitted to Congress in January 1969. NASA now planned to launch the first in 1972 and the second the following year. Each would fly at an altitude of 570 miles in a near-polar orbit, providing repeated Earth coverage every 17 days. The two main sensors were a Return Beam Vidicon (the television camera with an improved resolution of 80 meters now) and a Multispectral Scanner (a sensor that covered four spectral bands at the same improved resolution). Since a video tape recorder to store imagery onboard had not been developed, yet NASA initially intended to operate the sensors mostly over the United States and relay the data directly to one of three U.S. ground stations when within range.[49]

After receiving the approval of the SACC and MSFPC, NASA issued the Requests for Proposals in May 1969. SACC's NASA members asked their NRO counterparts for assistance on any technologies that could be useful to their satellites, and the latter believed that there was some largely unclassified hardware being explored that might be useful. They added that some major developments were near in electronic sensing and that NASA might even be able to help the NRO in this regard. The NRO members offered an observer to assist NASA in evaluating proposals, even though they recognized that it might be detrimental to NASA's image of engaging only in peaceful and scientific activities.[50] It is not known what, if any, NRO technologies were shared for possible use in ERTS-A and B and whether any NRO personnel helped in evaluating contractor submissions.

NASA quickly selected RCA to develop the Return Beam Vidicon and Hughes Aircraft to build the Multispectral Scanner. In 1970, it awarded General Electric the contract for system design and development and the firm became the prime contractor.[51]

NASA tested a multispectral camera in space for the first time during the Apollo 9 mission in March 1969. It was similar in many respects to the Multispectral Scanner ERTS-A and B would carry. Each of its four Hasselblad cameras simultaneously imaged test sites principally in the United States and Mexico—three using black-and-white film to record imagery in three different wavelengths and the fourth employing infra-red color film responsive to all these wavelengths. At the same time as the multispectral camera was in operation, the astronauts imaged the site with a handheld 70-mm camera and aircraft (including CIA U-2s) photographed it with 35-mm cameras to establish "ground truth." The U-2s also photographed the test sites in subsequent months with a multispectral camera system using the same film-filter combinations as Apollo 9. Researchers determined that the multispectral photography from Apollo 9 was beneficial for agricultural and geological studies, par-ticularly when analyzed together with the aerial photography taken at the same time or in subsequent months.[52]

President Richard Nixon established the U.S. objective of increased in-ternational cooperation in the remote sensing field in a September 1969 speech at the United Nations. He pledged to share NASA's data with the world and stated that the United States would offer several proposals to the world body on sharing the benefits of remote sensing. An inter-agency group made a number of recommendations to the president in December to carry out these commitments.[53]

Although NASA was prohibited from installing sophisticated image-forming sensors on its satellites, it still wanted to fly them at some time in the future when the restrictions were relaxed. Along these lines, NASA approached the NRO in 1969 about the possibility of utilizing CORONA hardware in its remote sensing program. CORONA was a very reliable op-erational system by that time, producing resolutions as high as 6 feet. It would satisfy most if not all of the long-standing requirements of NASA and the other federal civilian agencies for high-quality photography. The NRO was interested in the proposal because it would ensure that the satellite's manufacturing capability would be preserved in the event that the HEXAGON, CORONA's successor, did not become operational when scheduled to do so in 1970. However, the Office of Management and Budget refused to include in NASA's FY 1972 budget any funds for acquiring CORONA hardware. Homer Newell, associate administrator for space science and applications, advised John McLucas, the NRO director, of this in March 1970 and asked him to preserve CORONA production capability in the event NASA obtained funds in the future. McLucas replied that the NRO would try to make any surplus CORONA vehicles available to NASA. In the end, however, there was no excess hardware as all of it was flown due to delays in developing the HEXAGON.[54]

The national security authorities in 1969 examined but did not approve relaxing the restrictions on image-forming sensors that NASA could use. In May, Lee DuBridge, Nixon's science adviser, wrote the other members of the NRO's Executive Committee, Deputy Secretary of Defense David Packard and DCI Richard Helms, whether changes in the world political situation and advancements in unclassified technologies since the NSAM 156 Committee's 1966 report warranted such action. On the first point, he cited the widespread State Department support for remote sensing on international political goodwill grounds, the criticism of NASA's slow progress in developing an operational system by key legislators, the support of the National Academy of Sciences for establishing a program quickly, the lack of negative reaction to Gemini and Apollo Earth photography, and the recommendation of the pre-inaugural "Report of the Task Force on Space" that NASA's space-based Earth resources survey program should be greatly expanded immediately. With respect to the state of unclassified technologies, DuBridge noted that many unclassified systems allowed optical and ground resolutions that exceeded the limits of 0.1 milliradian and 20 meters from low-Earth orbit and that these restrictions have caused "considerable program and administrative difficulties." He recommended that the intelligence agencies first "redefine what technology, equipment, and fabrication techniques need to be classified to protect critical technology" and then ask the NSAM 156 Committee to consider changing the limits.[55]

The NRO's Executive Committee took up DuBridge's request at least in part at an August 1969 meeting that focused on the security and policy implications of the Strategic Arms Limitation Treaty negotiations on the National Reconnaissance Program. In discussions about the NASA proposal to the National Security Study Memorandum 28 Committee to openly operate a satellite to verify any agreement (which was soon rejected), DuBridge suggested the 20-meter resolution limit could be relaxed. Helms agreed, but evidently he and Vance believed it premature to go the NSAM 156 Committee at that point.[56]

The NRO director prepared the report on guidelines for NASA's Earth-sensing activities called for by DuBridge and sent it to the CIA. It remains classified, and the only information on the contents comes from a short memo from the DCI to the NRO's director concerning it. The memo stated that the report recommended that "there would be no further limitation on the type of optical systems used by NASA as long as they were pointed upward for astronomical purposes and not used to photograph the earth." It did not mention any change in the restrictions on sensors imaging the Earth, and the 20-meter limit for those used in NASA's Earth observation programs remained in place until 1978.[57]

Project Argo enjoyed continued progress in 1969. The user agencies completed their detailed studies on how they would use TALENT-KEY-HOLE photography and began employing it in their various fields. Of the nearly 10 civilian agencies involved in the project, the U.S. Geological Survey was the biggest user. It built a Sensitive Compartmented Information Facility in northern Virginia at which it and several other civilian agencies analyzed the film. The U.S. Intelligence Board approved furnishing CORONA photography of the United States (except for Hawaii) to the U.S. Geological Survey for updating its 1:250,000 scale maps, but much of this was over three years old and thus could not be utilized. The Board soon ordered that CORONA photograph 400,000 square miles of the United States each year and that the U.S. Geological Survey receive imagery from the high-resolution GAMBIT 1 and GAMBIT 3 systems. Although CORONA fell far short of meeting this requirement in the remaining missions, the U.S. Geological Survey nonetheless was able to revise 145 out of the 450 1:250,000 scale maps by 1971. It also started revising the larger-scale 1:24,000 maps of selected urban areas. The organization also used imagery from an unspecified satellite in 1970 to help make a preliminary route survey for the Alaskan pipeline.[58]

The Office of Emergency Planning and other civilian agencies often requested pre- and post-disaster coverage through the Argo Standing Committee. U-2s photographed tornado damage in Lubbock, Texas, damage from Hurricanes Camille and Celia, and earthquake damage in Los Angeles. The aircraft quickly provided the Department of the Interior with color imagery of the Santa Barbara oil spill. The Environmental Science Services Administration asked for satellite photography of the snow pack in three Sierra Nevada watersheds to determine spring run-off, but CIA U-2s on training flights from Edwards Air Force Base ended up providing the coverage for several consecutive years. An unidentified platform met the U.S. Coast Guard's requirement for photography to support the sailing of the SS *Manhattan* through the Northwest Passage to the Alaskan oil fields. The Army Corps of Engineers utilized imagery from an unspecified vehicle to determine the feasibility of the proposed Lake Michigan–Wabash River Barge Canal. CIA U-2s photographed two areas off Baja California for the Office of National Marine Fisheries of the new National Oceanic and Atmospheric Administration to determine how many whales were present. National Photographic Interpretation Center personnel analyzed the film and found them in both regions. The Agency for International Development, the only known organization to receive imagery of foreign nations, used photography from unidentified satellites to help assess drought conditions in Chile and West Africa, earthquake damage in Nicaragua, and war damage in Pakistan.[59]

Project Argo ended in 1973 when Nixon abolished the White House Office of Science and Technology. Donald Steininger, then the CIA's assistant deputy director for science and technology, gave it a mixed assessment. He said it definitely proved that TALENT-KEYHOLE photography had numerous civilian applications, but that the government had been unsuccessful in promoting its wide use within the civilian agencies. Among other things, Steininger recommended that the DCI become more active in this regard and TALENT-KEYHOLE clearances be made easier to obtain. After Project Argo ended, the DCI continued furnishing TALENT-KEYHOLE materials to the civilian agencies on an ad hoc basis until the Civilian Applications Committee was formed in 1975. This new body, chaired by an official of the Department of the Interior and with representatives of 11 civilian federal agencies, assumed the coordination of requests for classified imagery in the following decades.[60]

NASA Finally Launches ERTS-A and Obtains U-2s

By 1970 all that remained of the once-ambitious AAP program were plans to launch a Saturn S-IVB stage as a workshop in 1973 and have astronauts occupy it for three different periods over the next nine months. The new effort was designated Skylab, and NASA planned extensive solar astronomy, biomedical, and Earth resources survey experiments. In the Earth Resources Experiment Package, it intended to use a multispectral camera similar to that flown on Apollo 9 and the same Hycon lunar topographic camera flown on the Apollo 13 and 14 missions. At the recommendation of the NSAM 156 Committee, NASA officials briefed Soviet scientists on the remote sensing experiment and invited them to participate.[61]

ERTS-A and B continued moving forward in 1970 and 1971. RCA's Return Beam Vidicon system was successfully tested onboard a Lockheed U-2 in mid-1970. Over 600 domestic and foreign investigators met with NASA in February 1971 to learn about the data the satellites would generate and how they could obtain it. (National Photographic Interpretation Center personnel also attended.) NASA froze the design of the two satellites the following month. Later in 1971, it accepted delivery from RCA of the video tape recorders to store imagery onboard that would enable much greater coverage of the Earth.[62]

The possibility that NASA might be able to use higher-resolution cameras on the satellites in the future arose briefly in 1971 when the NRO director surprisingly proposed that the current resolution limit of 20 meters be reduced to 5 meters in 1975. The NRO's Executive Committee and the U.S. Intelligence Board apparently reviewed but did not approve any reduction, as the limit was not lowered until 1978.[63]

The national security community was uncertain about the potential intelligence value of ERTS-A and B during this period. Although the ground resolution of both the Return Beam Vidicon and Multispectral Scanner were far less than the cameras flown in the National Reconnaissance Program, it was believed that the imagery might have some usefulness nevertheless because it was multispectral and included various infrared frequencies. Of more interest was the equipment being developed in the program to process the near real-time photography (the NRO was then developing the first digital return imagery intelligence satellite, the KH-11, and launched it in December 1976). Along these lines, the National Photographic Interpretation Center stated that it would maintain close contact with NASA's program to benefit from its experience in this area.[64]

NASA successfully launched ERTS-A in July 1972 and later renamed it Landsat 1. It operated until 1978.[65] As will be seen in chapter 8, the multispectral imagery from it and the subsequent satellites proved to be very valuable for some national security applications.

NASA's Earth Resources Survey Aircraft Program received more capable equipment in June 1971 when the NRO loaned two U-2C reconnaissance aircraft to it. They joined NASA's existing fleet of a P-3A, C-130B, and RB-57F.

The NRO director had initiated this process in the spring of 1970 when he offered NASA several U-2Cs or U-2Gs. The aircraft were in "flyable storage" to replace any more advanced U-2Rs then in service that were lost, but since the latter's attrition rate was much lower than predicted, they were now surplus. NASA and the NRO reached an agreement later that year under which two U-2Cs would be loaned for domestic flights only. The NRO's Executive Committee (Deputy Secretary of Defense David Packard, DCI Richard Helms, and Special Assistant to the President for Science and Technology Edward E. David Jr.) approved the arrangement. NASA planned to use the U-2Cs initially to overfly test sites in the United States with sensors similar to those of ERTS-A and B. They would subsequently carry astronomy, atmospheric, and geophysics experiments. The two planes would not carry any classified hardware, and NASA would release all the data acquired. The maximum performance characteristics and sensors and hardware common to the U-2Rs would not be declassified.[66]

David wrote to Assistant to the President for National Security Affairs Henry Kissinger in December 1970 requesting that Kissinger and the 40 Committee examine the possible use of the U-2Cs or the RB-57F over foreign nations pursuant to bilateral agreements. (The 40 Committee was the subcommittee to the National Security Council, which, among other things, approved covert operations and aircraft and satellite reconnaissance missions over denied areas. Its members were the special assistant to the president for national security affairs, DCI, attorney general, deputy secretary of defense, chairman of the Joint Chiefs of Staff, and the undersecretary of state for political affairs.) Kissinger's staff forwarded the materials to the 40 Committee in early 1971 and posed the following questions: Would the program have any adverse effect on classified programs? What international problems might result from cross-border photography if the adjacent countries had not entered into an agreement with the United States? What, if any, limitations could be placed on access to the data collected? Would it be possible to selectively declassify satellite imagery to support the aircraft survey program? What are the implications of using a known spy plane in a civil program? NASA made it clear that it initially would use the U-2C only over test sites in the United States and that flights over foreign nations would await the results of these tests and agreements with their governments. The 40 Committee reviewed the issue at a March 1970 meeting. It evidently decided that NASA could use the RB-57F internationally but that it could not do the same with the U-2C until it had demonstrated the plane's usefulness in test flights. Any subsequent overtures to use the U-2C internationally would first have to be made to allies such as Canada. They based their decision on the concern that using the aircraft over foreign nations might deter its use for covert reconnaissance and that international political problems might result from taking photographs of cross-border areas of nations that were not parties to any bilateral agreements.[67]

NASA's Office of DoD and Interagency Affairs contacted the CIA's Steininger in March 1971 about obtaining a loan for NASA's U-2Cs of a multispectral photographic system and related equipment that remains classified today. Although the U-2Cs carried two different multispectral cameras during their flights over U.S. test sites in late 1971 and early 1972, their exact origin is unknown.[68]

The CIA's Office of Special Activities used its "black" contracting procedures to arrange for Lockheed support of the aircraft (primarily pilots and ground crews) during their first year of operation. NASA had requested this because it believed that the CIA could finalize a contract faster and cheaper. NASA asked the CIA to continue this arrangement past July 1972, but the CIA was reluctant on the basis that the "black" contracting procedures required that the DCI certify to Congress that the funds were expended for confidential purposes (which was not the case here). Although the CIA was comfortable in bending the rules for the first year of operation on the grounds that this action satisfied a high priority national need and saved millions of dollars, it did not want to continue doing so.[69]

Summary

NASA's remote sensing of the Earth from space held the promise of tremendous scientific benefits and increased prestige for both it and the United States. However, it also posed major threats to the National Reconnaissance Program. As a result, the defense and intelligence agencies monitored and limited NASA's program to an unprecedented extent. They imposed restrictions on the image-forming sensors NASA could use so it could not acquire any imagery approaching the quality of that from the classified programs, established mechanisms to monitor all of NASA's activities in this area, and delayed the development of its robotic remote sensing satellite. To compensate for these restrictions, they created a mechanism under which cleared federal civilian scientists gained access to some classified overhead imagery, but the results were mixed.

The next chapter describes the national security community's extensive involvement in experiments in the human spaceflight programs, including the restrictions placed on those involving photography. It also examines NASA's use of classified technologies in its lunar photography program and the security restrictions accompanying it. The chapter concludes with a review of the national security community's monitoring of NASA's space-based astronomy programs and the known and probable support it gave to them.

T. Keith Glennan, NASA's first administrator (1958–61). Under his leadership, NASA developed close ties with the national security community in a number of areas. Courtesy of the National Aeronautics and Space Administration (NASA).

First-generation Tiros weather satellite. Later NASA polar-orbiting weather satellites began providing data to meet some national security weather requirements, especially for tactical operations. The fourth-generation *TIROS-N* was the first to include a device to prevent unauthorized commands. Courtesy of the National Aeronautics and Space Administration (NASA).

James Webb, NASA's second administrator (1961–68). Webb oversaw tremendous growth in the size and stature of NASA and increased inter-action with the defense and intelligence agencies. One of the most im-portant areas was intelligence on the Soviet space program, particularly whether it had a manned lunar landing program competitive with Apollo. Courtesy of the National Aeronautics and Space Administration (NASA).

Lunar Orbiters photographed the Moon in 1966–67. In NASA's first known use of classified technologies, they used a modified E-1 camera from the cancelled SAMOS reconnaissance satellite program. Classified agreements between NASA and the DoD governed their acquisition and restricted the information that could be publicly disclosed about them. Courtesy of the National Aeronautics and Space Administration (NASA).

Robert Seamans, NASA associate administrator from 1960 to 1965 and deputy administrator from 1965 to 1967. He was the key contact with the national security agencies on a wide range of issues. Seamans later served as secretary of the Air Force from 1969 to 1973. Courtesy of the National Aeronautics and Space Administration (NASA).

Landsat 1, NASA's first remote sensing satellite, was launched in 1972. Classified agreements limited the types of sensors the satellite could carry, but the national security agencies soon began using the multispectral imagery to help estimate foreign agricultural production. Courtesy of the National Aeronautics and Space Administration (NASA).

Test launch of a Poseidon submarine–launched ballistic missile in the 1970s. Several NASA ground stations provided unclassified and classified support to interconti-nental and submarine–launched ballistic missile tests beginning in the late 1960s. Courtesy of the Department of Defense (DoD).

James Fletcher, NASA administrator from 1971 to 1977 and 1986 to 1989. During his first tenure, he obtained White House and congressional approval for the Shuttle and oversaw its early development. Courtesy of the National Aeronautics and Space Administration (NASA).

Skylab, America's first space station, was crewed for three periods in 1973–74. NASA received permission from the national security agencies to fly the Earth Terrain Camera, but they restricted what it could photograph and reviewed the imagery before public release. It was the first camera NASA employed to image the Earth that exceeded the resolution limits originally established by a 1965 NASA-NRO agreement. Courtesy of the National Aeronautics and Space Administration (NASA).

SEASAT-A, NASA's first oceanographic satellite, launched in 1978. The radar altimeter acquired geodetic data for civilian applications and for improving the accuracy of submarine–launched ballistic missiles. The national security agencies approved the use of the Synthetic Aperture Radar, the first openly flown in space, but limited the acquisition and dissemination of the imagery. Courtesy of the National Aeronautics and Space Administration (NASA).

Hans Mark, NRO director from 1977 to 1979, secretary of the Air Force from 1979 to 1981, and NASA deputy administrator from 1981 until 1984. He was a strong supporter of the Shuttle and was partially successful in accelerating the DoD's transition to it during the Carter administration. Along with Secretary of Defense Harold Brown, Mark helped persuade President Carter not to drastically cut the program when it faced serious technical and fiscal problems. Courtesy of the National Reconnaissance Office (NRO).

Tracking and Data Relay Satellite-5 shortly after its deployment from the Shuttle in August 1991. Launched beginning in 1983, these spacecraft relay data between the Shuttle and other satellites in Earth orbit and ground controllers. The defense and intelligence agencies have reportedly used them extensively. Courtesy of the National Aeronautics and Space Administration (NASA).

James Beggs, NASA administrator from 1981 to 1986. He helped bring the Shuttle to operational status, but unsuccessfully opposed the DoD's efforts in 1983–85 to resume acquisition of a small number of expendable launch vehicles for critical national security payloads. Courtesy of the National Aeronautics and Space Administration (NASA).

Polishing the *Hubble*'s primary mirror at the Perkin-Elmer plant in Danbury, Connecticut. The NRO permitted the firm to fabricate the Optical Telescope Assembly at this highly classified facility dedicated to building the HEXAGON's optical system but restricted NASA's access to it. Other known DoD assistance included supplying antenna booms, testing the rate gyros, and providing a container to ship the assembled spacecraft. Courtesy of the National Aeronautics and Space Administration (NASA).

X-37 after landing at the end of its first classified mission in 2010. Although originally developed by NASA, the Air Force, and Boeing as an unmanned reusable launch vehicle to test key technologies, the DoD assumed control of the project in 2004 and changed the focus to building a reconnaissance platform. NASA continued to give the project unspecified support. Courtesy of the U.S. Air Force.

Concerns over Human Spaceflight Program Experiments and Lunar and Astronomy Program Technologies

NASA conducted a wide range of scientific activities in space in addition to those of its applications satellites. It performed many experiments in the human spaceflight program, sent probes to explore the Moon and planets, and launched spacecraft to conduct astronomy. Although almost all data acquired from these efforts had only civilian applications, there was still considerable NASA-DoD interaction concerning them.

The experiments in the Mercury through Apollo-Soyuz programs were sponsored by civilian and military scientists within the federal government and civilian scientists from universities and other institutions from around the world. They produced a wealth of valuable data in many disciplines and generated tremendous goodwill for NASA and the United States. The DoD's first substantial involvement was in the Gemini program when it proposed many unclassified experiments and a small number of classified ones designed to evaluate the ability of humans to perform reconnaissance from space. Some officials at NASA strongly objected to the latter because they clashed with the policies of openness and nonparticipation in defense matters. However, NASA flew them under a compromise under which their existence would be acknowledged but their data could be withheld. The defense and intelligence agencies also became very concerned early in the Gemini program with the potential of the civilian photographic experiments capturing sensitive domestic or foreign sites. As a result, in late 1965 the DCI established a formal screening procedure of the imagery before public release. However, there are very few details available on what photographs were withheld or released in degraded form.

The intervention increased during Apollo when NASA was also required to submit the proposed photographic experiments before each mission. There is no evidence that any were rejected or modified, except for those which were part of the Earth-orbital contingency flights for the last five Apollo missions. Post-mission review of the imagery continued, but there is no information available on whether any of it was withheld or released in degraded form. The national security agencies also began testing some of the Apollo cameras before missions to ensure they did not exceed the resolution limits of the 1965 NASA-NRO agreement. They also conducted classified experiments in connection with the last four Apollo missions, although it appears as if DoD ground facilities conducted them and the astronauts were not involved.

NASA fought hard and finally received permission from the national security agencies and the White House to fly a camera for Earth resources purposes on Skylab that exceeded the resolution threshold established in the 1965 NASA-NRO agreement. It had to submit the proposed photographic experiments for approval and operate under strict constraints on what could be imaged and when with this and the other cameras. At least one photograph was withheld after post-mission review. With respect to Apollo-Soyuz, the available evidence suggests that the national security community did not make any changes to the proposed photographic experiments or withhold any imagery.

NASA used several classified cameras during the Apollo program to obtain the high-quality imagery of the Moon needed to select landing sites. This was the first known instance in which classified technologies were employed in one of its spaceflight programs. The national security agencies initially wanted to perform the missions employing them, but NASA firmly rejected these plans. NASA then selected and received permission to use a classified camera from the Air Force SAMOS reconnaissance satellite program in the Lunar Orbiters. It worked closely with the intelligence agencies on plans to fly a high-resolution GAMBIT 1 camera in a crewed mission around the Moon if the Lunar Orbiter photography proved deficient, but in the end this was not the case and the plans were cancelled. In order to use these cameras, NASA had to enter classified agreements tightly controlling the acquisition procedures and the information that could be released concerning these procedures, the origin of the cameras, and certain technical details about them. It also used several other classified cameras from the CORONA and strategic reconnaissance aircraft programs for lunar photography in the last five Apollo missions, and in all likelihood their procurement and information that could be released concerning them were governed by these existing agreements as well.

Beginning in the 1960s, there was considerable interaction between NASA and the defense and intelligence agencies regarding its space-based astronomy programs because some of the optical systems or pointing and stabilization systems in the Apollo Telescope Mount, Orbiting Astronomical Observatory, and other programs exceeded the limits established under the 1965 NASA-NRO agreement. The DCI and director of the NRO soon decided that high-quality photography of celestial bodies did not endanger the National Reconnaissance Program and removed the restrictions in 1969. However, they insisted that the security of the sophisticated optical systems and pointing and stabilization systems still had to be protected, and they recommended establishment of a new special security control system to govern their acquisition and use. Whether it was established or the existing procedures applicable to the classified cameras in the lunar photography program were employed to provide the necessary protection for the systems flown that exceeded the limits is unknown.

NASA had started planning in the 1960s for what was eventually designated the *Hubble Space Telescope*, which was to be the most capable astronomical observatory flown in space. In the early 1970s, the NRO offered NASA unspecified "highly refined optical manufacturing techniques" for the project. A newly established high-level, NASA-DoD-CIA board reviewed the sensitive technologies planned for the *Hubble* beginning in 1975, but it is not known exactly what these technologies were or what decisions were made concerning them. The NRO permitted Perkin-Elmer to use its highly classified facility in Danbury, Connecticut, to fabricate the *Hubble*'s Optical Telescope Assembly. It had been specifically constructed in the late 1960s for manufacturing the HEXAGON's optical system. NASA's access to the facility was severely restricted at first and then gradually relaxed. What is still unclear is whether any classified manufacturing or testing processes were employed on the Optical Telescope Assembly. Lockheed received the contract for the Support Systems Module, and Eastman Kodak got the contract for a backup primary mirror. If either firm employed classified facilities in this work they undoubtedly also received permission from the NRO to do so. The national security agencies provided other support to the *Hubble*. The Defense Advanced Research Projects Agency funded two antenna booms and the Air Force tested its rate gyros at a special laboratory normally used for testing inertial guidance systems for missiles. NASA did use an NRO shipping container to transport the *Hubble* from California to Florida for its launch.

The NRO offered a classified facility for testing the completed Optical Telescope Assembly, but it is still unclear as to why NASA rejected it. Such a test would have revealed the flaw in the primary mirror before launch.

Mercury and Gemini Experiments

All seven unmanned Mercury flights carried 16-mm motion picture cameras, some of which used color and others black-and-white film. Astronauts imaged the Earth and weather in color with fixed 70-mm cameras in the two suborbital flights in 1961. In the four orbital missions during the next two years, they photographed the daylight horizon, meteorological phenomena, and selected areas in North America and Africa in color with handheld 35-mm or 70-mm cameras with 80-mm lenses. All the imagery was disseminated to the public. The astronauts received pre-mission briefings from NASA personnel on the photographic experiments, including which areas they should and should not image. The latter included sensitive U.S. and allied military installations. There is no evidence that the national security agencies were involved in the planning of these experiments or were concerned about the public distribution of the imagery.[1] All four orbital flights flew at an inclination of 32.5 degrees, thus any possible photography of the USSR was very limited.

NASA did, however, provide copies of all the imagery from the orbital flights to the CIA's National Photographic Interpretation Center. The CIA used the stellar photography in its development of the stellar camera that it soon placed on the CORONA imagery intelligence satellites. The Center was very interested in the Earth photography to determine the benefits and disadvantages of space-based color photography of the Earth and to evaluate its possible employment in the classified reconnaissance programs. (CORONA only employed black-and-white film at the time.) For example, what was its ground resolution compared with black-and-white imagery, and what surface features showed up better in color than in black-and-white? Although there were not a large number of Earth photographs from the Mercury program and only three types of color films were used, undoubtedly the Center obtained at least some preliminary answers to these questions.[2]

The DoD's involvement in Gemini was much greater. Because of its growing interest in the possibilities of humans performing reconnaissance from space, McNamara proposed to Webb in 1961 that the DoD manage all human spaceflight programs and operate all crewed flights in Earth orbit, with NASA operating all manned lunar missions. Not surprisingly, Webb rejected this brazen power grab. McNamara then proposed to manage Gemini jointly, but again Webb rebuffed him. NASA and DoD signed an agreement in January 1963 that resolved the dispute. Among other things, it established the Gemini Program Planning Board to ensure that the project met the objectives and requirements of both organizations. Seamans and Brockway McMillan, undersecretary of the Air Force and (covertly) NRO director, were cochairs. The other NASA members were the heads of the Office of Manned Space Flight and Office of Defense Affairs, and the other DoD members were the special assistant (space) under the DDR&E and Gen. Bernard Schriever, head of the Air Force Systems Command.[3]

The board first met in February 1963. It focused on overseeing the planning and conduct of experiments and the programs to man-rate the Air Force Titan II launch vehicle. The board declared that Gemini was to first satisfy the objectives of the Manned Lunar Landing Program and then other NASA and DoD requirements (with the DoD experiments having priority over NASA's). Although the board expressly provided that the DoD would determine the security classification of its experiments, its demand that certain experiments be classified upset some NASA officials because they believed doing so would compromise the open nature of NASA's programs and damage its peaceful image. This was, of course, basically the same issue that arose with the ANNA geodetic satellites during the same time period but with one difference. While there were civilian applications for the ANNA data the DoD wanted to classify, the data from classified Gemini experiments had little or none. In any event, NASA's top management compromised on the matter. The official history of NASA's Office of Defense Affairs states, "A plan was worked out under the aegis of the GPPB whereby DoD classified experiments were installed prior to flight and removed after recovery without exposure to the public, and results were screened by a joint committee who decided what should and should not be released. The arrangements worked out very satisfactorily."[4] In short, the existence of all DoD experiments would be acknowledged but some or all of the data could be withheld.

The board and NASA's leadership approved approximately 50 experiments for Gemini. NASA's included synoptic Earth and weather color photography using handheld 70-mm cameras with 38-mm or 80-mm lenses, a handheld 35-mm camera with a 250-mm lens, and general-purpose color photography with a handheld 16-mm motion picture camera.

Fifteen of the 50 experiments were DoD sponsored and were largely focused on determining the effectiveness of humans performing reconnaissance from space. They included measuring the radiation in the capsule, determining the usefulness of star occultation measurements and certain tools for spacecraft navigation, testing the ability of the crew to discriminate small objects on Earth, and the usefulness of image intensification equipment in viewing the Earth at night. They also included three photographic experiments—Basic Object Photography (D-1), Nearby Object Photography (D-2), and Surface Photography (D-6)—all of which were scheduled for Gemini V, which flew in August 1965. A prelaunch NASA press release described D-1 and D-2 as testing the ability to acquire, track, and photograph other objects in space, both nearby and at a distance. The objects included the Titan II second stage that placed them in orbit, a rendezvous evaluation pod to be released from the Gemini capsule, the Moon, and "other objects of opportunity." The last set of targets was not explained any further, but it probably referred to Soviet spacecraft or space debris. D-6 was to test the ability to acquire, track, and photograph targets on Earth, which included "selected cities, rail, highways, harbors, rivers, lakes, illuminated night-side [sic] sites, ships, and wakes" within the United States and Africa. All three DoD photographic experiments utilized a handheld 35-mm camera with a 250-mm or 1,200-mm telephoto lens, the latter of which produced much higher quality imagery than the civilian experiments.[5]

As in the Mercury program, the astronauts received pre-mission briefings on what they should and should not photograph. There was very little overflight of the USSR, since the Gemini spacecraft flew at an average inclination of about 29 degrees. Using their 70-mm cameras, astronauts in the first two crewed missions—Gemini 3 and IV—acquired approximately 175 Earth and weather photographs. Most were of the United

States, Mexico, Pacific islands, northern Africa, and parts of the Middle East, but the Gemini IV astronauts also took four of China. There were no reported photographs of the Soviet Union, Eastern Europe, or Cuba. That the astronauts photographed China at all was surprising, since it presumably risked provoking a reaction. However, along with the other reported imagery, it was all evidently publicly disseminated.[6]

On 9 August 1965, NRO director Brockway McMillan wrote Seamans regarding the possible compromise of the National Reconnaissance Program from astronaut photography. He noted that the image of Cape Kennedy taken by the Gemini IV astronauts and published in a NASA newsletter had "minor intelligence or political significance," but that a similar image taken of "certain domestic and most foreign locations could create serious problems." McMillan was particularly concerned with the upcoming Gemini V photography of the Earth in the DoD and NASA experiments. He did not want it to go to the Manned Spacecraft Center's Photographic Laboratory in Houston for processing prior to public release, which had been the case with all the previous Gemini imagery. Instead, he recommended that it go to the CIA's National Photographic Interpretation Center where its personnel, Seamans, and he could review it. Mindful of NASA's reluctance to appear involved in any classified activities and wanting to better conceal the NRO's involvement, McMillan suggested that the fully cleared Col. Floyd Sweet of the Office of Defense Affairs pick up the DoD and NASA films on the recovery aircraft carrier and bring them to the Center. This was done, and after processing, its personnel worked with Seamans and McMillan on which frames could be released.[7]

It is impossible now to reconstruct what exactly took place with the Gemini V photography brought to the National Photographic Interpretation Center. A NASA index listed that the astronauts acquired about 250 images under the synoptic Earth and weather experiments. These included more than 20 of China and 2 of Cuba, but none of the USSR or Eastern Europe. NASA released all the approximately 250 photographs.[8]

Post-mission NASA press releases and briefings stated that DoD experiments D-1 and D-2 produced only a few images of the Moon and other celestial bodies, but these were apparently not released. These sources also indicated that because of electrical power problems in the capsule, the astronauts only took a few photographs of the rendezvous evaluation pod with the 16- and 70-mm cameras. NASA released several of these frames. The astronauts stated that they did not observe or photograph any other spacecraft. However, Air Force records indicated that the North American Air Defense Command (which operated the Space Detection and Tracking System) provided NASA with a series of eight-hour forecasts of satellites passing within visual range of the capsule during the mission. Pursuant to them, one of the astronauts sighted an adapter piece from a prior U.S. launch at a range of 45 kilometers within 10 seconds of the forecast's predicted time of sighting. The astronaut did not photograph the debris because it was on the side of the capsule occupied by the other astronaut, who was asleep. With respect to the DoD's D-6 experiment, NASA publicly reported that one of the five acquisition and tracking modes was very successful in obtaining images of "preselected terrestrial objects," but also that cloud cover prohibited coverage of some targets. However, it did not provide any count or inventory (the Johnson Space Center states today that it holds about 225 D-6 images, but there is no index). Only a few were released to the public in the months following the mission, probably in degraded form. For example, the 11 October 1965 issue of *Aviation Week & Space Technology* had photographs of Dallas, Tibet, and Kashmir taken with the 35-mm camera. It is clear that the astronauts were very impressed with the 1,200-mm telephoto lens. They reported to ground controllers several times during the mission that the detail seen through it was remarkable.[9]

The NRO's role in examining the Gemini V photography concerned the CIA. The deputy director for science and technology (DDS&T) wrote the deputy director for intelligence in September 1965, "It is important that the intrusion of NRO into NASA programs not exceed the limits established in the DoD/CIA–NASA Agreement of 1963. NRO action on GEMINI V photography clearly usurps the responsibility of the director and could be embarrassing to him, since the NRO is to receive policy guidance on security from the Director of Central Intelligence."[10]

Gemini VI and VII flew in early December 1965, and the astronauts took more than 200 synoptic Earth and weather photographs. A NASA index from the latter mission included 3 of a Polaris submarine–launched ballistic missile test firing off the coast of Florida, 3 of China, and more than 10 of Cuba. NASA described almost all the photographs of Cuba as "sufficiently degraded to be considered useless, or nearly so," but it released the remaining ones and those of China. Once again, there was no known photography of the Soviet Union or Eastern Europe. Later that month, DCI Raborn wrote to Webb concerning the review of Gemini photography. Based on his responsibilities to protect intelligence sources and methods and to establish the security policies of the National Reconnaissance Program, he informed NASA's administrator that he had designated the chairman of the U.S. Intelligence Board's Committee on Overhead Reconnaissance to be in charge of policies and procedures in this area. Raborn assured Webb, "We are attempting to be helpful to you in this matter, and not play the role of censor." He noted that the DDS&T had already contacted Seamans about the review of Gemini VI and VII imagery.[11]

Under the review procedures established, the film from the astronauts' cameras was taken to NASA's Photographic Technology Laboratory. NASA developed it, and a small team of representatives from the National Photographic Interpretation Center and apparently several other intelligence organizations reviewed each frame. It made masters of any photographs that could not be released and took them back to Washington. The originals stayed in a vault at the NASA facility. NASA was not informed why the particular images were not releasable.[12]

Astronauts brought back over 2,000 additional synoptic Earth and weather photographs during the remaining Gemini missions (Gemini VIII through Gemini XII). The United States, South America, and parts of Africa were the most frequent targets. The only photographs of communist countries that the NASA indexes list are of China and North Vietnam during Gemini X and Cuba during Gemini XII, all of which were released.[13]

Unfortunately, apart from the DoD photographic experiments in Gemini V there is little information on what photographs were withheld or possibly released in degraded form during the Gemini program. The only available record is a March 1968 memo from the NRO director to the deputy secretary of defense in which he stated, "Among the photographs withheld were those illustrating the capability to show airfields (one withheld photograph of Bergstrom Air Force Base approximated 20–30 foot resolution, at which it was possible to identify and count the B-52s on the base)."[14]

Apollo Experiments

Apollo photography from 1967 to 1972 was more extensive and complex, and during the lunar orbital and lunar landing missions, it targeted the Moon mostly. All the missions flew at an inclination of roughly 33 degrees when orbiting the Earth, precluding imaging most of the USSR.

An automated 70-mm camera took more than 700 photographs of the Earth from an altitude of about 10,000 miles during Apollo 4, an unmanned Earth orbital flight in November 1967. Apollo 5, another un-manned mission launched in January 1968, did not carry any cameras.[15]

Pre-mission review of NASA's proposed photographic experiments by the intelligence agencies began with the unmanned Apollo 6 launched in April 1968. It is not known what prompted this step. In February 1968, NASA's David Williamson sent the photographic plans to the head of the U.S. Intelligence Board's Committee on Imagery Requirements and Exploitation (the new name of the former Committee on Overhead Reconnaissance). Among other things, NASA planned to employ the automatic 70-mm camera in the command module "to record engineering data on spacecraft orientation." Williamson did add, however, that "it is expected that the engineering film data will also be used by the earth resources survey program participants to provide an insight into the value of this class of photography for each of their disciplines." The camera would only operate during the spacecraft's initial two orbits from an altitude of about 100 miles. The anticipated best resolution was 125 feet. At an inclination of 32.5 degrees, most of the imagery would be of Africa, some Pacific islands, and the southern United States. There would be overlapping coverage between frames, and thus stereo analysis would be possible. Williamson forwarded the plans the same day to the CIA's DDS&T and remarked, "In our brief discussion with him [the head of the committee], he expressed the view that there would be no difficulties raised concerning the mission as planned." He added, "Following your suggestion, we will work directly with [redacted] office as representing all interested parties in the review of the actual photograph prior to its release, continuing the pattern established during the Gemini program." NASA also intended to bring the Apollo 6 photographic plans to the 303 Committee for approval.[16] (The 303 Committee was the subcommittee to the National Security Council which, among other things, approved covert operations and aircraft and satellite reconnaissance missions over denied areas. Its members were the DCI, national security advisor, secretary of defense, and the undersecretary of state for political affairs.)

NASA's proposal did not sit well with the intelligence agencies, particularly the NRO. In a 3 March letter to Deputy Secretary of Defense Paul Nitze, the NRO director conceded that NASA's plans did not violate the resolution limit of 20 meters from low-Earth orbit. However, he added that the plans had not been examined by the two DoD-NASA coordinating bodies established to review all of NASA's efforts in this area under the September 1966 Top Secret/BYEMAN "DoD-NASA Coordination of the Earth Resources Survey Program" agreement—the Survey Applications Coordinating Committee (SACC) and the Manned Space Flight Policy Committee (MSFPC). The NRO director strongly recommended that this be done, that a screening procedure be adopted before public release of any imagery, and that the 303 Committee look at the political sensitivity of the photography. Nitze wrote to the 303 Committee, and the issue was heard at its 19 March meeting. The 303 Committee, among other things, decided that the NASA proposal should be examined by the SACC and that "all photography, either domestic or foreign, should be screened by the COMIREX mechanism prior to public release."[17]

These steps were soon taken. At the 21 March SACC meeting, the DoD members expressed concern that the Maurer 70-mm camera might exceed the 0.1 milliradian limit originally established by the August 1965 NASA-NRO agreement. The group directed that technical experts from DoD and NASA review all the data and report back. They concluded that the camera would operate at an angular resolution of 0.3 milliradians and thus did not exceed the threshold.[18]

A memorandum from NASA's associate administrator for space science and applications to the associate administrator for manned space flight the day before the launch detailed the screening process. It stated, "The authorization NASA has received to obtain the Maurer photography is predicated on an interagency review prior to the release of any photography" and "These procedures, while oriented primarily towards Apollo 6 Maurer photography, will provide an outline for subsequent interagency review of future missions." NASA's Office of DoD and Interagency Affairs, in consultation with the other NASA offices and the intelligence agencies represented on the newly designated Interagency Review Team, had established these procedures. They were similar to those implemented during the Gemini program. NASA's Photographic Technology Laboratory was to give the team the original and two duplicates of all the imagery from a mission. The team was to inform the director of the Manned Space Flight Center in writing of their decisions concerning the release or classification of the photography. If an image could not be sufficiently protected by being classified at the Secret level, it would be taken to Washington. Access by NASA personnel to the laboratory was to be limited during the team's presence to those with a "need-to-know." No public disclosure of the fact of the screening would be made. However, in the event of direct inquiries, selected NASA personnel were authorized to state that a review had been done with other government agencies, that not all the imagery was disseminated because of cloud cover or poor illumination, and that when "national interests so require," photography would not be released.[19]

NASA launched Apollo 6 on 4 April, and the 70-mm camera took over 760 pictures during the brief period it operated. The Interagency Review Team screened them and determined that all the imagery could be released. (Because of darkness and cloud cover, only photography of Baja California to the East Coast of the United States during one orbit had any usefulness.) NASA soon reported to SACC that it had already released five frames to the public and would inform the body of further releases.[20]

NASA undoubtedly submitted the proposed Earth photography experiments in the next few Apollo missions to the intelligence agencies for review. Whether they mandated any changes is not known.

The handheld Hasselblad 70-mm camera to be used in Apollo 7 underwent testing by NASA personnel to determine whether it exceeded the 0.1 milliradian threshold. The National Photographic Interpretation Center reviewed the results and concluded that the performance was "equivalent to an angular resolution of approximately 0.45 milliradians," well below it. The astronauts onboard Apollo 7, the program's first manned flight, which was placed in Earth orbit in October 1968, acquired more than 500 synoptic Earth and 500 synoptic weather photographs with the Hasselblad. Almost all were of the United States, Mexico, Central and South America, Africa, the Middle East, Australia, Indonesia, Philippines, and Japan. There were only a few frames of China.[21] The Interagency Review Team examined the imagery and determined that all of it could be released to the public. However, before this was done, United Press International and *Aviation Week & Space Technology* queried NASA's Office of Public Affairs about holding back photography from release. NASA consulted with the DoD members of SACC and gave a somewhat vague response.[22]

Apollo 8 was a lunar-orbital mission. The astronauts used two handheld Hasselblad 70-mm cameras with either an 80-mm or 250-mm lens. Although no photography was planned during the Earth orbits on the first day of the flight, they were specifically instructed not to use the

250-mm lens to image the Earth in this phase, since the photography would likely exceed the 20-meter limit. As it turned out, the astronauts did not image the Earth at all during these orbits.[23]

The next Earth photography took place during Apollo 9, launched in March 1969. Astronauts acquired almost 800 images of the Earth and weather with a handheld Hasselblad 70-mm camera. These were mostly of the United States, Mexico, Central and South America, Africa, the Middle East, South Asia, Australia, and Japan. However, there were a few frames of China and Cuba. The mission also carried the first multispectral camera in space in an experiment designed to evaluate crops, rangeland, and forests. Seamans had originally contacted the NRO director in June 1966 concerning using a multispectral camera in a future Apollo Applications Program flight. The director responded in September that the multispectral camera did not involve "sensors of reconnaissance quality" and thus he had no objections to its use. However, he did note that because of the political sensitivities, no DoD organizations could participate in this or any other open NASA space experiment employing optical sensors, and he requested that all NASA press materials and related items delete the participating DoD organizations. With the drastic cutbacks in the Apollo Applications Program in the following years, NASA decided to fly the instrument on Apollo 9. Each of its four cameras simultaneously imaged test sites located in the United States or Mexico—three using black-and-white film to record imagery in three different wavelengths and the fourth employing infrared color film responsive to all these wavelengths. At the same time as the multispectral camera was in operation, the astronauts imaged the site with the handheld 70-mm camera and aircraft (including CIA U-2s) photographed it to establish "ground truth." Each of the four cameras produced about 130 images during the mission.[24]

The Interagency Review Team examined all the Apollo 9 imagery at the Photographic Technology Laboratory in Houston. It concluded that the Hasselblad 70-mm color images "are of little technical/intelligence value to the Intelligence Community." The team was most interested in the multispectral photography and noted in this regard that "most of the scenes are domestic from an equatorial track from [two lines redacted] is 50/75 percent cloud covered." It commented that "the various emulsions and filter/emulsion combinations . . . provide depth perception and could excite some interest within the Community." The team further noted that the multispectral imagery would be made available to the "IEG [Imagery Exploitation Group] to decide if the various emulsion combinations warrant an extensive study to determine their intelligence value."[25] The results of any IEG study are unknown.

The interest in the multispectral photography was not surprising, since Soviet and Chinese agricultural production was an intelligence objective of growing importance. During the early 1960s, the CIA's Office of Research and Development began investigating the use of aerial photography to evaluate the state of crops and their yield. Shortly thereafter, it formally established a "crop yield" project. As part of this effort, U-2s with multispectral cameras flew over wheat-growing areas in North Dakota, and the CIA maintained regular contact with the Department of Agriculture, which was also studying multispectral imaging. NASA's airborne multispectral sensor program (which overflew U.S. test sites) was undoubtedly followed closely.[26]

The photographic plans for Apollo 10 through 12 did not include imaging the Earth, and presumably there was no liaison with the defense and intelligence agencies concerning them. However, NASA planned to photograph the Earth during any Earth-orbital contingency missions of Apollo 13 through 17 and had to obtain permission from the White House to do so. These missions would occur if after launch the Apollo spacecraft could not continue on to the Moon but could safely remain in Earth orbit. NASA's Willis Shapley submitted the Apollo 13 contingency mission proposal to the 40 Committee (the successor to the 303 Committee) in early April 1970, and it was soon approved. Myron Krueger, from NASA's Office of DoD and Interagency Affairs, submitted the Apollo 14 proposal to the 40 Committee in November 1970, and that too was approved quickly.[27] (Both Apollo 13 and 14 carried the classified Hycon camera originally developed for an unknown strategic reconnaissance aircraft, which NASA had obtained permission to use for lunar photography.) Neither Earth-orbital contingency mission was flown.

NASA Administrator James Fletcher wrote Assistant to the President for National Security Affairs Henry Kissinger in May 1971 concerning the Apollo 15 contingency mission. Fletcher noted, "This contingency would offer a unique opportunity to acquire earth survey photography in support of the NASA Earth Resources Survey Program." All the onboard cameras would be utilized to image the United States and contiguous areas, but the key one would be the Itek 24-inch focal length panoramic camera which was first carried on this flight. (This was a classified camera from the U-2 program which NASA had obtained permission to carry for photographing the Moon during the last three Apollo missions.) The contingency mission would be flown at an altitude of 230 nautical miles and at an inclination of 40 degrees (higher than the normal Apollo mission and resulting in greater overflight of the USSR). It was expected that the ground resolution would be from 25 to 35 feet, thus exceeding the 20-meter limit. The Interagency Review Team would examine all the photography before public release. NASA's proposal was approved by the 40 Committee on 22 June under the conditions that the photography would emphasize the United States and Western Hemisphere, no photography of the Sino-Soviet Bloc would be acquired, and the U.S. Intelligence Board's Committee on Imagery Requirements and Exploitation would review all the photography before public release. Fletcher wrote to Kissinger in January 1972 requesting approval for the Apollo 16 contingency mission photographic experiments, and the 40 Committee gave its approval under the same restrictions.[28] Once again, neither contingency mission was flown.

NASA planned to fly the Apollo 17 Earth-orbital contingency mission also at an altitude of 230 nautical miles and an inclination of 40 degrees. In the fall of 1972, it proposed greatly expanding the quantity and quality of the photography by imaging the Sino-Soviet Bloc and other previously prohibited areas at a resolution as high as 10 meters.[29]

NASA ran the contingency mission proposal by the NRO, CIA, Office of Secretary of Defense, State Department, and Arms Control and Disarmament Agency before submitting it to the White House. All but the State Department strongly opposed the plans on the basis that openly acquiring and disseminating higher-quality imagery might provoke some nations to argue that their consent was needed to be imaged and that this might eventually lead to international restrictions on U.S. reconnaissance satellites. An internal NRO memorandum also complained that NASA was "going out of channels again." The State Department favored the proposal in the interests of promoting an "open skies" environment and the principle that a nation's agreement was not necessary to photograph it from space. Despite the opposition, NASA included the expanded coverage in its submission to Kissinger on 17 November. The 40 Committee rejected it and only approved the contingency mission under the same conditions as before.[30] NASA did not end up flying the Apollo 17 contingency mission.

The DoD conducted some classified tests in connection with the Apollo 14, 15, 16, and 17 missions, and NASA publicly disclosed this fact in the post-flight mission reports. The *Apollo 14 Mission Report* set forth the following six tests performed for the DoD and Kennedy Space Center: "Chapel Bell (classified Department of Defense test), Radar Skin Tracking, Ionospheric Disturbance from Missiles, Acoustic Measurement of Missile Exhaust Noise, Army Acoustic Test, Long-Focal-Length Optical System." The *Apollo 15 Mission Report*, *Apollo 16 Mission Report*, and *Apollo 17 Mission Report* listed the same six tests and a seventh ("Sonic boom measurement").[31]

There is no information available on any of the tests other than Chapel Bell. However, by their descriptions the others seem to be related to detecting and tracking vehicles in the atmosphere and space. Chapel Bell was the codename for an Office of Naval Research R&D project from the late 1950s into the 1970s involving over-the-horizon-radars. It was one of several DoD projects that investigated the use of high frequency radars (whose signals bounce off the ionosphere) to detect and track missiles and aircraft at long ranges. Located at Muirkirk, Maryland, the Chapel Bell radar attempted to detect and track every DoD and NASA vehicle launched at Cape Canaveral and Kennedy Space Center. The facility would receive the countdowns for DoD launches by encrypted communications sent by an International Telephone and Telegraph station at the Florida launch complexes and the countdowns for NASA launches through a NASA communications network. Buoys in the Atlantic served as range and azimuth markers and the International Telephone and Telegraph station in Florida would measure the strength of the Chapel Bell radar before a launch and relay the information to Muirkirk. NASA did not change its missions or flight plans to accommodate Chapel Bell.[32]

Skylab and Apollo-Soyuz Experiments

Skylab was the next human spaceflight program after Apollo. George Low, NASA's acting administrator, wrote to Henry Kissinger in April 1971 concerning plans to use the high-resolution Earth Terrain Camera as part of the Earth Resources Experiment Package. (The Earth Terrain Camera was the new name of the classified Hycon camera that NASA had received permission to use for photographing the Moon during Apollo 13 and 14.) He explained that the maximum resolution was expected to be around 30 feet but that most frames would have a resolution less than that. NASA wanted to use the camera to obtain high-resolution photos in interpreting the data from the Earth Resources Experiment Package's multispectral camera and to assist in understanding the relative importance of spatial and spectral resolution in remote sensing. Skylab was to fly at an inclination of 50 degrees, higher than any previous human spaceflight mission. Most of the photography would be of the Western Hemisphere and the contiguous oceans and would take up only 70 hours of the entire time Skylab was occupied. Low asked the 40 Committee to review the possible use of the Earth Terrain Camera. He maintained that there were no technological security issues and that "international sensitivity toward space imagery at useful resolutions significantly grosser than those required for intelligence purposes seems to us virtually to have disappeared."[33]

Kissinger asked the NSAM 156 Committee to examine the matter and forwarded to them a letter he had recently received from NASA's new administrator, James Fletcher. In the fall of 1971, the committee recommended conditional approval of NASA's use of the Earth Terrain Camera as an exception to the limit established by the August 1965 NASA-NRO agreement and not as a revision, provided that the photographic plans were reviewed beforehand to preclude photography of sensitive areas and the imagery was screened before public release. It also strongly recommended that the 40 Committee review the possible international reaction to the experiment immediately prior to launch. Kissinger approved the recommendations.[34]

The national security implications do not appear to have been examined closely again until the NSAM 156 Committee prepared a formal report for the 40 Committee approximately five weeks before the launch of the unmanned workshop in May 1973. It recommended that the 40 Committee approve the following NASA photographic plans and safeguards to protect the classified reconnaissance programs. The Earth Resources Experiment Package's multispectral camera would operate during only 65 of the 2,000 crewed orbits and the Earth Terrain Camera in only 40. They would image sites selected by domestic and foreign investigators, but under no circumstances would they photograph the USSR, China, North Korea, the Israeli border areas, the Suez Canal, and certain "sensitive" areas within the United States. Although the border between India and Pakistan was also considered "sensitive," it would be photographed because investigators from both nations had requested complete coverage of the subcontinent. The Earth Terrain Camera, the lower-resolution multispectral camera, and the hand-held 35-mm and 70-mm handheld cameras would only photograph the United States, Mexico, and Canada during the first crewed period. Representatives of the CIA, Defense Intelligence Agency, State Department, and "other interested agencies" should screen the photography before public release. NASA would release imagery acquired by the lower-resolution cameras first, followed by Earth Terrain Camera photography of the United States. If there were no negative international reactions to it, the imagery of Canada and Mexico would then be released. During the second and third crewed periods, the astronauts would image other foreign sites as well. NASA would release Earth Terrain Camera photography of other foreign sites only if there continued to be no adverse reactions.[35] The 40 Committee undoubtedly reviewed and approved the NSAM 156 Committee's report.

NASA launched the unmanned workshop on 14 May and the three-astronaut crew shortly thereafter for the Skylab 2 mission. During the 28-day flight, the Earth Resources Experiment Package's multispectral camera and Earth Terrain Camera were activated on only 13 passes and they acquired about 5,300 photographs. In general conformance with the NSAM 156 Committee's recommendations, they only imaged areas in the Western Hemisphere. The astronauts took over 650 photographs of targets of opportunity with the two handheld cameras. The second crew reached the workshop on 28 July to begin the 59-day Skylab 3 mission. Operating during 48 passes, the multispectral camera and Earth Terrain Camera took almost 13,500 images of the Western Hemisphere, Western Europe, Africa, Southeast Asia, and Australia. The astronauts acquired almost 1,100 photographs with the handheld cameras. NASA launched the third and last set of astronauts in November on the 84-day Skylab 4 mission. The multispectral camera and Earth Terrain Camera operated during 49 passes and produced over 17,000 images of the same areas as in Skylab 3. Over 2,500 photographs were acquired with the two handheld cameras. NASA reported that the highest resolutions produced by the Earth Terrain Camera were 17 meters with high-definition black-and-white film, 21 meters with high-resolution color film, 23 meters with high-resolution infrared color film, and 30 meters with infrared color film.[36]

An interagency group reviewed all the imagery after each mission before public release. In the spring of 1974, it withheld a photograph of Area 51 in Nevada (it was probably taken with the Earth Terrain Camera during Skylab 4). Although all the crews had been instructed not to photograph this site, one astronaut did so inadvertently. The Air Force, NRO, Joint Chiefs of Staff, and the Office of the Secretary of Defense believed it should be withheld, while NASA and the State Department took the position that it should be released. They set forth their views in a position paper submitted to DCI William Colby for the final decision. Colby initially questioned the withholding of it because the Soviets had undoubtedly photographed Area 51 with their own satellites, the image did not actually reveal details of the site, and the activities there could always be explained by simply stating that the Air Force performed "classified work." Nevertheless, in the end he ordered the photograph withheld.[37]

The last human spaceflight program before the Shuttle was Apollo-Soyuz in 1975, the first joint mission with the Soviets. Soyuz 19 with two cosmonauts flew into space on 15 July 1975, while an Apollo command/service module with three astronauts was launched later that same day. From the 17th to the 19th, the capsules docked and undocked twice. The astronauts returned to Earth six days later. The Apollo command/service module flew at an inclination of almost 52 degrees, higher than even Skylab. This, of course, resulted in considerable overflight of the USSR.[38]

Photography of the Earth was one of the many science experiments during the mission. Astronauts used a 70-mm Hasselblad camera on a fixed mount for mapping photography. A timing mechanism triggered the shutter every 10 seconds for the 60-mm lens and every 6 seconds for the 100-mm lens, providing stereoscopic coverage. They employed handheld 70-mm Hasselblad and 35-mm Nikon cameras to image preselected visual observation sites and targets of opportunity for geologic, oceanographic, and meteorological studies. Almost 2,000 photographs were obtained, of which approximately 75 percent were of excellent quality. Although it is not known whether the intelligence agencies approved the photographic plans before the mission, they reviewed the photography before public release. However, there is no evidence that any images were withheld or released in degraded form.[39]

Lunar Photography Program

NASA established its first lunar program when it took over the Air Force's lunar probe project in October 1958 and designated it Pioneer. All four launches of the vehicle, which carried scientific instruments and not cameras, were unsuccessful. The Jet Propulsion Laboratory established the next program, Ranger, in December 1959. It originally envisioned lunar near-miss missions to measure solar radiation and conduct other experiments, followed by impact missions to photograph the lunar surface.

In May 1960, the Laboratory initiated the Surveyor program to photograph the Moon with an orbiter and to perform surface experiments with a lander. President Kennedy's announcement in May 1961 that the nation would land humans on the Moon by the end of the decade made these programs essential for selecting landing sites and gathering other critical data for the Apollo missions.[40]

An NRO history states that several unspecified NASA personnel contacted the NRO's immediate predecessor in April 1961 for permission to examine and use the E-1 film readout system in the lunar photography program. It had been built by Kodak for the Air Force's SAMOS reconnaissance satellite. The E-1's first launch in October 1960 was unsuccessful, but three months later the second reached orbit and sent back photographs with a ground resolution of approximately 100 feet. This proved the feasibility of a readout system, but all work on the E-1 was soon cancelled because more capable SAMOS photographic payloads were under development. In any event, the NRO gave NASA permission to deal with Kodak, but there are no details available on any contacts until 1963.[41]

Both Ranger and Surveyor quickly experienced major problems and delays. Because of this and the increasing demand for accurate information on lunar landing sites for the Apollo program, the orbiter portion of Surveyor was cancelled and the orbiter project reassigned to the Langley Research Center in early 1963. Langley soon came up with a spacecraft using different hardware and another launch vehicle in the newly designated Lunar Orbiter program.[42]

John Rubel, deputy director of defense research and engineering, met with Seamans at the end of 1962 and recommended that Air Force–NRO launch vehicles, spacecraft, instrumentation, and data reduction and processing equipment could be used with some modification by NASA to photograph and map the Moon. Rubel subsequently contacted Brockway McMillan, the new NRO director, to voice his concerns that NASA not waste resources to independently develop and acquire hardware that was already available in the classified reconnaissance programs and that if NASA were to do that it might compromise or damage these programs. McMillan replied that with respect to NASA's lunar reconnaissance effort, the NRO had already loaned personnel to help in the preparation of work statements, evaluation of contractor proposals for lunar photography and mapping hardware, and transferred some unspecified equipment. He added that advanced data handling techniques and systems developed for the classified programs could easily be used by NASA in its effort. McMillan recommended that the NRO be given more specific roles in this area.[43]

Rubel considered McMillan's response weak and continued trying to establish a mechanism under which NASA's lunar photographic and mapping program would not compromise or damage the classified reconnaissance programs. In an early April 1963 meeting, Seamans indicated to Rubel that NASA was currently studying two types of cameras for both the unmanned and manned lunar reconnaissance of the Moon and that NASA would be in contact when it was done. He did not commit NASA to using any image-forming technologies developed for the National Reconnaissance Program. Rubel quickly contacted the secretary of defense about the lack of progress, and McNamara wrote Webb in late April suggesting that the DoD perform the lunar reconnaissance mission for NASA. Webb did not expressly reject the offer, but replied that NASA's requirements had not been firmly established and that in any event it would follow all security guidelines.[44]

Joseph Shea led a NASA delegation in early May 1963 that met with Eugene Fubini, the new deputy director of defense research and engineering, and others. Shea described plans to photograph the Moon from both manned and unmanned vehicles prior to any landing, stated that the cameras would have to be tested in Earth orbit first, and asked for the NRO's assistance. Fubini informed McNamara and McMillan. In the coming weeks, Webb and Seamans discussed with DCI McCone and others the possibility of NASA utilizing unclassified contracts for sensitive cameras with either the CIA or NRO covertly managing development. McCone, Fubini, and others were skeptical that NASA could protect the classified programs and their technologies under such procedures.[45]

The CIA's new deputy director for science and technology (DDS&T), Albert Wheelon, met with unknown NASA, DDR&E, and NRO personnel in early July 1963 to explore what National Reconnaissance Program cameras and contractors NASA could use. Their discussions led to McNamara and Webb signing the Top Secret/BYEMAN "DoD/CIA–NASA Agreement on NASA Reconnaissance Programs" the following month. Its purpose was to match NASA's needs for manned and unmanned lunar reconnaissance with NRO capabilities. Among other things, it provided that if NRO-developed reconnaissance technologies classified above the Confidential level met NASA's specifications, the NRO would select one of its contractors to build the hardware in a secure or "black" fashion. NASA would concurrently enter into an open or "white" agreement with the contractor. One fully cleared NASA employee and one NRO employee would jointly administer the contract. The former would manage the open contract in such a manner as to provide sufficient information to NASA and contractor personnel without the requisite clearances to enable them to design circuitry, telemetry, and other systems. The NRO employee would manage the development of the actual payload. If NASA also wanted to employ an NRO-developed spacecraft, the NRO employee would manage its development. NASA would reimburse the NRO for the costs of both the "black" and "white" contracts and was obligated to follow NRO security practices at all times. Any lunar reconnaissance program that did not require the equivalent performance of classified NRO-developed hardware was not subject to the agreement and could be conducted openly if both NASA's associate administrator and DDR&E agreed.[46]

Seamans and the NRO director signed a six-page Top Secret/BYEMAN appendix on security at the same time. It listed the classification level of approximately 40 types of information and activities related to the agreement. For example, the existence and text of the agreement between NASA and the NRO on lunar reconnaissance was classified Top Secret/BYEMAN, while the fact of such an agreement between the DoD and NASA was unclassified. NASA could publicly disclose that the Air Force was developing the camera, the contractor's name, and the fact that the details of the camera were classified. NASA's official statement of lunar reconnaissance requirements and the NRO analyses thereof were Top Secret/BYEMAN. General statements of design objectives for lunar reconnaissance were Confidential. The Requests for Proposals, contractor replies, and Source Selection Board deliberations and results were Top Secret/BYEMAN. Initially processed lunar photography was Secret, and in its sanitized form it was unclassified. Any Earth photography taken by the camera was Top Secret/TALENT-KEYHOLE.[47]

NASA, NRO, and CIA officials met with Frederic Oder, a Kodak executive formerly associated with the SAMOS program, in late July 1963. Oder saw no problems with modifying either the E-1 or E-2 film readout system to NASA specifications. In either case, it would weigh approximately 120 pounds and be capable of seven-foot resolution from low-lunar orbit.[48]

Langley discussed details of the Lunar Orbiter system with potential contractors during the summer of 1963. It prepared a Request for Proposal inviting firms to submit bids on the project and released it on 30 August 1963. Although not known for certain, its release may have been delayed until after Webb and McNamara had signed the "DoD/CIA–NASA Agreement on NASA Reconnaissance Programs." Five firms—Hughes, TRW/Space Technology Laboratories, Martin, Lockheed Missiles and Space, and Boeing—submitted bids that the Langley Source Evaluation Board evaluated. Boeing proposed the smallest spacecraft, weighing just less than 800 pounds, which used a great amount of proven hardware. It would carry a modified E-1 camera with high-resolution and medium-resolution lenses that could take pictures simultaneously and photograph more of the Moon and at a higher resolution (about 1 meter) than any other proposed camera system. Additionally, the E-1 was the only system that had flown in space and used low-speed film that could be protected much more easily from radiation.[49]

The Langley Lunar Orbiter Program Office made it clear to the Source Evaluation Board that it favored Boeing's proposal. The board recommended to Seamans that Boeing be chosen as the contractor over Hughes, the second-highest ranked bidder. Although Boeing's bid of over $83 million was more than twice that of Hughes, Seamans ultimately decided in favor of Boeing based on the strength of its technical proposals and management arrangements, and a contract was signed in early May 1964 for delivery of the first two vehicles two years later.[50]

Problems quickly arose with the August 1963 "DoD/CIA–NASA Agreement on NASA Reconnaissance Programs." McMillan now believed that it would be a mistake for NASA to enter any unclassified contracts because they would probably reveal NRO's participation and would likely lead to security breaches. Seamans soon agreed, and in late March 1964 the two signed a supplemental agreement. It directed that the NRO enter into the contracts for any camera for the lunar reconnaissance program and provide it as classified government furnished equipment to NASA. A NASA-NRO management team would continue to oversee all the details, and NASA would reimburse the NRO for all costs.[51]

NASA planned that the Surveyors' low-resolution imagery of the surface of the Moon and tests of its composition in conjunction with the Lunar Orbiter photography would enable the agency to identify and certify landing sites. If they did not, however, NASA planned manned circumlunar missions with a classified camera to photograph potential locations under what was designated as the Lunar Mapping and Survey program. Along these lines, McMillan and Seamans signed an unclassified DoD/

NASA Agreement on the NASA Manned Lunar Mapping and Survey Program in April 1964, which had codeword annexes. The procedures established were almost identical to those in the August 1963 "DoD/CIA–NASA Agreement on NASA Reconnaissance Programs" as amended by the March 1964 supplemental agreement. NASA would submit a statement of performance and delivery requirements, and the NRO would select a contractor in coordination with NASA. The NRO would contract for the hardware, provide it to NASA as classified government furnished equipment, and NASA would reimburse the NRO for the costs. Contractors began studies the following month to establish requirements and evaluate which NRO system could best meet them. Among other things, NASA wanted ground resolution even better than the Lunar Orbiter's 1 meter. NASA transferred over $800,000 to cover the costs of the studies and support. The project was codenamed UPWARD.[52]

The contractors concluded that the GAMBIT 1 high-resolution system, which first flew in 1963, was the best choice. It could produce higher-resolution imagery of the Moon from low altitudes and cover a broader area from higher altitudes. The final plans were to place a modified camera in an orbiting control vehicle that would be carried atop the Saturn V rocket in the compartment built for the lunar module. The command and service module would detach from the stack, turn around 180 degrees, dock with the orbiting control vehicle, and pull it free (just as would be done if a lunar module were carried). It is not known whether the vehicle would operate independently from the command/service module or how the astronauts would retrieve the film once exposed. Lockheed won the contract in June 1965 to modify the camera and install it in the orbiting control vehicle. The contract called for delivery of the first unit in July 1967 and a flight test in Earth orbit in December.[53]

Five of seven Surveyors successfully soft-landed on the Moon from May 1966 to January 1968 and sent back thousands of low-resolution pictures and data on the lunar surface. NASA launched five Lunar Orbiters from August 1966 to August 1967, and each successfully transmitted high-quality images of the Moon. Since the first three missions largely satisfied Apollo requirements, the last two focused on increasing scientific knowledge. NASA had identified 23 possible landing sites by December 1966 based on photography from *Lunar Orbiter I* and *II*, of which eight underwent further study.[54] As a result of this success, the Lunar Mapping and Survey System was no longer necessary, and NASA soon cancelled UPWARD.[55]

Apollo missions continued to photograph the Moon, but now primarily for various scientific investigations. Both Apollo 13 and 14 carried a Hycon lunar topographic camera. It was a modified Air Force KA-7A camera that flew on an unknown reconnaissance aircraft. The best lunar surface resolution the camera achieved during the Apollo flights was 6 to 9 feet. Whether NASA's acquisition of it was done under the procedures established by the August 1963 NASA/CIA–DoD agreement as amended or the April 1964 DoD/NASA agreement is not known.[56] The flights also carried a 70-mm still camera, a 70-mm electric camera, a 16-mm data acquisition camera, and a 35-mm lunar surface stereoscopic close-up camera.[57]

The Apollo 15, 16, and 17 service modules carried the more capable Itek panoramic camera and a Fairchild mapping and stellar camera. The former was a modified Air Force KA-80A optical bar camera, which Itek started manufacturing in 1966 for use in U-2s (it was designated IRIS II in the U-2 program). As best as can be determined, the existence and design of the camera were unclassified, and the only classified information regarding it was the vehicle use and the customer. While the best ground resolution of the KA-80A in the U-2 was less than 1 foot from an altitude of 70,000 feet, as modified for Apollo it achieved a lunar surface resolution as high as 3 feet from its orbit roughly 70 miles above the Moon. The combined mapping and stellar camera was a modified Fairchild Dual Image Stellar Index camera flown on CORONA satellites from 1967 to 1972. It had a maximum lunar surface resolution of approximately 80 feet from its lunar orbit.[58] Once again, there are no accessible records shedding light on whether NASA's acquisition of these cameras was subject to the August 1963 or April 1964 agreements. These flights also carried three 70-mm still cameras, a 70-mm electric camera, a 16-mm data acquisition camera, and two television cameras.[59]

Astronomy Program

NASA engaged in a number of solar and stellar astronomical programs beginning in the Apollo era. It launched dozens of small Explorer satellites, some of which conducted gamma and x-ray astronomy. Four automated Orbiting Astronomical Observatories were placed in orbit to observe many portions of light emitted by stars which could not penetrate the Earth's atmosphere to ground-based telescopes. NASA launched seven automated Orbiting Solar Observatories. It also developed the

Apollo Telescope Mount, a solar observatory with telescopes to be flown several times in the Apollo Applications Program. After being launched, it would rendezvous and dock with a crewed capsule. The astronauts would operate the instruments and then collect and return the film to Earth at the end of the mission. Due to severe budget cutbacks in the Apollo Applications Program, the Apollo Telescope Mount only flew on Skylab in 1973–74.[60]

NASA launched the first *Orbiting Astronomical Observatory (OAO-1)* in April 1966, but the mission ended after three days due to a power failure. The Survey Applications Coordinating Committee (SACC) and Manned Space Flight Policy Committee (MSFPC) shortly thereafter began examining some of the astronomical programs. Donald Hornig, the special assistant to the president for science and technology, wrote to the NRO director in the fall of 1966 expressing concern that the Apollo Telescope Mount might jeopardize the National Reconnaissance Program in some unspecified manner. SACC's DoD members then reviewed NASA's long-range Physics and Astronomy Program at the second meeting in November 1966. Based on a preliminary examination, they determined that the planned projects did not pose any conflicts or problems.[61]

The NSAM 156 Committee's July 1966 Top Secret/BYEMAN report on NASA's remote sensing program requested that the DCI, in consultation with the NRO director, establish "security restrictions on cameras and other sensing apparatus and equipment which can be made available for NASA's program of non-military applications of satellite earth-sensing." DCI Helms sent the requested report to the committee in September 1967. Although the committee's July 1966 report did not address the issue of NASA's use of sensitive hardware in its lunar or astronomical programs, the DCI's report focused extensively on the subject. It noted, "The matter of lunar or extra-planetary exploration poses certain problems, in that the use of high as well as medium range resolution optics is in all probability necessary to obtain any meaningful scientific data." Although space-based photography of celestial bodies would not produce any adverse international reaction, either from it or the equipment could be "extrapolated the probable capability of the same equipment operating against the earth's surface from earth orbit." Accordingly, the procedures established under the April 1964 "DoD/NASA Agreement on the NASA Manned Lunar Mapping and Survey Program" should be followed when NASA had demonstrated a need for using sensors capable of resolutions greater than that established by the August 1965 NASA-

NRO agreement. With respect to Earth-orbiting spacecraft with optical telescopic devices for astronomical observation, the report concluded that "optical and spacecraft technology and technical requirements for telescopes for stellar and solar observations are of a degree of significant similarity with NRP high resolution systems as to warrant the development of a security guide for such activities similar to that prescribed under Project UPWARD."[62] However, the NSAM 156 Committee did not act upon the DCI's report.

The MSFPC and SACC examined NASA's astronomy programs again in late 1967 and early 1968. NASA planned to launch *OAO-2* in late 1968 and *OAO-B* in 1969. Each would carry progressively larger and more powerful ultraviolet telescopes and have increasingly greater pointing accuracy and stability. At the November 1967 MSFPC meeting, Seamans stated that *OAO-2*, *OAO-B*, *Voyager* (the planned Mars and Venus probe), and the large astronomical telescope under study were all designed to exceed the pointing and stabilization limits originally established by the August 1965 NASA-NRO agreement. He argued that a restrictive interpretation of the above-described September 1967 DCI report to the NSAM 156 Committee "could seriously inhibit NASA's activity." The NRO director remarked that the existing guidelines might need to be modified with respect to astronomical systems and that a distinction should be made between astronomical and Earth-oriented systems. He stated that he would assist NASA in preparing a position paper for the DCI on the pointing and stabilization requirements for astronomical payloads.[63]

Whether such a paper was prepared and sent is unknown. However, the NRO director quickly conducted a study in consultation with the DCI on the appropriate restrictions on NASA's astronomical programs. It initially noted that there was a fundamental difference between the NRO and NASA's use of optical sensors. While the NRO had to conceal their presence, NASA did not, and this affected which technologies should be given special security protection and the system of security controls. The possible negative international reaction to imagery of celestial bodies was very slight, regardless of the resolution. However, the "provocation value of research, development, and fabrication activities related to such activities will in all probability be significant, inasmuch as the equipments produced will be comparable in most respects to those high performance equipments used for satellite reconnaissance purposes." As a result, there should be strict security controls over research, development, and fabrication activities, but "some allowance can and in all probability should be made with respect to data reduction, compilation, and analysis activities associated with the resulting imagery, even to the extent of providing the basic geometry related to the optical sensor."[64]

The study specifically recommended the following for special security controls in systems for astronomical use that were designed to produce resolutions greater than the 0.1 milliradian threshold established in the August 1965 NASA-NRO agreement: (1) all data regarding the internal composition of the image-forming sensor, not essential to data reduction, compilation, and analysis of the resulting imagery, (2) all processes related to image-forming sensor manufacture, related lens grinding and polishing, mirror construction, and film manufacture, (3) command and control processes that reveal data in the first category, such as image motion compensation techniques and focus, filter, and exposure controls, (4) non-image-forming sensor data that directly relates to sensor targeting, timing, and attitude controls and temperature and other environmental requirements critical to sensor operation. It recommended a new security control system to be managed by NASA but subject to CIA oversight. This system would not involve semicovert security procedures as those used by the NRO, and it was intended to be separate from but complement the BYEMAN system. Among other things, it would require Top Secret clearances, physical security protection consistent with BYEMAN standards, and prior coordination with the NRO through the SACC before negotiations with contractors involving covered activities.[65] It is not known whether such a security control system was adopted and, if so, when.

NASA launched *OAO-2* with 11 ultraviolet telescopes in December 1968, and it operated successfully until early 1973. This was the first time astronomers had been able to observe in the ultraviolet spectrum.[66] As best as can be determined, the pointing accuracy was greater than the threshold, and the NRO and others must have agreed to NASA obtaining and using the technologies to achieve this.

The NRO director sent the CIA a proposed new set of guidelines regarding NASA's Earth-sensing programs in October 1969. Although the guidelines remain classified, a declassified memo from DCI Helms to the NRO's director in response reveals some information concerning them. Helms wrote that the CIA had no problems with the proposed changes. However, he noted that the report stated "that there would be no further limitation on the type of optical systems used by NASA as long as they were pointed upward for astronomical purposes and not used to photograph the earth. That still leaves open the question of making available in unclassified form detailed design data on optical systems which could be readily adapted for reconnaissance purposes." The DoD and NASA had apparently already agreed to the new guidelines, but Helms suggested that they be referred to the NSAM 156 Committee because of its long-standing role in the matter and because "it is clearly a subject of interest to more than these two organizations." The available records of the NSAM 156 Committee after this date do not indicate that it ever made any such changes to the limitations originally established by the August 1965 NASA-NRO agreement and that its July 1966 report reaffirmed.[67] All that can be concluded from the fragmentary evidence is that it appears that there was some still-classified NASA-DoD agreement reached during 1969 which eliminated the resolution restrictions concerning optical systems used in NASA's astronomy programs. Whether the new agreement modified the limitation on pointing and stabilization systems too is unknown.

NASA launched *OAO-B* with a 38-inch ultraviolet telescope in November 1970, but the booster failed to separate and the satellite soon burned up as it reentered the atmosphere. *OAO-3* was placed in orbit in August 1972 and operated until 1981. Carrying a 32-inch telescope, its pointing performance exceeded the threshold.[68] In all likelihood, NASA had to obtain permission from the NRO or other national security agencies before acquiring and using the pointing and stabilization system.

During this period, the first known transfer of classified hardware to NASA for use in its astronomy programs took place when it received an unknown number of "large glass blanks" from the NRO for use in "optical projects." The items were probably 72-inch mirror blanks designed for the KH-10 camera to be flown on the Manned Orbiting Laboratory, a program the Nixon administration had cancelled in the summer of 1969. There are some reports that some or all ended up in the Multiple Mirror Telescope at the University of Arizona.[69]

NASA's astronomy program during the early years of the Shuttle era centered on the *Hubble Space Telescope*. (Designated the *Large Space Telescope* until 1983 and *Hubble Space Telescope* thereafter, it will be referred to as the *Hubble* for the sake of simplicity.) It had started studying the possibility of developing a massive space telescope in the early 1960s. As mentioned previously, a telescope in orbit could detect the many types of light that stars emit which the atmosphere blocks from ground-based telescopes.[70] It is very likely that the SACC reviewed the preliminary studies for the project when it examined NASA's astronomical programs at several meetings during the late 1960s.

NASA awarded Boeing, Martin-Marietta, and Lockheed small design contracts for the telescope's Support Systems Module in the early 1970s. Of the three, only Lockheed is known to have had a long history of building spacecraft for the National Reconnaissance Program. It was the prime contractor for CORONA from 1958 to 1972 and specifically developed the Agena upper stage for it. Lockheed also furnished the Agena for the GAMBIT program's two high-resolution satellites, which flew from 1963 to 1984. It provided the Satellite Basic Assembly for the HEXAGON launched from 1971 to 1986. At 60 feet long and 10 feet in diameter, the HEXAGON was considerably larger than NASA's planned space telescope and certainly gave the company considerable expertise in building large spacecraft.[71]

Itek and Perkin-Elmer received small design contracts for the project's Optical Telescope Assembly during the same period. Both had extensive experience supplying image-forming sensors (some of which had mirrors) for the National Reconnaissance Program. Itek built the IRIS II and DELTA III cameras flown on U-2s beginning in the 1960s, the panoramic cameras carried by CORONA, the 240-inch telescope and its strip camera flown once on a CORONA in 1963, and the mapping camera carried on 12 HEXAGONs from 1973 to 1980. It submitted a bid to the CIA in 1966 to build the panoramic cameras for the HEXAGON, but Perkin-Elmer's proposal was approved, and it ended up making them. Perkin-Elmer also manufactured the Type I camera flown on all 29 CIA A-12 strategic reconnaissance aircraft missions conducted in 1967–68.[72]

The two firms also had supplied image-forming sensors to NASA. Itek supplied the panoramic cameras (derived from the U-2's IRIS II), which flew in the service modules on the last three Apollo missions. Perkin-Elmer made the 12-inch telescope on the Stratoscope I balloons flown in 1957 and 1959, the 36-inch telescope carried on Stratoscope II balloons launched from 1963 to 1971, the mirror for the 38-inch ultraviolet telescope carried on the unsuccessful *OAO-B* in 1970, and the mirror for the 32-inch ultraviolet telescope on *OAO-3* which reached orbit two years later. Both companies also did extensive consulting work for NASA.[73]

The plans during the early 1970s were for the *Hubble*'s optical and ultraviolet telescope to have a 120-inch primary mirror and a much smaller secondary mirror. The former was larger than any that had flown on a NASA or NRO spacecraft. (The larger the mirror that collects light, the sharper the image of the object from which the light emanates if the mirror is very smooth.) Fabricating the primary mirror posed immense technical challenges. NASA and the astronomical community had agreed that the telescope should be diffraction limited, i.e., the deviation from perfection would by imposed by light and not by hardware. The technology did not exist at the time to achieve this degree of optical perfection with a large mirror. The mirror had to weigh less than a ton (in contrast, the 200-inch mirror at the Mt. Palomar observatory weighed more than 14.5 tons). It had to be invulnerable to the expansion and contraction in the extreme temperatures of space and had to be securely mounted to protect it from damage during launch. The *Hubble* was to be the most precisely pointed astronomical instrument ever. NASA's objective was that the telescope could be pointed to a position with an accuracy of 0.01 arc second or less and lock on a target for up to 24 hours without deviating more than 0.007 arc second. Consequently, designing the fine guidance sensors and pointing and control system to achieve these goals posed another major engineering challenge.[74]

The project faced strong opposition in Congress when NASA requested $6.2 million in planning funds during 1974. In the end, it appropriated a little less than half that sum and directed NASA to design a less expensive instrument and to obtain the participation of other nations. NASA and its contractors compared the proposed cost and scientific performance of the 120-inch primary mirror with a 94-inch and 71-inch primary mirror. The 94-inch resulted in more modest cost savings but considerably less degradation in scientific performance, while the 71-inch was the least expensive but resulted in a major loss of scientific performance. Perkin-Elmer and Itek stated that the two smaller primary mirrors could be built and tested at existing manufacturing facilities, but the 120-inch would require the expensive construction of a new plant. Weighing all the factors involved, NASA decided in May 1975 to proceed with the 94-inch.[75]

NASA had also decided that the telescope would be designed to permit all the major components except the optics to be replaced in orbit by Shuttle astronauts. It claimed that this would save money by not having to develop long-lived components and by not having to subject them to extensive ground testing. The national security agencies had studied the possibility of the Shuttle performing on-orbit repairs of their satellites and had expressed interest in the potential benefits.[76] However, it is not known whether they had a role in NASA's decision to build the telescope in this manner. At the very least, they must have considered the *Hubble* as a test case in this regard and closely followed the new and complex design and fabrication that enabled repairs on orbit.

The interagency mechanisms created in the 1960s to establish guidance and oversee NASA's use of sensitive technologies in its space-based imaging programs—the NSAM 156 Committee, Manned Space Flight Policy Committee, and Survey Applications Coordinating Committee— had disbanded by the early 1970s. Nevertheless, just as there was on issues related to Earth observation, there must have been extensive informal liaison between NASA and the NRO and other national security agencies after their disbandment on at least certain aspects of the *Hubble*. Along these lines, the NRO director wrote to the secretary of defense in late 1972 that his agency was offering NASA "highly refined optical manufacturing techniques applicable to their Large Space Telescope."[77] However, no details are available on this assistance.

The informal nature of the contacts regarding the *Hubble* and other space-based imaging programs ended when the secretary of defense, DCI, and NASA's administrator signed the Secret/TALENT-KEYHOLE "Memorandum of Agreement for the Conduct of Intelligence and Civil Space Programs" in the summer of 1975. Its objectives were to assure that the inadvertent disclosures of technology in the civil programs did not jeopardize the covert programs, data and information release policies recognized national security considerations, the design and management of overt programs avoided conflicts with the classified ones, and appropriate use was made by both sectors of technology, data, and information generated by the other. The agreement established several interagency groups to replace the defunct Manned Space Flight Policy Committee and Survey Applications Coordinating Committee. The Program Review Board was responsible for resolving interagency issues, coordinating the two sectors' Earth-sensing programs, recommending to agency heads changes in program direction that were mutually agreed upon, and reporting to the agency heads issues that could not be resolved. Its members were the DDR&E, NASA's deputy administrator, the NRO director, and the CIA's deputy director for science and technology. The board would refer unresolved interagency issues relating to national policy interpretation or recommended changes in national policy to the National Security Council. It would supervise two new interagency groups under it. The Technology Review Committee, chaired by a senior DoD official, was charged with assessing civil and military technology flows into the public domain and coordinating the development, transfer, and public release of new technologies. The Data and Information Release Committee, chaired by a senior NASA official, was responsible for reviewing policies and procedures concerning Earth-oriented science and applications data and information release and trying to resolve interagency disputes.[78]

An October 1975 memo from the NRO to the Office of the Secretary of Defense noted that the Program Review Board examined the *Hubble* project at its first meeting and that the two committees were following up on various unspecified action items.[79] However, what decisions they made are not known.

Despite scaling back the project and reducing its cost, NASA agreed with the Office of Management and Budget to not request any monies in the FY 1977 budget to begin development due to the higher priority of the Shuttle and other factors. After getting the European Space Agency's agreement to participate in the project, planning to use the Shuttle to launch the *Hubble*, and taking other actions NASA received the enthusiastic support of the White House and Congress the following year for FY 1978 funding.[80]

NASA issued the Requests for Proposals for the Support Systems Module and Optical Telescope Assembly in January 1977. Boeing, Martin-Marietta, and Lockheed submitted bids on the former, and NASA selected Lockheed. Using its highly classified facility in Sunnyvale, California, where National Reconnaissance Program satellites were assembled, the firm would integrate the *Hubble*'s many components into a flight-ready vehicle. It planned to construct a new Vertical Assembly and Test Area in which to work on the *Hubble*. Lockheed presumably received permission from the NRO to use part of the facility for the project.[81]

Perkin-Elmer and Kodak/Itek submitted bids on the Optical Telescope Assembly. The NRO had previously given permission to Perkin-Elmer to use part of its Danbury, Connecticut, facility for work on the Optical Telescope Assembly. After the CIA awarded it the contract to build the HEXAGON's cameras in late 1966, Perkin-Elmer constructed the plant to make them. It was primarily a BYEMAN facility, with access requiring that high-level clearance with only a few exceptions. Among other things, Perkin-Elmer proposed in its bid to use a new polishing method controlled by computers and to verify the specifications of the primary and secondary mirrors in separate tests. It is not known whether these methods were employed in making the mirrors for the HEXAGON's cameras or were classified in any respect.[82]

Kodak was a key contractor in the National Reconnaissance Program. It supplied the films used in the cameras on satellites, aircraft, and drones. The company built the cameras (which had much smaller mirrors) for the GAMBIT 1 and 3 high-resolution satellites. Kodak also reportedly made the KH-10 camera for the Manned Orbiting Laboratory before the project's cancellation in 1969 and the optical system for the KH-11, the first digital return system, which was initially launched in late 1976. It also had a long-standing relationship with NASA, supplying the Lunar Orbiter cameras and the films used in the human spaceflight programs. Kodak/Itek planned to use conventional polishing methods and to verify the specifications of the two mirrors with a full-assembly test.[83] Whether the two firms were going to use any of their facilities in which classified work was done is unknown. If so, they presumably received permission from the NRO.

NASA awarded Perkin-Elmer the contract. Most of the manufacturing would be done at the Danbury plant, but some preliminary work would be conducted at the nearby Wilton facility. The contract stated that Perkin-Elmer was authorized to use on a "rent-free, non-interference basis" a wide range of government-owned equipment accountable under an existing Air Force contract (undoubtedly for the HEXAGON's optical system). It also provided that the firm could modify the government-owned Thermal Vacuum Chamber and its support equipment. Kodak also received a contract to provide a backup 94-inch primary mirror in case Perkin-Elmer encountered problems with its mirror.[84]

The Marshall Space Flight Center had overall responsibility for the *Hubble*. It also directly oversaw the manufacture of the Optical Telescope Assembly, including the primary and secondary mirrors and the pointing and control systems (the Goddard Space Flight Center was responsible for its scientific instruments such as the Wide Field/Planetary Camera). Corning supplied the primary and secondary mirror blanks, and Perkin-Elmer started the rough grinding of the former at its Wilton facility in December 1978. In June 1980, the firm completed it and moved the primary mirror to Danbury for polishing, cleaning, and measuring. It is not known whether NASA personnel accessing the Danbury plant needed any clearances. To minimize the exposure of the BYEMAN-controlled work there for HEXAGON, the DoD and NASA agreed in the beginning to limit the number of NASA personnel with access to the Danbury plant, but the project manager at Marshall opposed the limitation and at some unknown date DoD agreed that additional NASA personnel could be assigned to Danbury. Most of the polishing and cleaning work was presumably done in non-BYEMAN-controlled areas where NASA personnel could be present. Along these lines, NASA released several photos of these operations. However, Perkin-Elmer's Optical Operations Division prohibited any access by NASA officials and even Perkin-Elmer quality assurance personnel to the metrology area when certain activities were under way, including the design, building, and testing of special hardware (such as the null correctors used to determine the shape of the *Hubble*'s primary mirror). The prohibition apparently resulted from some parts of the *Hubble*'s metrology operations involving either proprietary functions or classified technologies or processes.[85]

Perkin-Elmer completed the fine polishing of the primary mirror in April 1981. (Meanwhile, Kodak had finished the grinding and polishing of the backup primary mirror in 1980.) As often as several times a day during the process, the mirror was moved from the polishing machine to the null correctors to take measurements to determine where more polishing was required. One of the null correctors had been incorrectly assembled, and this led to the defect in the mirror that was not discovered until after the *Hubble*'s long-delayed launch in 1990. After completion of the polishing, cleaning, and measuring, Perkin-Elmer applied a new type of coating that was transparent to both ultraviolet and visible light.[86]

The company finished fabricating the Optical Telescope Assembly with the primary and secondary mirrors in September 1984. At the end of October, the entire structure was moved on a NASA aircraft from a nearby Air Force base in New York to Lockheed's facility in Sunnyvale, California, for integration with the Support Systems Module.[87]

There was DoD support to the building of the *Hubble* separate and apart from permitting Perkin-Elmer to utilize the Danbury facility and perhaps classified mirror manufacturing or testing processes. The Defense Advanced Research Projects Agency funded the design, fabrication, delivery, and installation of the two antennas booms made of a graphite-fiber and aluminum, which resulted in considerable weight savings and increased stiffness and dimensional stability. The Air Force tested the rate gyros in its Advanced Inertial Test Laboratory before installation. This was a seismically quiet facility that normally tested inertial guidance systems for missiles.[88]

One area in which the DoD apparently offered support but was turned down was testing the primary and secondary mirrors together (what Kodak/Itek had stated they would do in their bid and what was evidently done with all the mirrors built for imagery intelligence satellites). Such a test would have detected the flaw in the primary mirror that was discovered only after the *Hubble* reached orbit. Some senior NASA personnel involved with the project stated in 1990 after the defect had been detected that they were unaware such an offer had been made. This contradicted NASA's written responses to questions from a subcommittee of the Senate Appropriations Committee that same year, though. In them, NASA noted that it "considered using Department of Defense (DoD) optical test facilities to validate the primary optics. However, there was no facility that could satisfactorily verify the adherence to specification of a 2.4 meter, diffraction limited mirror. Without such a facility, the ability to meet the performance requirements of the HST program could not be confirmed until after its initial focusing while in orbit." NASA maintained in its responses that to have built such a facility would have cost $100 million and that the plans to test and verify the primary and secondary mirrors separately were approved by the contractors and the academic community.[89]

NASA announced in early January 1986 that the Shuttle would carry *Hubble* into orbit during an October mission, nearly three years after the original launch date. The Challenger tragedy later that month, of course, imposed another delay. The assembled *Hubble* remained at Lockheed in Sunnyvale and underwent several upgrades and further testing to improve it. In October 1989, NASA employed a shipping container used for reconnaissance satellites to move the *Hubble* from Sunnyvale to Kennedy Space Center on an Air Force C-5A. There was no publicity given to the flight. NASA finally launched the *Hubble* in April 1990.[90]

The national security community's assistance to the *Hubble* project did not end with its launch. Automated surfacing and metrology techniques originally developed by Tinsley Labs under a Strategic Defense Initiative Organization program were used by the firm to build and test the corrective optics for the space telescope and the main optical components for the new Wide-Field Planetary Camera.[91] As best as can be determined, these processes were unclassified.

Summary

Despite the fact that almost all of NASA's varied scientific pursuits in space outside of the applications satellite programs were conducted for only civilian purposes, there was considerable interaction with the defense and intelligence agencies. Photography of the Earth in the human spaceflight program received intense attention with the establishment of procedures to review proposed photographic experiments and to screen imagery before public release to ensure no political threat to the National Reconnaissance Program developed. The national security agencies began testing some of the cameras employed to ensure they did not exceed the resolution limits established in the 1965 NASA-NRO agreement. They granted NASA the first waiver of these limits for one camera flown in the Skylab program.

The DoD's insistence on flying a small number of classified experiments in Gemini revived the dispute which originally arose with the ANNA geodetic satellites over NASA conducting such activities in one of its spaceflight programs. In this case, NASA reached a compromise with the DoD under which the existence of the experiments was publicly acknowledged but selected data could be withheld. The DoD conducted several classified experiments in the last four Apollo missions, but little information is available on them.

The intervention in the lunar photography and astronomy programs was very different. In the former, the national security agencies permitted NASA to use classified cameras in the Apollo program under agreements governing their acquisition and the information that could be disclosed concerning them. They closely monitored the astronomy program beginning in the late 1960s and, as best as can be determined, quickly exempted it from the restrictions on NASA's use of classified technologies in imaging the Earth. However, their acquisition and the data that could be released about them were apparently also governed by the existing agreements applicable to lunar photography cameras or entirely new agreements. The national security community's involvement in the *Hubble* project was extensive. It shared data with NASA on relevant optical manufacturing techniques in the early 1970s, permitted an NRO contractor to use a highly classified plant and perhaps classified manufacturing and testing processes to manufacture the Optical Telescope Assembly, and imposed restrictions on NASA's access to the facility. It also provided antennas, a container for transporting the completed spacecraft, and testing facilities.

The next chapter discusses the early years of the Shuttle, the spaceflight program in which NASA had the closest and longest interaction with the national security agencies. It focuses on their massive influence on the project and NASA's abandonment of its policies of conducting spaceflight programs openly and for civilian objectives only.

CHAPTER 6

The Shuttle

NASA's Radically New Partnership
with the National Security Agencies

NASA's search for a major post-Apollo program was lengthy and frustrating. It began developing the Apollo Applications Program in the mid-1960s as the successor, but the political climate was changing rapidly, and only projects that demonstrated practical and tangible benefits would receive any support. Since the Apollo Applications Program did not have many, the White House and Congress soon drastically reduced its funding. NASA's alternative proposals for a manned mission to Mars and a space station promised even fewer such benefits and never went beyond the study stages.

NASA then advanced a reusable Shuttle on the grounds that it would be a cheaper and more reliable launch vehicle for all U.S. government payloads. To ensure the critical political support of the defense and intelligence agencies for the project, it formed an unprecedented partnership with them and incorporated their requirements for both a larger vehicle and performance specifications beyond what it needed. The project received White House and congressional approval in 1972, with an expected operational date of 1978. During the following years, NASA worked more closely with the defense and intelligence agencies than previously, designed and operated a spaceflight program to meet their more exacting needs, and sacrificed its guiding principles far beyond what it had done before.

While the DoD planned to fly about 25 percent of the missions, it consistently refused to pay for any orbiters, which were the single most expensive item. However, it did agree early on that it would construct a launch complex at Vandenberg Air Force Base, develop the Interim Upper Stage to boost civilian and national security payloads into higher orbits, and for its classified missions install a secure command and control system at NASA designated Controlled Mode.

The Shuttle was facing severe budget and technical problems by 1977, and it was clear that the first flights would not take place the following year. Only the strong support of the secretary of defense, the DCI, and the NRO director at the White House saved the program from drastic reduction that year. President Jimmy Carter soon directed that the system become the exclusive launch vehicle for all U.S. government payloads, with priority being given to national security missions. He would continue to strongly support the Shuttle during the remainder of his administration.

Major budgetary and technical problems continued, and NASA postponed the first orbital fight several times to 1981. It was clear that the Shuttle would meet neither the projected flight rates nor the original performance specifications required for selected future national security payloads. As a result, NASA began studying several performance-enhancing measures.

The DoD's transition plans from expendable launch vehicles (ELVs) to the Shuttle moved slowly, as many of its officials were highly skeptical of the unproven system. However, firm schedules were finally developed during the Carter administration under pressure from the White House and the civilian leadership of the national security community. Initially, these directed that selected satellites would begin transitioning in 1981 at Kennedy Space Center and in 1983 at Vandenberg. The DoD would have ELVs as backups for most of these satellites during the first two years of flights from each complex, and it would continue to use ELVs as the primary launch vehicle for other payloads during this period. If the Shuttle were successful, production of ELVs would end in 1982, and the inventory of ELVs would be used up by the end of 1985. Thereafter, the Shuttle would launch all national security payloads. Payloads scheduled for launch on the Shuttle during the transition had to be specially designed and tested to make them dual-compatible, which was a costly and complex process. Satellites designed and built for launch only on the Shuttle were Shuttle-optimized. Because of their size, weight, and other factors, they either could not be launched at all on existing ELVs or could only be after expensive reconfiguring. The transition plans would undergo many changes as the Shuttle program continued to experience various delays and technical problems.

During the Carter administration, the defense and intelligence agencies began planning for their own Shuttle mission control center in Colorado to provide them with complete control of their classified flights and a higher level of security than Controlled Mode. They also undertook an unprecedented review of all the proposed civilian scientific experiments for the early Shuttle missions and imposed restrictions on some.

Despite all the turmoil in the Shuttle program, the civilian leaders of the national security community maintained their strong support for it at the White House and Congress and concluded that the problems had not yet caused any degradation in any national security space program.

During the Shuttle era, NASA experienced many leadership changes. President Nixon selected James Fletcher to succeed Thomas Paine as administrator in 1971, and he served in that position until 1977. Fletcher had previously worked at the Guided Missile Division of the Ramo-Wooldridge Corporation and Aerojet and was a cofounder of the Space Electronics Corporation. After leaving NASA, he returned to the private sector and also served on several key government advisory boards, including the one that helped develop the Strategic Defense Initiative. Fletcher served as administrator again from 1986 to 1989. George Low, deputy administrator from 1969 until 1976, had worked at the National Advisory Committee on Aeronautics for many years before joining the newly established NASA in 1958.[1]

Robert Frosch was chosen by President Carter in 1977 to replace Fletcher. He had been director of nuclear test detection at and then deputy director of the DoD's Advanced Research Projects Agency before becoming the assistant secretary of the Navy for research and development in 1966. From 1973 to 1977, Frosch had worked in a United Nations environment program and then at the Woods Hole Oceanographic Institution. Alan Lovelace, deputy administrator from 1976 until 1981, had held several Air Force scientific and technical positions for many years before coming to NASA in 1974 as the associate administrator for aeronautics and space technology.[2]

President Ronald Reagan selected James Beggs to be the new NASA administrator, and he served in that position from 1981 to 1986. Beggs had extensive experience in the federal government at NASA and the Department of Transportation and in the private sector at Westinghouse and General Dynamics. Hans Mark was deputy administrator from 1981 to 1984. Prior to this, he had been a physicist at the Atomic Energy Commission's Lawrence Livermore National Laboratory during the early 1960s, director of NASA's Ames Research Center from 1969 to 1977, NRO director from 1977 to 1979, and secretary of the Air Force from 1979 to 1981. Founder of a defense consulting firm and member of a number of government advisory boards dealing with national security issues, William Graham succeeded Mark as deputy administrator in 1984. Dale Myers, a longtime aerospace executive and consultant, replaced Graham in 1986 and served as deputy administrator until 1989.[3]

The Decision to Build the Shuttle

NASA's search for a post-Apollo human spaceflight program turned to a space station and a manned mission to Mars once it was evident that budget cuts in the Apollo Applications Program would prevent it from meeting the original objectives. It began examining the possibility of a reusable launch vehicle to supply the station, awarding small study contracts to Lockheed, North American Aviation, General Dynamics, and McDonnell Douglas in January 1969. An advisory group headed by Charles Townes, a Nobel Prize–winning physicist, submitted its report on space policy to the new Nixon administration later that month. Among other things, the majority recommended against developing a space station or a Shuttle or attempting any manned landing on Mars.[4]

The president soon established the Space Task Group primarily to give him another set of recommendations on the nation's future space programs, including NASA's post-Apollo programs. Chaired by Vice President Spiro Agnew, its other members were Thomas Paine, NASA's administrator, A. Lee DuBridge, Nixon's science advisor and head of the Office of Science and Technology, and Robert Seamans, formerly NASA's deputy administrator and now secretary of the Air Force. The Space Task Group quickly called for a new study on space transportation, and Paine and Seamans agreed to a joint committee to conduct it. George Mueller, NASA's associate administrator for manned space flight, and Grant Hansen, assistant secretary of the Air Force for research and development, were the cochairs.[5]

The committee's classified June 1969 report to the Space Task Group enthusiastically endorsed building a Shuttle and greatly expanded the grounds for it. Although the primary purpose was still to supply a space station, the committee also concluded that it could launch satellites, repair spacecraft in orbit, retrieve and return satellites back to Earth, and inspect foreign spacecraft. National security payloads greatly influenced the capacity and dimensions of the proposed vehicle. The report set forth a requirement for a 50,000-pound payload capability to accommodate "advanced surveillance missions, propulsive stages and payloads for synchronous orbits or escape trajectories, mission equipment for crisis information or attack assessment, and cargo and/or passengers in a passenger module." A cargo bay diameter of 15 feet was needed for "space station logistics support, propulsive stages, and satellites such as advanced versions of TACSAT or of surveillance systems." The cargo bay had to be 60 feet long for "ocean surveillance spacecraft, stages-plus-payloads for synchronous missions, or two medium altitude surveillance satellites." The largest and heaviest known classified spacecraft under development during this period was the HEXAGON imagery intelligence satellite (the successor to CORONA), which was to be launched in 1970 from Vandenberg into low-Earth, near-polar orbits. By 1968, its configuration had been fixed at 10 feet in diameter, almost 59 feet long (with the mapping camera module), and an estimated weight of 25,000 pounds at liftoff. There is no information available on any of the other satellites. The report also included a requirement for a cross-range of 1,500 nautical miles. This was driven by the need to potentially fly a one-orbit mission from Vandenberg and permit the vehicle's safe return to a landing strip on the West Coast after one orbit (because of the Earth's rotation, the Shuttle would be around 1,200 miles west of California). None of NASA's designs incorporated all of these requirements, and to ensure continued national security agency support for a Shuttle, NASA specifically agreed to include the Air Force, NRO, and others in the further design work and policy decisions.[6]

After receiving considerable input from a number of government agencies and outside organizations such as the American Institute of Aeronautics and Astronautics, the Space Task Group submitted its report to the president in September 1969. It basically set forth three options, each of which contained the same major new programs of a Shuttle, a series of space stations, and a crewed expedition to Mars. They only differed with respect to the timetable. Nixon did not commit to any option and gave only a cautious endorsement of NASA's proposed human spaceflight programs in March 1970. Along with budget cuts imposed by the White House and severe criticism of NASA's proposals in Congress, this resulted in NASA dropping plans for a Mars expedition and space stations that year. If there was to be any post-Apollo human spaceflight project, it would be the Shuttle.

Paine and Seamans signed an agreement in February 1970 that stated the proposed Space Transportation System would consist of the Earth-to-orbit space shuttle and that its goal was "to provide the United States with an economical capability for delivering payloads of men, equipment, supplies, and other spacecraft to and from space by reducing operating costs and order of magnitude below those of present systems." The agreement established the NASA/USAF Space Transportation System Committee to continually review the program and recommend steps "to achieve the objectives of a system that meets DoD and NASA requirements." NASA's associate administrator for manned space flight and the assistant secretary of the Air Force for research and development were the cochairs. Although the agreement expressly stated that the project would be "generally unclassified," the agency heads had agreed separately that justifications for DoD performance requirements could be treated as classified information. Secretary of Defense Melvin Laird, Seamans, and others repeatedly stated to NASA that the DoD did not have funds for the Shuttle's development and that they expected NASA to fully pay for it.[7]

The Space Transportation System Committee proved to be an important body. In 1970, it published a general security guide for the program. DoD representatives gave a classified briefing (undoubtedly at the codeword level) to the committee on their requirements for the Shuttle. Their current and projected payloads still mandated that it have a 60 by 15 foot cargo bay but an increased capability of carrying up to 40,000 pounds into low-Earth, near-polar orbits from Vandenberg and 65,000 pounds (including an upper stage) into low-Earth, low-inclination orbits from Kennedy. The cross-range ability was now reduced to 1,100 nautical miles. DoD personnel cautioned that if the Shuttle did not meet these requirements, then it could not transition completely from ELVs to the new vehicle as planned and that it would have to keep and improve its inventory of ELVs. NASA and the DoD presented their separate studies on the joint requirement for a space tug, orbit-to-orbit Shuttle, or upper stage that would boost payloads from the Shuttle to higher orbits or permit the repair or retrieval of satellites in higher orbits. The committee oversaw the creation by the Air Force's Space and Missile Systems Organization of a Shuttle office in Los Angeles and the beginning of detailing personnel from that organization to the three NASA centers with major responsibilities for the program. It approved "technology sensitivity guidelines" regulating the disclosure of Shuttle data to foreign industry and governments.[8]

NASA convened a meeting with Shuttle study contractors and DoD personnel in January 1971 for the purpose of defining the vehicle's requirements to guide the work of the contractors. Although the DoD requirements presented to the Space Transportation System Committee were far more than what NASA needed and made the Shuttle considerably more complex and expensive, it agreed that they would be met. NASA, of course, realized that if it did not do so, the national security agencies would not support the project and it would never get approved.[9]

The Office of Management and Budget (OMB) during much of 1971 advocated for a reduced NASA FY 1973 budget that would essentially terminate the Shuttle project. James Fletcher, NASA's new administrator as of June 1971, and other top NASA officials repeatedly argued against the cuts and in the end persuaded the OMB to at least present an option for a less expensive and smaller Shuttle (with a 30 by 10 foot cargo bay and a payload capacity of 30,000 pounds from Kennedy) to President Nixon in early December. He quickly approved it. However, NASA continued to strenuously argue for a larger Shuttle that met the DoD requirements of a 60 by 15 foot cargo bay and a payload capacity of 65,000 pounds from Kennedy or, as a less preferable alternative, one with a 45 by 14 foot cargo bay and a payload capacity of 45,000 pounds from Kennedy.[10]

Edward David, Nixon's science advisor, met with top NASA and OMB officials on 3 January 1972. NASA was hoping to at least obtain approval of the smaller of the two vehicles it had lobbied for, but the OMB director surprised everybody by announcing that the president had approved the full-sized Shuttle. Several factors influenced Nixon's decision: the need to maintain U.S. leadership in human spaceflight, the Shuttle's potential value to the DoD, and the positive effect the program would have on the aerospace industry. He publicly announced the decision on 5 January. Among other things, Nixon stated that the Shuttle would likely replace all but the largest and smallest ELVs (the remaining Saturn Vs and Scouts that were still being produced, respectively) and might bring operating costs down to as low as one-tenth of those of conventional rockets. He also set forth the goals of the first crewed flight in 1978 and an operational system shortly thereafter.[11]

The Shuttle's payload capacity was far greater than any NASA or Air Force ELV then in use for robotic spacecraft. NASA's Scouts could only boost very small payloads and its Deltas medium payloads into low-Earth orbits. Employed only at Cape Canaveral, the Air Force Atlas Agena could launch 8,500 pounds into low-Earth orbits and 625 pounds into geosynchronous orbit and the Atlas Centaur could put 11,650 pounds into low-Earth orbits and 2,050 pounds into geosynchronous orbit. Its Atlas E could place 2,300 pounds into low-Earth orbits from Vandenberg. Several versions of the Titan III were used by the Air Force as a heavy-lift vehicle. The Titan IIIB Agena, launched only at Cape Canaveral, could boost 10,200 pounds into low-Earth orbits and 1,035 pounds into geosynchronous orbit. Titan IIICs with a Transtage, also used only at Cape Canaveral, could place payloads 10 feet in diameter, 15 feet long, and 29,200 pounds into low-Earth orbits and 3,220 pounds into geosynchronous orbit. Titan IIIDs, employed only at Vandenberg, could boost spacecraft 10 feet in diameter, 60 feet long, and weighing 25,000 pounds into low-Earth orbits.[12]

The payload capacity also greatly exceeded that of any subsequent ELVs developed by NASA or the Air Force before the Shuttle became operational. Improved Atlases, Deltas, Scouts, Titan IIIBs, Titan IIICs, and Titan IIIDs entered service, but the increases in their capacities were small. The new Titan IIIEs used for NASA's interplanetary missions from 1975 to 1977 could launch spacecraft 14 feet in diameter, 55 feet long, and weighing 12,000 pounds into an escape orbit. The new Titan 34D, which the Air Force began developing in 1978 and initially launched in 1982, could place satellites 10 feet in diameter, 60 feet long, and weighing 27,600 pounds into low-Earth orbits from Vandenberg. With the most advanced upper stage, it could launch payloads of 32,900 pounds into low-Earth orbits and 4,000 pounds into geosynchronous orbit from Cape Canaveral.[13]

Shuttle Development Begins, and the DoD Examines Its Potential Uses

The final design for the Space Transportation System (Shuttle) included the reusable orbiter, two reusable solid rocket boosters, an expendable external tank to supply fuel to the orbiter's main engines, and associated ground support facilities. NASA had overall management authority for the system and was responsible for the development and acquisition of the orbiters and other flight hardware; launch, integration, and landing facilities at Kennedy; flight planning and mission control at Johnson Space Center; and communications with the orbiters through its Spaceflight Tracking and Data Network controlled at Goddard Space Flight Center. These, of course, represented the vast majority of costs associated with building and operating the system. This huge disparity between NASA and the DoD's financial contributions to the program would continue. NASA awarded the contract for the orbiters in July 1972. It was planned that the first two would be built during the development phase for testing and then be refurbished for spaceflight. The last three orbiters would be built during the production phase when the vehicle was declared operational.[14] It should be noted, however, that despite the plans for five orbiters, NASA's budget only included funding for three.

NASA planned to complete the first orbiter in mid-1976 for use in atmospheric flight testing, the initial crewed orbital flight in 1978, and the first operational flight shortly thereafter. With the DoD, it developed a revised mission model with 445 flights in the first 11 years of operations—6 in 1978, 15 in 1979, 24 in 1980, 32 in 1981, 40 in 1982, 60 annually from 1983 through 1987, and 28 in 1988. The DoD would conduct 34 percent of these missions. Around 60 percent of the national security payloads required some sort of an upper stage to place them in higher orbits.[15]

NASA and the DoD decided in April 1972 that Kennedy and Vandenberg would be the launch sites. Both would use Kennedy for placing payloads into geosynchronous, escape, or other low-inclination orbits and Vandenberg for satellites in low-Earth, high-inclination orbits. NASA was responsible for all facilities and launch and recovery operations at Kennedy for all civil and military missions, the first of which was scheduled for 1978. The Air Force was responsible for doing the same at Vandenberg for all civil and military missions, the first of which was scheduled for 1982.[16] This became the DoD's first major cost in the Shuttle program.

Gen. Jacob E. Smart, NASA's assistant administrator for DoD and Interagency Affairs, wrote Carl Duckett, the CIA's deputy director for science and technology (DDS&T), in May 1972 concerning the effect of the Shuttle on new space systems the CIA was developing. Duckett accepted the offer of having a NASA engineer brief his people.[17] There is no further information available on how this liaison evolved.

Both NASA and the DoD initiated detailed studies on exactly how the Shuttle should be used. John Foster, director of defense research and engineering (DDR&E), requested in May 1972 that Frank Lehan, co-founder with James Fletcher of Space Electronics Corporation in the late 1950s and a longtime NRO consultant, to prepare a study of future DoD space operations using it. Lehan submitted his Top Secret/BYEMAN "Space Shuttle Implications on Future Military Space Activity" within a few months. Although the report remains classified, partially released comments on it by NASA's administrator, the CIA's director of special projects, and the CIA's DDS&T shed some light on it.

Lehan's general recommendations included the necessity of NASA keeping Shuttle operating costs at the projected level or even lower, reducing payload costs, and building an upper stage to place payloads into higher orbits. Although the Shuttle had the capability to launch larger and heavier payloads into polar orbits from Vandenberg, the Titan IIID's payload capacity and packaging flexibility were sufficient for most of the future national security satellites envisioned for launch from that complex. If the Shuttle achieved considerable reductions in launch costs, it might change the current policy from deploying a limited number of signals intelligence satellites into geosynchronous orbit from Cape Canaveral to launching a larger number into low-altitude orbits. Its increased capacity to place payloads such as communications satellites in geosynchronous orbit from that facility, if accompanied by reduced launch costs, made it very attractive for this mission. However, these flights needed an upper stage that had not been developed yet.

Lehan noted that there were no unique reconnaissance or intelligence missions that the Shuttle's ability to recover satellites in space and return them to Earth made feasible. It did provide the opportunity to perform three potentially important tasks—inspecting and testing satellites before their deployment, repairing satellites in orbit, and building antennas in space. With respect to on-orbit repairs, however, the design compromises needed to permit them were unknown and the reliability of components was constantly improving, which might render such repairs unnecessary. Since large antennas were most needed on satellites in geosynchronous orbits, a means of moving them to higher orbit after their construction in low-Earth orbit had to be built.[18]

Fletcher soon proposed to Foster that NASA and the DoD establish a joint user committee to determine the practical civil and military uses of the Shuttle. It would complement the joint Space Transportation System Committee, which oversaw the development of the vehicle. Foster rejected the proposal because of "the nature of military space activities." The DoD instead formed the DoD Shuttle User Committee. It was chaired by an Air Force officer, and the other members came from the Office of the Secretary of Defense, the Joint Chiefs of Staff, and the three military services. NASA was not completely shut out of the Committee, however, as one of its officials sat as an observer.[19]

The August 1973 "DoD Space Mission Model (FY 1980–FY 1991)" listed all the current and future national security payloads planned to be launched during that period and their dimensions, weights, orbital elements, and expected transition dates to the new vehicle. (The Navy's Transit navigation satellite program was the only one at the time that would not transition.) Reconnaissance satellites were referred to only as Support Mission I through IV, while all others were listed by their actual name. The only spacecraft with a diameter greater than 10 feet was Ocean Surveillance II, which was to transition to the Shuttle at Vandenberg in 1983 for placement in a low-Earth, near-polar orbit. The heaviest to be boosted from that complex was a Support Mission I satellite with a diameter of 10 feet, length of 60 feet, and weight of 24,000 pounds. This was the HEXAGON, scheduled to transition to the Shuttle in 1984. Other spacecraft to be boosted by the new vehicle from Vandenberg beginning in 1983 or 1984 included the Satellite Data System, Defense System Application Program, Defense Navigation Satellite System (concept no. 2), and Support Mission IV.[20]

The largest satellite to be launched from Kennedy was Support Mission III (10 feet in diameter and 35 feet long), while the heaviest were Support Mission II and Defense Support Program at 3,200 pounds each. These and the other satellites to be launched from Kennedy—Defense Satellite Communications System, Fleet Satellite Communications, Survivable Satellite Communications System, Ocean Surveillance III, Defense Navigation Satellite System (concept no. 1), Spacetrack Augmentation Satellite, and two Space Test Program experiments—required an upper stage to place them in higher orbits. All were to transition to the Shuttle from 1980 to 1984.[21]

The Lehan report led to the formation of a subpanel of the Space Transportation System Committee to speed up the development of a geosynchronous orbit capability. NASA and the Air Force had been jointly studying a space tug for some time, but by 1973 they concluded that it could not be developed until the mid-1980s, and they began examining using existing upper stages as an interim measure. Deputy Administrator George Low asked DDR&E Malcolm Currie in September 1973 about the Air Force funding and managing the development of the orbit-to-orbit stage to reduce the pressures on NASA's budget and to give the DoD a greater stake in the Shuttle program. They tentatively agreed that the DoD would fund development of a reusable Interim Upper Stage by modifying an existing expendable upper stage. If this could not be achieved for less than $100 million, the vehicle would be expendable. After a favorable endorsement by the NRO director and the deputy secretary of defense, Defense Secretary James Schlesinger approved the agreement in early 1974.[22]

The Fourth and Fifth Orbiter Issue during the Ford Administration

NASA and the DoD had agreed for some time that, to fully meet civilian and national security requirements, at least one and possibly two additional orbiters were required beyond the three included in NASA's budget. However, there was no agreement on who would pay for them, and this issue surfaced several times beginning in 1974. Certain elements within the DoD were skeptical of the Shuttle's value and had persuaded Secretary Schlesinger to proceed slowly in committing DoD resources to it. He met with Fletcher in July 1974 to discuss the program and, although expressing support for it, he rejected any DoD purchase of orbiters.[23]

The joint NASA-DoD Aeronautics and Astronautics Coordinating Board addressed the question in early 1976, and the DoD representatives continued to refuse to pay for the orbiters. Its cochairs, NASA's deputy administrator, George Low, and the DDR&E, Malcolm Currie, jointly recommended to the secretary of defense and NASA's administrator in May that a fourth and fifth orbiter be acquired and that NASA pay for them. To prepare the FY 1978 budget, James Lynn, director of the OMB, soon requested that NASA and the DoD jointly examine the matter. Fletcher met with Secretary of Defense Donald Rumsfeld in August to request again that the DoD pay for the two new orbiters, but Rumsfeld refused to do so on the basis that if it did, using the Shuttle would no longer be cost-effective and it would have to greatly cut back critical weapons system improvements to procure them. As it had done many times before in the

Shuttle program and would continue doing, NASA acceded to the DoD. Submitted in September, the joint report concluded that on the basis of the projected number of 560 flights (50 percent NASA, 20 percent DoD, and 30 percent other users), a fourth and fifth orbiter were needed, and it stated that NASA had agreed to procure them at an estimated cost of $1 billion.[24]

DCI George Bush wrote Lynn in October 1976 that the intelligence agencies supported the report's call for the fourth and fifth orbiters based on the current mission models, but noted that they were concerned about the estimated high transition costs from ELVs to the Shuttle. He stated that the NRO director was planning in most cases to transition to the Shuttle at the same time as implementing major satellite system upgrades (thus no existing satellite designs would be changed solely to permit their launch on an orbiter). Bush concluded by stating that the Shuttle "will provide a significant increase in capability to the National Foreign Intelligence Program," but to do so it "must maintain an adequate launcher and facilities capability and keep its user costs close to current projection."[25]

DDR&E Currie wrote to Bush concerning an ad hoc high-level interagency group that examined transition and backup ELVs. It concluded that backups were needed for both military space systems (to transition beginning in 1980) and intelligence space systems (to transition beginning in 1982) and that payload design must be flexible to permit launch on either an orbiter or ELV. The Shuttle should launch between four and eight priority military satellites before any intelligence satellites. Backup ELVs should be acquired for all DoD payloads scheduled for launch on the Shuttle in the first two years of its operations at Kennedy and Vandenberg. ELV production was currently scheduled to continue through 1982, and a decision could be made at that time on whether to maintain or end it depending on how many milestones the Shuttle had met.[26]

Shuttle Operations' Pricing and Security

While discussions were under way in 1976 on the fourth and fifth orbiters, NASA and the DoD began negotiations on the issue of what the latter would pay for utilizing the Shuttle. They ended in March of the following year when the two organizations signed an initial agreement entitled "Basic Principles for NASA/DoD Space Transportation Launch Requirements." Under it, the DoD was obligated to provide launch support at Vandenberg for all non-DoD users in return for NASA furnishing all launch support at Kennedy and flight operations support for all DoD missions. The DoD would pay NASA $12.2 million in FY 1975 dollars per dedicated DoD flight from either Kennedy or Vandenberg during the first six years of Shuttle operations. After this period, the price was to be adjusted annually. NASA would charge non-DoD users roughly $20 million in FY 1975 dollars for each flight from Kennedy.[27]

The survivability of U.S. military and intelligence spacecraft, including the Shuttle, against anti-satellite weapons and other threats received a great deal of attention after the Soviets resumed testing of their co-orbital interceptor in 1976. Brent Scowcroft, the president's assistant for national security affairs, directed the National Security Council to review near-term measures to reduce the vulnerability of U.S. satellites, the necessary steps to afford protection over the next 15 years, and the options to develop a U.S. anti-satellite capability. The NSC Survivability Panel, whose members were civilian experts, issued interim and final reports during 1976. Although they remain classified, other documents shed some light on them. The reports concluded that the Soviets probably already had a "limited operational capability with their non-nuclear interceptor against U.S. low altitude satellites," but there was no evidence of any capability against U.S. high-altitude spacecraft. There were numerous near-term countermeasures the United States could use to minimize the impact of Soviet anti-satellite weapons, and the technology existed to provide the protection when a decision was made to do so.[28]

President Ford issued National Security Decision Memorandum 333 in July 1976 on the subject. It directed the secretary of defense and the DCI to prepare an action plan to implement a series of still-classified short- and intermediate-term steps "to enhance the survivability of critical military and intelligence space capabilities against Soviet non-nuclear and laser threats at low altitudes and Soviet electronic threats at all altitudes." They were also requested to develop long-term measures to provide protection for both ground- and space-based elements of critical satellite systems from these threats and others such as radiation from nuclear detonations in space.[29] The action plan remains inaccessible, and it is not known what it recommended with respect to the Shuttle.

The related issue of security for preflight planning and operations, launches, and command and control of classified missions also began to receive intensive study in 1976. NASA had always conducted these activities itself, but with the Shuttle program it shared responsibility with the DoD to shift some of the financial burden and to provide security for classified missions and payloads. NASA would prepare the Shuttle for launch, process and integrate the upper stages and payloads, and install cargo in the orbiter bay for missions from Kennedy (the Air Force would do the same for missions from Vandenberg). Johnson would conduct flight planning, flight readiness, and flight control for all missions. Goddard would track and communicate with the Shuttle during all missions through its Spaceflight Tracking and Data Network ground stations and the planned constellation of Tracking and Data Relay Satellites designed to replace almost all of them. The Air Force Satellite Control Facility, the principal command and control network for DoD spacecraft, would apparently assist NASA's network in tracking and communicating with the Shuttle when needed. Its Satellite Test Center in Sunnyvale, California, would be responsible for controlling DoD payloads in the Shuttle's cargo bay, controlling the Interim Upper Stage when it was utilized to boost them into higher orbits, and controlling all DoD payloads once deployed.

A joint NASA-DoD study completed in 1977 concluded that Secret was the highest classification level needed to protect at least the initial DoD missions with military payloads such as communications and navigation satellites. Only after the Shuttle proved itself would reconnaissance satellites be launched that required a command and control system that could operate at the Top Secret or codeword level. The study recommended that the two organizations work to satisfy their individual requirements within existing and programmed facilities to the maximum extent possible. Along these lines, NASA's deputy administrator and DDR&E created an ad hoc Shuttle Security Group under the Aeronautics and Astronautics Coordinating Board to arrive at a mutually acceptable low-cost approach to allow NASA to support classified missions through the Secret level. Its members came from the Air Force, NASA, and CIA. NASA administrator Robert Frosch and the new secretary of defense, Harold Brown, approved its recommendations to implement an interim measure designated Controlled Mode at the Kennedy, Johnson, and Goddard Centers beginning in December 1982 and for the DoD to pay for it. Under Controlled Mode, there would be a separate DoD work area in the Mission Control Center at Johnson. This, along with hardware changes in shared equipment and other steps, permitted Secret-level flight planning, simulation, training, flight control, and post-mission analysis. DoD security guidelines and operational requirements would determine the manner in which common facilities were to be employed during national security missions. Not addressed was the extent to which NASA's Spaceflight Tracking and Data Network would be used to receive and send classified data considering its technical and political limitations on doing so.[30]

DoD's Review of Proposed Civilian Experiments

The planned experiments for the Shuttle were far greater in number and complexity than during any previous U.S. human spaceflight program. This was primarily due to the vehicle's large size, which enabled it to carry many more experiments and the much larger number of missions flown. Among others, the experiments involved communications, materials processing, astronomy, medicine, and remote sensing of the land, oceans, and weather.

The recently established joint NASA-NRO-CIA-DDR&E Data and Information Release Committee began examining proposed civilian experiments for the early Shuttle missions in 1976. Most of the experiments did not raise any national security concerns and were quickly approved, but there were a few that did. NASA's John Naugle, its chair, stated at an August 1976 meeting that the geological community and oil industry were pressing for a film camera (soon designated the Large Format Camera) capable of 10 meters ground resolution in color and producing stereo photography to image the entire world. The other members were initially opposed to flying the sensor based on the probable negative reaction of many foreign countries to having this economic intelligence acquired without their permission. As best as can be determined, the fact that the resolution exceeded the 20-meter limit from low-Earth orbit originally established by the 1965 NASA-NRO agreement was not a significant point. (As it turned out, Presidential Directive/NSC-54 issued in November 1979 lowered the maximum permissible resolution for space-based civil remote sensing to "at or better than ten meters . . . under controls and when such needs are justified and assessed in relation to civil benefits, national security, and foreign policy.") By the end of 1976, however, the defense and intelligence representatives on the Data and Information Release Committee no longer opposed the Large Format Camera provided that the photographic plans were reviewed by the national security community before it was flown.[31] NASA awarded Itek Corporation the contract in 1978 to develop the camera, which was capable of achieving 10 meters ground resolution with black-and-white film.[32]

The national security representatives of the Data and Information Release Committee also had reservations about the Shuttle Imaging Radar-A experiment. This instrument, similar to the *SEASAT-A* Synthetic Aperture Radar to be flown in 1978, would be the second radar imaging device openly flown in space by the U.S. government. Its purpose was to evaluate its potential in geologic mapping and to determine the value of radar imaging in conjunction with Landsat imagery for Earth resources investigations. The instrument's maximum ground resolution of 38 meters was less than that of the Synthetic Aperture Radar planned for *SEASAT-A*. In contrast to the Synthetic Aperture Radar's data, the Shuttle Imaging Radar-A's data would not be downlinked to ground stations but instead would be recorded on film for recovery and processing once the Shuttle landed. As with the Large Format Camera, the Committee's NRO, CIA, and DDR&E representatives did not object to the device being flown but insisted in late 1977 that NASA submit the data acquisition and dissemination plans to the Program Review Board, which oversaw it.[33] It is not known whether this was done and, if so, what the Board directed.

Finally, these members had concerns regarding the Electromagnetic Emanation Experiment planned for the third flight of Spacelab, a reusable laboratory to be built by the European Space Research Organization and carried in the Shuttle cargo bay on several missions. Sponsored by NASA, the experiment was designed to measure and characterize the actual and potential interference to spacecraft communications caused by Earth-based emitters operating over the frequency range of 0.4–40 Gigahertz. These frequencies were allocated by the International Telecommunications Union and used by many nations for satellite communications, terrestrial microwave links, and radars. The data collected would be analyzed by NASA and then shared with other federal agencies and the International Telecommunications Union.[34]

Although the exact national security concerns are not known, they were probably related to the possibility that the information might reveal the ability to intercept these emitters from space and the existence, location, and frequencies of certain U.S. and foreign microwave circuits and radars. Along these lines, the national security representatives on the Data and Information Release Committee had strongly objected in 1976 to the analysis and dissemination of the radio frequency interference data collected by NASA's *ATS-6* satellite the prior year over the United States and the ongoing collection of it by the spacecraft over the Indian Ocean. NASA agreed to stop the acquisition, analysis, and dissemination of the data and turn over to the NRO all materials connected to the experiment. Whether NASA was permitted to release any information subsequently is not clear.[35]

NASA informed the committee in late 1976 that a Phase B study of the Electromagnetic Emanation Experiment had been completed in close coordination with the NRO, and it was believed that the collection plan would not create any national security concerns. The study specified that at an inclination of 57 degrees, the experiment would acquire almost 59 hours of information over all the inhabited continents during a six-day mission. At an August 1977 meeting, the NRO reported to the committee that the experiment would not acquire "voice data nor intelligible communications and would have a ground transmitter location capability of no better than 50 nm radius." The committee then decided that NASA would prepare a memo for the Program Review Board regarding information acquisition, processing, analysis, dissemination, and interagency review procedures.[36] No information is available on whether this was done and, if so, what action the Board took. In any event, the experiment was never flown on the Shuttle, and it cannot be determined at this point why.

Shuttle Problems, Their Impact, and DoD's Critical Support during the First Years of the Carter Administration

The Shuttle program faced mounting technical and financial problems when Jimmy Carter became president. James McIntyre, the new director of the OMB, Frank Press, the new science advisor and director of the Office of Science and Technology, and Vice President Walter Mondale favored a drastic reduction. McIntyre developed two proposals to resolve funding shortfalls in the FY 1978 and 1979 budgets, the first of which was to acquire the three orbiters budgeted for and construct only the Kennedy launch complex and the second to procure four orbiters but build both the Kennedy and Vandenberg launch complexes. Secretary of Defense Harold Brown wrote to the president in early 1977 and noted that under either option the DoD would have to maintain ELVs for an extended period of time and that "we would probably opt to drop out as users of the Shuttle program." Brown strongly urged Carter to retain the current plan for five orbiters and two launch sites or, if this were not done, terminate the program.[37]

McIntyre then came up with a third option of keeping the planned five orbiters and two launch sites but with the DoD paying for the fourth and fifth vehicles. Hans Mark, the NRO director, and others advised Brown that what was important was not that a fifth orbiter be purchased but that the major components be bought so that it would be available within two years of a decision to proceed with assembly (an option designated "four-plus").[38]

At the end of November 1977, McIntyre convened a meeting attended by Brown, Mark, Deputy Secretary of Defense Charles Duncan, DDR&E William Perry, and DCI Stansfield Turner. McIntyre presented his office's original two options in slightly modified form and the more recent one for five orbiters and two launch sites, with the DoD purchasing the fourth and fifth vehicles. As before, he favored three orbiters and one launch site at Kennedy. Mark, with the support of Brown and Perry, argued strongly for the two launch complexes and five orbiters. However, they adamantly opposed the DoD paying for the final two orbiters. No decisions were made at the meeting.

A follow-up meeting attended by McIntyre, Brown, and Turner was held in December. The secretary of defense, supported by the DCI, maintained that the two launch sites and at least four orbiters and production of major components for a fifth were needed to meet national security requirements. They advanced a new argument that the fourth and fifth vehicles were needed since the first two operational ones would weigh more than the later ones and could not carry the very heaviest projected national security payloads. It is unclear whether NASA's new administrator, Robert Frosch, attended these November and December meetings. However, he presented NASA's position in correspondence and several separate meetings with McIntyre and Carter during this period. In the end, the president decided that NASA would build four orbiters—two full-capability and two overweight—with the provision that one overweight vehicle could be modified to a full-capability one if necessary and that a fifth could be procured if projected flight rates or the loss of an orbiter warranted it. He also affirmed that both launch complexes were to be constructed and operational by 1984, NASA and the DoD were to transition completely from ELVs, and the Interim Upper Stage was to be available by mid-1980.[39]

President Carter gave another major endorsement to the Shuttle program in May 1978 when he issued the Top Secret Presidential Directive/NSC-37 entitled "National Space Policy." The declassified portions concerning the Shuttle reaffirmed that NASA with the DoD's cooperation would build and operate the Space Transportation System to service all authorized users—domestic and foreign, commercial and governmental. They were to provide "launch priority and necessary security to military and intelligence missions while recognizing the essentially open character of the civil space program." National security missions could be dedicated. The agency conducting a mission was responsible for its control.[40]

Presidential Directive/NSC-37 also mandated the NSC Space Policy Review Committee to organize task forces on a number of issues related to the nation's civilian and defense space programs and submit their reports to the president by 1 September. One of the seven task forces organized was designated "Strategies to Utilize the Shuttle." Chaired by Hans Mark and John Yardley, NASA's associate administrator for space flight, its other members came from the CIA, OMB, Department of Commerce, National Security Council staff, and the Office of Science and Technology Policy.[41]

One of the key issues for the task force was the DoD's transition from ELVs to the Shuttle. During the FY 1979 budget reviews in the spring of 1978, Carter had expressed concern that this was occurring too slowly. This prompted a letter from Secretary of Defense Harold Brown to the president in June. He stated that the Shuttle would begin launching military satellites in 1981 and reconnaissance satellites in 1984 from Kennedy and would initially carry military satellites in 1983 and reconnaissance satellites in 1985 from Vandenberg. During this transition, ELVs would remain the primary launch vehicle for selected payloads until the beginning of 1985. Brown noted that spacecraft to be launched by the Shuttle in this period must also be capable of launch on ELVs so they could reach orbit if the former encountered delays or problems (dual compatibility). If the Shuttle were successful, Titan III production would end in early 1982 and the backup launch capability would be terminated at the end of 1985. He pointed out that beginning in 1985 there would be completely new spacecraft that would require the additional capability of the Shuttle because ELVs could not launch their increased weight and size (Shuttle-optimized). Additionally, there might be one or two new reconnaissance satellites recoverable from orbit. Brown stated that the DoD was also studying how a manned presence in space could contribute to future systems, particularly the erection of large structures on orbit, and assured Carter that he was continually reviewing the transition plans with DCI Turner and would keep him apprised.[42]

Achieving dual compatibility was complex and expensive. Titan IIICs with a Transtage placed many DoD payloads into higher orbits, but the Transtage was not compatible with the Shuttle. To provide a backup for these payloads and to achieve greater reliability and payload capacity overall during the transition period to the Shuttle, the Air Force began developing the Titan 34D in 1978 to replace both Titan IIICs and Titan IIIDs. Titan 34Ds could use the Inertial Upper Stage being built for the orbiters and boost heavier payloads—with the Inertial Upper Stage almost 800 more pounds than Titan IIICs into geosynchronous orbit from Kennedy and 2,700 more pounds than Titan IIIDs into low-Earth orbits from Vandenberg. However, satellites could not simply be removed from an ELV and launched on a Shuttle or vice versa. The acceleration, vibration, and acoustic environments were very different, and a payload built and tested for launch on one had to undergo extensive requalification before it could be launched on the other.[43]

A declassified memorandum to Carter from Frank Press is the only source of information on the recommendations of the Shuttle task force and the Space Policy Review Committee's review of it. The memorandum stated that all the agencies agreed that five unspecified military and intelligence systems should become candidates for accelerated transition to the Shuttle. No additional systems should be considered for this until it had demonstrated its flight readiness. The DoD's policies on backup ELVs as described in Brown's recent letter to the president were prudent and should be approved. Regarding other Shuttle issues, NASA, the DoD, and the DCI were to submit a report on the issue of joint versus separate NASA and DoD mission control centers before the FY 1980 budget review. An interagency task force was to examine by the same date improvements in the Shuttle such as extending the on-orbit time and giving it the capability to fly to geosynchronous orbit. Concerning the matter of survivability, the Shuttle for the time being should be limited to existing maneuvering capabilities and encryption of uplinks and downlinks. NASA, the DoD, and DCI were to prepare a report by August 1979 on what additional steps might be necessary to increase survivability. All these recommendations were essentially incorporated in full in the Secret Presidential Directive/NSC-42 from October 1978 entitled "Civil and Further National Space Policy."[44]

The president's science advisor soon directed the NRO director and NASA's deputy administrator to cochair a new interagency task force to study means of increasing the performance of the Shuttle. At this point, NASA estimated that the two overweight orbiters (the refurbished Structural Test Article, STA-099, and 102) could boost close to the required 65,000 pounds from Kennedy but less than 24,000 pounds from Vandenberg. As a result, it planned to use them only at Kennedy. NASA estimated that the last two orbiters (103 and 104) could launch 65,000 pounds from Kennedy but less than 27,000 pounds from Vandenberg. Consequently, it would employ them only at the latter site. The heaviest satellite then being boosted from Vandenberg was the HEXAGON, which now weighed about 27,000 pounds. However, there were plans to launch a new 32,000-pound imagery intelligence satellite from there beginning in the mid-1980s (subsequently designated Mission 4), and none of the orbiters could carry this spacecraft. Press also separately directed NASA, the DoD, and the DCI to prepare a report on whether a joint mission control center or separate NASA and DoD centers should be used in the later years of the Shuttle program. The DoD was also to address the need for a backup command and control center to its Satellite Test Center in Sunnyvale, California, for all of its space programs including the Shuttle.[45]

The DoD published a revised version of its "DoD Space Shuttle Transition Plan (FY 1977–1991)" in March 1979, which covered 20 existing and planned space systems. It was based on the projected initial flight of the Shuttle from Kennedy in 1980 and from Vandenberg two years later. All the transitions were "dependent on satellite development status and demonstrated Shuttle performance—technical and schedule. If the Shuttle schedule or technical performance changes, transition planning is impacted." The phase-out of ELVs was "related to the demonstrated Shuttle performance and the number of expendable boosters required to back up Shuttle missions," but this "is a complicated issue involving assessment of potential Shuttle slippages and of Shuttle unavailability due to possible grounding in the outyears."

Eleven current systems used Cape Canaveral. The DoD planned to transition four to the Shuttle at Kennedy in 1982: Defense Satellite Communications System, Special Mission V, Space Test Program, and

Transit. Between 1983 and 1985, six others would transition: Satellite Data System, Defense Support Program, Global Positioning System, Special Mission II, Special Mission III, and NATO III/IV. Fleet Satellite Communications would not transition. Two new ones (Nuclear Forces Communications Satellite and Deep Space Surveillance System) would exclusively use the Shuttle beginning with their initial launches in 1988. All 12 systems to be carried on the Shuttle would require the newly designated Inertial Upper Stage to place them in higher orbits, except for Special Mission V and some Space Test Program experiments. None required the entire 60 by 15 foot cargo bay or weighed near 65,000 pounds. Only one version of Special Mission V and several Space Test Program experiments had a diameter greater than 10 feet, and the longest payload was Satellite Data System and its Inertial Upper Stage at 43 feet. The heaviest payload was Special Mission III and its Inertial Upper Stage at 49,000 pounds.

Five existing systems used Vandenberg (at which the Shuttle launch complex was now expected to be completed in 1983). The DoD planned to transition all five in 1984 and 1985: Defense Meteorological Satellite Program, Special Mission I (one version of which was the HEXAGON), Special Mission IV, Space Test Program, and Transit. The Shuttle would exclusively launch two new systems (Clipper Bow and the Deep Space Surveillance System/LASS) beginning in 1988. All seven flew in low-Earth orbits and thus did not require the Inertial Upper Stage. Once again, none required the entire 60 by 15 foot cargo bay or weighed near 32,000 pounds. Only one Space Test Program experiment had a diameter larger than 10 feet, and just one version of Special Mission I (HEXAGON) was close to 60 feet long. The heaviest payload was HEXAGON at 27,000 pounds. For unknown reasons, the document did not list the new Mission 4 satellite weighing 32,000 pounds, which was scheduled to be launched in the mid-1980s as discussed above.

The Defense Support Program, Defense Satellite Communications System Block II, and Transit systems were the furthest along in the transition process, and the DoD estimated the total transition costs for all three through FY 1984 to be nearly $138 million. The original spacecraft designs for nine other systems were being "tailored for the Shuttle capabilities" and did not require backup ELVs.

The backup ELV policy now mandated that only the requisite number of Titans be available to launch the highest-priority national security payloads from Kennedy for two years after the initial Shuttle flight there in 1980 and one year from Vandenberg after the first Shuttle flight there in 1983. To achieve this, four complete Titan IIIs and two sets of long-lead Titan III materials were needed in addition to the Titan IIIs already in the inventory or coming off the production line. Decisions could be made in 1982 to expand the backup capabilities if the Shuttle experienced further delays or other problems.[46]

Those payloads with backup ELVs were, of course, dual-compatible. Existing systems at the time of their transition and completely new systems to be launched only on the Shuttle were undoubtedly Shuttle-optimized. For those which flew in high orbits that the Shuttle could not reach, this probably meant that they were shorter, wider, or heavier than if designed for launch on ELVs. For those which flew in low-Earth orbit, this possibly meant that they too were shorter, wider, or heavier but also capable of being serviced on-orbit. Shuttle-optimized satellites either could not be launched on existing ELVs at all or could be only after expensive reconfiguring.[47]

Despite these comprehensive plans, there were some who were concerned that the transition was moving too slowly. Hans Mark, who left the NRO and became secretary of the Air Force in 1979, informed the secretary of defense and DCI late that year that a major long-term problem at the NRO was the conversion of space systems to the Shuttle. He acknowledged that there were legitimate concerns over the Shuttle, including its technical and economic viability and the fear that a common launch vehicle would lead to loss of control over programs and endanger national security. Nevertheless, Mark urged that the NRO and other national security agencies maintain a strong commitment "to take advantage of the unique properties of the Shuttle" and "to ensure that proper organizational arrangements are developed so that the national security community retains adequate control over Shuttle operations."[48]

NASA's administrator informed the White House during early 1979 that the Shuttle program was continuing to experience major cost and scheduling problems and that without additional funding the first orbital flight and deliveries of the follow-on orbiters would be delayed. Additionally, problems with the main engines and thermal protection system were growing. The first orbital test flight was soon postponed from September 1979 to late 1980, with a good chance of a further delay.

Frosch testified before Congress in April 1979 that the program faced a cost overrun of $600 million over the next four years, only some of which could be covered by switching monies from the production budget to the development budget. He and McIntyre told the president in early September that the shortfall had now grown close to $1 billion. They added that the OMB, National Security Council staff, Office of Science and Technology Policy, and the DoD concurred with NASA's assessment that the schedule adjustments would not "affect the important initial SALT related launch scheduled for the Shuttle in early 1983, and all subsequent national security related missions."[49] The early 1983 launch was of some still-classified reconnaissance satellite.

Carter ordered an intensive review of the Shuttle program and met with Frosch, McIntyre, Mark (who was attending in the absence of Brown), Press, Zbigniew Brzezinski (the president's national security advisor), and their staffs on 14 November 1979. At the outset, he announced that he had decided to support the Shuttle program and emphasized its importance in performing national security missions. All the principals expressed their agreement with this course of action. Carter soon directed McIntyre to find the roughly $1 billion in extra funds to continue the program, which he did by taking it from an Air Force program to install new engines on KC-135 tankers.[50]

During the fall of 1979, NASA, DoD, and the DCI submitted their report on whether a joint Shuttle mission control center or separate facilities were needed and whether a backup site to the Satellite Test Center was needed for all of DoD's space programs. The report recommended that a Consolidated Space Operations Center be constructed at an Air Force base in New Mexico or Colorado to provide DoD with its own Shuttle mission control center and a backup to the Satellite Test Center for its robotic satellites. It argued that the new facility would afford growth potential for future space operations (the Satellite Test Center was near capacity), reduce the vulnerability of the Satellite Test Center, eliminate Johnson as the single facility for Shuttle mission control, provide security for mission control at a level higher than Secret, and permit complete DoD control of flights. The estimated cost was over $500 million to complete the new facility by 1985. Approval of the concept was given by the White House, and funding to begin construction was included in the Air Force's FY 1981 budget.[51]

In early 1979 the Air Force chief of staff (with the concurrence of the NRO director) directed the establishment of the Manned Spaceflight Engineer program under which active duty officers would be trained to serve as payload specialists on Shuttle missions flying classified DoD payloads. One Navy and 12 Air Force officers were selected and trained from early

1980 until late 1981. A second group of 14 Air Force officers was selected in 1982 and finished their training in early 1984. The Air Force picked a third group of five officers in 1985. Although NASA wanted to train the Manned Spaceflight Engineers as it had done with all of its astronauts, the Air Force opposed this and conducted the training. With the reduced number of DoD Shuttle flights due to various problems, the opportunities for employing the Manned Spaceflight Engineers were limited. Nevertheless, each of the two dedicated DoD flights in 1985 carried one. After the *Challenger* accident, the program came under increasing pressure and was finally cancelled in 1989.[52]

The Shuttle Program at the End of the Carter Administration

In early 1980 NASA and the DoD executed a new "NASA/DoD Memorandum of Understanding on Management and Operation of the Space Transportation System." Mark met with Frosch and Deputy Administrator Alan Lovelace several times during the negotiations for it. He insisted that the agreement include a provision that gave the DoD absolute priority in launch operations because of the critical importance of certain space systems to the nation's security. Frosch and Lovelace protested that this would result in the "militarization" of NASA. Their opposition was puzzling, since Presidential Directive/NSC-37 expressly provided for this. In any event, they dropped their opposition when Mark pointed out the vital support the DoD had given to the program and stated that without such a provision it would be impossible to resist the ongoing pressure within the DoD to retain an independent ELV capability. In addition to the provision that the DoD would enjoy launch priority, the agreement established two classes of national security missions. The first were those conducted by NASA in which its personnel would be working under the ultimate direction of DoD mission directors who had the final responsibility of achieving mission goals. The second were "Designated National Security Missions" in which an Air Force flight director would be responsible for "operational control, including flight vehicle and crew safety, through the Air Force chain of command."[53]

Throughout 1980 Mark and Brown continued to argue forcefully for the required extra funding for the Shuttle before Congress and emphasized the program's importance to national security. Although virtually no details are available, the CIA apparently also strongly supported the program at congressional hearings and various interagency forums. Two months before the November elections, Mark took the unusual step of meeting with Richard Allen, who was expected to become national security advisor if Ronald Reagan won. The meeting was arranged by Albert Wheelon (formerly the CIA's deputy director for science and technology and now a senior vice president of Hughes Aircraft). Over four hours, Mark discussed the Shuttle with Allen and believed he conveyed its vital role in satisfying defense and intelligence requirements.[54]

The Shuttle program encountered further delays during the last years of the Carter administration. Continuing technical problems and additional funding constraints caused NASA to postpone until 1981 the initial orbital test flight of *Columbia* (which was delivered to NASA in 1979) and the operational availability of this orbiter to late 1982. The second orbiter for spaceflight, Orbital Vehicle-099 (*Challenger*), was now to be delivered in December 1983; the third, Orbital Vehicle-103 (*Discovery*), in December 1984; and the fourth, Orbital Vehicle-104 (*Atlantis*), in 1985.[55]

Along with major postponements in their availability, NASA now estimated that none of the orbiters could meet NASA's performance goals of launching 65,000 pounds into low-Earth, low-inclination orbits from Kennedy or 32,000 pounds into low-Earth, high-inclination orbits from Vandenberg. Although as discussed above no current or future system set forth in the March 1979 "DoD Space Shuttle Transition Plan (FY 1977–1991)" required these performance levels, NASA and the DoD were particularly concerned about the vehicle's inability to conduct the new Mission 4 from Vandenberg planned for 1986, which now involved a four-man crew over seven days launching a payload weighing 32,000 pounds into a 98 degree, 150 nautical mile orbit and retrieving a 25,000 pound spacecraft.[56]

NASA planned a series of improvements to *Columbia* and *Challenger* to achieve the 65,000 pound capacity at Kennedy by the end of 1983, including a lighter external tank, increasing the main engine thrust level from 100 percent to 109 percent, and changes to the solid rocket boosters such as increasing the burn rate and reducing the case weight. However, these would still render them incapable of performing Mission 4 from Vandenberg. NASA was examining a number of possible additional improvements to *Discovery* and *Atlantis* so they could conduct it, including building them lighter, increasing the main engine thrust level to 115 percent, and augmenting the thrust by strap-on solid motors attached to the solid rocket boosters or a liquid boost module (Titan stage 1 engines and tanks) mounted on the aft end of the external tank.[57]

Other delays in the Shuttle program included the Air Force moving back the initial operational date of the Inertial Upper Stage one year to 1981 due to developmental problems. Slippages in orbiter deliveries, dramatic cost overruns, and the need to redesign some facilities at the Shuttle launch complex at Vandenberg caused the Air Force to postpone the opening of the site one more time to mid-1984. The original operational date of late 1982 for Controlled Mode operations at Johnson was pushed back, and it was decided that it would be built to handle only one classified mission at a time instead of two.

NASA postponed the launch of NASA's first Tracking and Data Relay Satellite from 1980 to 1983 because of developmental problems and delays in the Shuttle from which it would be launched into geosynchronous orbit on an Inertial Upper Stage. This impacted the security that could be provided for classified missions. The complete Tracking and Data Relay Satellite System included two operational spacecraft in geosynchronous orbit (one over the mid-Atlantic and the other over the mid-Pacific), one spare in geosynchronous orbit, and the ground station in White Sands, New Mexico. The constellation was designed to provide tracking and communications for up to 100 spacecraft separately and 25 simultaneously, with 85 percent coverage of satellites at orbital altitudes between 100 miles and 600 miles (including the Shuttle) and 100 percent coverage of satellites in higher orbits. This was a vast improvement over the coverage provided by NASA Spaceflight Tracking and Data Network ground stations around the world that could only communicate with spacecraft in low-Earth orbit about 20 percent of the time because of line-of-sight limitations. The Tracking and Data Relay Satellite System would also bypass the technical and political limitations that restricted many Spaceflight Tracking and Data Network ground stations from supporting national security space programs. Work-around solutions such as utilizing the worldwide network of Air Force Satellite Control Facility ground stations during classified missions were possible, but these facilities too would only be able to communicate with the Shuttle about 20 percent of the time because of line-of-sight restrictions.[58]

The Shuttle delays and other problems caused the DoD to again revise its plans for transitioning from ELVs to the Shuttle and for backup ELVs. They resulted in DoD postponing the planned 1983 transition of two satellite programs to 1984 and 1985 (these were evidently the Defense Support Program and Defense Satellite Communications System). Although a new satellite would have to be placed in orbit to avoid an expected operational gap in one case, this would actually result in lower costs by not having to design it to be dual-compatible. However, in the other case the delay would greatly increase costs as it would require the acquisition of an older model satellite (reconfiguration of the new model for launch on an ELV would be prohibitively expensive), conversion of a backup Titan 34D into a Titan IIIB, and acquisition of an Agena upper stage. Additionally, there were two unknown Shuttle-optimized spacecraft scheduled for launch on the Shuttle in 1983 and 1984, and these could not be launched by the Titan 34D without extensive and costly work. If those launch dates could not be delayed, in one instance the Titan/Centaur launch capability would have to be restored (at a cost of over $250 million), and in the other a new payload fairing would have to be developed (at a cost of $60 million).

The DoD still maintained the policy that backup ELVs were needed for critical national security payloads during the first two years of operations at Kennedy and the first year of operations at Vandenberg. Shuttle delays forced the DoD to reschedule two unspecified critical missions from that vehicle to Titan IIIs and thus reduced the number of backup ELVs from seven (five complete Titan IIIs and the acquisition of components for two more) to five. It explored various options for increasing the number of Titan backups and keeping the production line open, including the acquisition of three more complete rockets or three complete rockets per year over several years, but no firm decisions were made at the time. There were 13 planned launches of noncritical spacecraft scheduled for the Shuttle between 1983 and 1985 for which there were no backup ELVs. Air Force officials believed if backups were needed they could probably use some of NASA's Atlases and Deltas for at least the smaller payloads.

Notwithstanding the growing number of problems with the Shuttle program, DoD officials believed at the end of 1980 that they had not caused any known operational degradations in any of their space programs. However, they acknowledged these might occur if the existing problems continued or new ones developed.[59]

Summary

NASA had very ambitious plans for the post-Apollo period, but with little political support for a space station or a manned Mars mission, it ended up with only the Shuttle. To obtain the critical support of the national security agencies for the project, it entered an unprecedented partnership with them to build the Shuttle as a cheaper and more reliable launch vehicle for all U.S. government civilian and national security payloads. However, this mandated a larger and more complex vehicle. NASA abandoned its guiding principles of openness and pursuit of only peaceful and scientific objectives in space to a far greater extent than before. For the first time ever, it would openly and repeatedly carry classified payloads and conduct classified experiments, employ secure command and control procedures, and withhold extensive information from the public.

A series of technical and financial problems plagued the program during the 1970s, and the Shuttle had not flown in space by the end of the decade. Nevertheless, no national security space programs had been adversely affected yet. Furthermore, despite the skepticism of some officials toward the Shuttle, the national security community was still planning on completely transitioning to the Shuttle by the middle of the next decade.

The next chapter discusses the use of the Shuttle by the defense and intelligence agencies once it began operations, the impact on them of its failure to meet the projected flight rate or payload capacity, the steps they took in 1984 to maintain an ELV capability over NASA's fierce opposition, the result of the *Challenger* accident in greatly accelerating their abandonment of the Shuttle, and the overall benefits and costs of their participation in the program.

CHAPTER 7

The National Security Agencies
Abandon the Shuttle

NASA finally launched the first Shuttle into space in April 1981, over two years after the originally scheduled date. President Ronald Reagan initially reiterated his predecessor's directive that the Shuttle become the exclusive launch vehicle for all U.S. government payloads and that national security missions receive priority. The program continued to experience major problems, however. These included numerous issues with the main engines and heat protection tiles, a much longer turnaround time than estimated, which made it unable to meet the projected flight rate, and the inability to satisfy the original performance specifications. There were also delays with the Vandenberg launch complex, the Inertial Upper Stage, the Tracking and Data Relay Satellite System, and the Consolidated Space Operations Center in Colorado with its secure mission control center for DoD Shuttle operations.

NASA began examining further performance-enhancing measures and, with the DoD, started to modify the Centaur upper stage to double the weight of spacecraft the Shuttle could place in higher orbits from Kennedy. Reflecting the growing concern of the defense and intelligence agencies about totally relying on the Shuttle, Reagan ordered in 1982 that they might require "special-purpose launch capabilities." NASA continued lobbying hard for a fifth orbiter, but the president only ordered in 1983 that production of structural and component spares continue.

The many setbacks caused major problems with DoD's transition to the Shuttle. Along with starting to shut down the production lines of expendable launch vehicles (ELVs) in 1983, it abandoned the policy of having a backup ELV for critical payloads during the first two years of Shuttle operations from both Kennedy and Vandenberg and scheduled each of the small number of remaining ELVs to launch a specific satellite. Over NASA's bitter opposition, the DoD also obtained approval from the White House and Congress in 1984 to acquire 10 new heavy-lift ELVs to begin launching in 1988 critical payloads it wanted to remove from the orbiters. They would be capable of launching satellites the same size and weight as the Shuttle's original performance specifications.

Although the DoD had to cancel several planned Shuttle missions prior to the *Challenger* accident in 1986, out of the 20 flights flown before that it placed classified experiments or satellites on 3 and unclassified experiments and satellites on several more. The Controlled Mode finally became operational at the Johnson, Kennedy, and Goddard Centers, which provided security at the Secret level for preflight activities and mission control for the classified experiments and satellites. The DoD probably had some role in selected civilian experiments as well, particularly those involving photography with handheld and fixed cameras and the two involving Synthetic Aperture Radars.

The *Challenger* accident in 1986 greatly increased NASA and the DoD's problems in accessing space. NASA had planned 15 Shuttle missions in 1986, of which the DoD would fly 2 from Kennedy and 2 from Vandenberg in the first use of the West Coast complex. The remaining three orbiters were originally scheduled to be grounded for only a brief period after the accident, but flights did not resume until 1988. DoD's problem of accessing space was compounded by the grounding for almost a year of the six remaining heavy-lift Titans after one exploded during its April 1986 launch, the impossibility of acquiring any additional current ELVs for several years due to the previous closure of their production lines, and the inability to advance the delivery date of the 10 new heavy-lift ELVs.

In the months following the accident, NASA cancelled further performance-enhancing measures and development of the more capable Centaur upper stage for the Shuttle. These actions increased the number of national security payloads it could not carry. The DoD mothballed the unfinished Vandenberg launch complex because the Shuttle's performance limitations rendered it unable to launch any reconnaissance satellites from there and cancelled the Shuttle mission control portion of the Consolidated Space Operations Center. It also received approval to acquire more of the new heavy-lift ELVs and two new types of medium-lift ELVs. The DoD had already paid for nine Shuttle missions and planned to launch classified payloads on them which, as best as can be determined, were Shuttle-optimized and either incapable of being launched on ELVs or prohibitively expensive to reconfigure to permit that. It ended up flying classified payloads on eight dedicated national security missions from 1988 to 1992. During this period, the DoD also used the last of the existing ELVs and took delivery of and began employing the new ELVs. Once again, these vehicles became the DoD's only means of accessing space, and the longest and most far-reaching partnership between NASA and the defense and intelligence agencies came to an early close.

DoD's Participation in Early Shuttle Flights and Continuing Program Problems

In 1981 James Beggs became NASA's administrator and Hans Mark the deputy administrator. Their initial goals for the Shuttle program were to achieve operational status, acquire at least a fifth orbiter, develop a usable upper stage, and get the Tracking and Data Relay Satellite System operational.[1] Despite continuing problems in many areas, these objectives were eventually achieved.

Columbia finally flew the first of four test flights in space in April 1981 (STS-1), more than two years after the originally scheduled date. It flew the second six months later, the third in March 1982, and the last four months later. Before the Challenger accident in January 1986, the four Shuttles (Columbia, Challenger, Discovery, and Atlantis) conducted another 20 missions. This was far less than even the 1981 NASA projection of 34 flights through 1985, which was part of a new schedule it made that year reducing the number of estimated missions through 1992 from 487 to 300.[2]

The National Security Council reviewed the Shuttle program shortly after the conclusion of STS-1. It directed George Keyworth, the new director of the Office of Science and Technology Policy, to examine as the highest priority the whole issue of transitioning from ELVs to the new vehicle and the timetable for it. This reflected the continuing concern

regarding the ability of the Shuttle to meet national security requirements, a concern that would increase in the coming years and cause considerable bitterness between NASA and the national security agencies.
For example, the NRO wrote Vice President George Bush's chief of staff
shortly before *STS-1*:

> Over the last several months there has been considerable discus-
> sion and recommendations to the effect that critical Department
> of Defense and NRP missions should not be totally depending on
> the Space Shuttle as a means of achieving orbit. It is our under-
> standing, and we certainly support, a reassessment of the total
> commitment to the Space Shuttle depending to some extent on
> the success of the initial Shuttle flight currently scheduled for next
> month.

In any event, Keyworth's report has not been released, and thus it is not
known what conclusions and recommendations he made.[3]

The Shuttle carried many more civilian and national security experiments than any previous human spaceflight program. Beginning in
1976, the joint NASA-NRO-CIA-DDR&E Program Review Board and the
two committees under it examined the Shuttle's many proposed civilian experiments for the early flights. Beyond this, the national security
agencies were very likely involved in more of these experiments than the
available records describe.

One candidate in this regard was the extensive handheld photography
conducted by the astronauts on virtually every Shuttle mission using
35-mm, 70-mm, and 125-mm cameras with a variety of lenses and films.
Just as in the prior human spaceflight programs, the crews received intensive training before each flight in photography and such subjects as
geology, meteorology, and oceanography. Scientists prepared a list of
photographic targets for each mission, and the astronauts received daily
instructions during the flight on which to photograph depending on
cloud cover and orbital conditions. The astronauts were also permitted
to image targets of opportunity. Each of the first seven Shuttle missions
between 1981 and 1983 brought back between 500 and 1,000 photographs. The astronauts on subsequent flights usually acquired between
1,500 and 3,000 photographs.[4]

Most of the Shuttle flights in the first 10 years were at inclinations of
28 degrees, but some were as high as 57 degrees. These higher-inclination
missions, of course, resulted in much greater flight over the USSR. There

is no evidence that the national security agencies reviewed the plans for handheld photography before any mission as they did in Apollo, Skylab, and Apollo-Soyuz, but at times they examined the imagery from these cameras and other image-forming sensors before public release as they had been doing since early in the Gemini program. Press reports indicated that NASA officials were upset with the crew of STS-9 (a December 1983 mission that flew at 57 degrees) for taking photographs of sensitive Soviet facilities. Following this mission, the director of the Defense Intelligence Agency wrote the deputy director of central intelligence expressing concern that the public distribution of Shuttle imagery "could lead to full disclosure of sensitive national security information, including force readiness, force disposition, potential targeting, and details of critical defense facilities." The deputy assistant to the president for national security affairs wrote NASA's administrator during the same period asking that it work with the defense and intelligence agencies to ensure that no sensitive imagery be released. Undoubtedly due to these high-level concerns, NASA and the national security agencies reviewed the handheld photography from the next three flights before any dissemination (STS-41B in February 1984, STS-41D in August 1984, and STS-41G in October 1984). There is no information available on whether any photographs were withheld.[5]

The second Shuttle mission, STS-2, in November 1981 carried a NASA Office of Space and Terrestrial Applications payload with various sensors, including the first Shuttle Imaging Radar (SIR-A). The national security agencies had approved the use of this instrument in 1977, but had apparently done so on the condition that they review the data acquisition and dissemination plans beforehand. However, the only available evidence of compliance with this is a NASA technical report which stated that foreign investigators would receive imagery only of their sites that had been approved for inclusion in the experiment, no other foreign distribution would be made without the approval of NASA's Office of International Affairs, and no imagery would be given to anyone other than the investigators and NASA headquarters without the permission of both. Numerous domestic and foreign investigators participated in the experiment. The planned coverage was of small portions of the United States, Central America, South America, Middle East, Africa, Indonesia, Philippines, Japan, and Australia. The majority of these areas were imaged during the mission, but for various reasons the actual coverage also included several swaths of the Pacific and Atlantic Oceans, Spain, South

Asia, Southeast Asia, and China. The national security agencies must also have taken an interest in the potential use of the SIR-A product for their own purposes because, among other things, it was able to image subsurface areas in the Sahara up to a depth of 16 feet.[6]

President Reagan issued National Security Decision Directive No. 8 on the Space Transportation System during the *STS-2* flight. It essentially reiterated the principles set forth in Carter's Presidential Directive/NSC-37, mandating that the Space Transportation System be the primary space launch system for both the U.S. military and civilian payloads, national security payloads be integrated into the system, and national security missions be given priority and could be dedicated.[7]

STS-4, a seven-day mission in the summer of 1982, was the last research and developmental flight and carried several civilian and three classified DoD experiments. Notwithstanding their classified status, the press reported considerable information on them. One was the Cryogenic Infrared Radiance Instrument for the Shuttle, whose goal was to evaluate the usefulness of employing extremely sensitive long-wavelength infrared surveillance systems in space. Because of a stuck telescope cover, the instrument was never activated during the flight. The other two were an ultraviolet sensor to evaluate its feasibility in space-based surveillance systems and a sextant designed to provide robotic spacecraft with navigation data in the absence of extensive ground support. The results of these two experiments are not known.[8]

The classified experiments mandated changes in NASA's long-standing policies of openness and transparency in its human and other spaceflight programs. Contrary to its earlier practices, it did not release any information at all concerning the three experiments. The mission flight plan had several long periods of time in which no activities were listed. At the request of the Air Force, NASA agreed to restrict the release of ground-air voice communications pertaining to the experiments and to not provide any television coverage of them.[9]

It is unclear what level of security the classified experiments required and how it was provided. Controlled Mode (which provided protection through the Secret level) was not operational at the Johnson, Kennedy, or Goddard Centers. Whether secure flight planning, simulation, training, and mission control took place in the absence of Controlled Mode and, if so, how cannot be determined.

Congress directed NASA in 1982 to modify the Centaur upper stage for its upcoming Shuttle-launched planetary exploration missions because they were too heavy for the Inertial Upper Stage. NASA and the Air Force soon agreed to share the project's management and costs. The Centaur G was designed to place NASA and DoD spacecraft up to 10,000 pounds in geosynchronous orbit (compared with the Inertial Upper Stage's 5,000 pounds). A stretched Centaur G would be built to launch even heavier NASA payloads such as *Galileo* and *International Solar Polar* (soon renamed *Ulysses*) into their escape orbits.[10]

President Reagan issued National Security Decision Directive No. 42 entitled "National Space Policy" at Edwards Air Force Base on 4 July 1982, the same day *Columbia* landed there to end *STS-4*. The provisions regarding the Shuttle in both the unclassified and Top Secret versions were essentially the same, except for one paragraph in the latter which is still classified. Among other things, the directive stated that the first priority was "to make the system fully operational and cost-effective in providing routine access to space," U.S. government satellites "should be designed to take advantage of the unique capabilities of the STS," and "transition to the Shuttle should occur as expeditiously as possible." Reflecting the growing concern in the NRO and some other national security agencies that the Shuttle could not be relied on to be the exclusive launch vehicle for their payloads, it provided that ELV operations be continued "until the capabilities of the STS are sufficient to meet its needs and obligations" and that "unique national security considerations may dictate developing special-purpose launch capabilities."[11]

National Security Decision Directive No. 42 also established the Senior Interagency Group (Space) to review and advise on changes to national space policy and refer issues to the president as necessary for decision. Chaired by the assistant to the president for national security affairs, its other members were the deputy undersecretary of state, deputy secretary of defense, deputy secretary of commerce, DCI, chairman of the Joint Chiefs of Staff, director of the Arms Control and Disarmament Agency, and NASA's administrator. Representatives from the Office of Management and Budget and the Office of Science and Technology Policy were observers.[12]

Beggs noted at the first meeting in September 1982 that the three main issues for NASA were acquisition of a fifth orbiter, approval of a space station, and international cooperation in remote sensing. The Space Launch Policy Working Group was established to examine the fifth orbiter issue prior to making a recommendation to the president on whether to include funding for the vehicle in the FY 1984 budget.

It set forth three options in its November report. The first was to not build another vehicle, the second was to maintain production of selected structural parts and major structural assemblies for up to two years after completion of the fourth orbiter, and the third was to start building a complete orbiter in FY 1984 for delivery in late 1988.[13]

The Senior Interagency Group (Space) reviewed the report in December 1982. However, the members held widely different views and did not reach a consensus. William Clark, the assistant to the president for national security affairs and the chair, requested that each agency submit its position in writing for him to use in reviewing the matter with the president. NASA and the Department of Commerce strongly endorsed the third option to begin building a complete orbiter, while the State Department and Office of Science and Technology favored the first option of not acquiring it. The CIA stated that national security programs alone did not require another orbiter, but that the second option should be approved to keep the production line open and thus provide a capacity to meet unforeseen contingencies or repair an orbiter. The positions of the other organizations represented on the Senior Interagency Group (Space) are not known, but they probably reflected the CIA's. Shortly after the agencies made their submissions, NASA indicated that it could support the second option.[14]

Reagan set forth his decision on the fifth orbiter in the February 1983 National Security Decision Directive No. 80. He approved the second option and directed that production continue of structural and components spares so that Shuttle operations could be maintained in case of "minor problems, modifications or other periods of extended Orbiter outages."[15]

STS-6 in April 1983 was the first flight of *Challenger* and the initial use of the Inertial Upper Stage on the Shuttle. (On top of the first Titan 34D launched in October of the previous year, it had successfully placed two Defense Satellite Communications System satellites in orbit.) The payload was the first Tracking and Data Relay Satellite, but problems with the Inertial Upper Stage placed it in the wrong orbit. Ground controllers slowly commanded the spacecraft to the desired geosynchronous orbital position off the northeast coast of Brazil. Once in place, however, one of the satellite's major links was lost. This meant that it could no longer transmit maps, schematics, and photographs between the Mission Control Center at Johnson and the orbiter and instead had to rely on a backup onboard Apollo-era system to transmit text-only instructions to the astronauts. Nevertheless, NASA declared the satellite operational in December 1984.[16] How this affected the satellite's ability to be used in classified DoD experiments and missions is unknown.

DoD Begins Moving Away from Total Reliance on the Shuttle

NRO Director Pete Aldridge and other key DoD officials in 1983 began trying to get approval for building a new ELV to launch a limited number of payloads they wanted to remove from the Shuttle. Their actions were not surprising given the very mixed record of the program to date. The Shuttle had flown its first mission in space nearly three years after NASA had originally planned. Turnaround times between missions were vastly greater than NASA had estimated. Only two orbiters had been built and flown in space, and even with the adoption of limited performance-enhancing steps these were incapable of launching the planned heaviest national security payloads. Although lighter in weight, the final two orbiters under construction also could not carry these payloads without additional performance-enhancing measures, which for the most part had neither been finalized nor implemented. Moreover, a series of problems from fuel line leaks to loss of heat protection tiles cast doubt on the overall reliability of the orbiters.[17]

The DoD was also responsible in whole or part for other problems in the program. The Inertial Upper Stage underwent major modifications and retesting after the malfunctions during STS-6 and would not be available for use again for several years. Development of the Centaur upper stage was proceeding slowly, and it too would not be available for some years. Both the Shuttle launch complex at Vandenberg and the Consolidated Space Operations Center in Colorado were experiencing large cost overruns and contracting delays. As a result, the initial operating date of the former was pushed back to late 1985 and the latter to 1986.[18]

It was clear that because of the above events NASA would not be able to meet the projected 34 flights through 1985 and would probably fall short of the planned 24 missions per year beginning later in the decade. They also greatly affected DoD's transition to the Shuttle. The DoD had already cancelled several planned missions, including the 1983 flight carrying the unknown reconnaissance satellite that had been described as essential for monitoring the SALT II agreement and two others in 1983 and 1984 which required the Inertial Upper Stage. At this time, the DoD had 13 Titan IIIs or 34Ds and 20 Atlases remaining in its launch vehicle inventory. The Titan production line had started to shut down in mid-1983, and the Atlases were the last of the long-obsolete Atlas E intercontinental ballistic missiles being modified to use as an ELV. All 33 vehicles were assigned to launch a specific satellite, with the last Titan scheduled for use in 1987 and the last Atlas in 1990. There were no longer any backup ELVs for DoD payloads on the Shuttle. As a result, an unknown number of the payloads scheduled to fly on the cancelled Shuttle missions were simply not placed in orbit. The available evidence does not indicate whether this caused any degradation of coverage in any national security space programs.[19]

The Joint Chiefs of Staff and Secretary of Defense Casper Weinberger received a briefing on U.S. and Soviet space capabilities in late December 1983, of which a portion was devoted to Air Force and NRO concerns about total reliance on the Shuttle. The following day Weinberger approved the Air Force and NRO plans to acquire 10 complementary expendable launch vehicles (CELVs) capable of boosting payloads the same size and weight as the original Shuttle performance specifications established in the early 1970s. Weinberger wrote Reagan in January 1984 stating that the national security agencies were very concerned about the complete dependence on the Shuttle and that this policy had been a mistake. He emphasized that the DoD remained committed to the Shuttle but strongly urged that the 10 CELVs be procured. The White House evidently approved the proposal, as Weinberger issued a "Defense Space Launch Strategy" directive in early February.[20] It stated, in part:

Existing Defense space launch planning specifies that DoD will rely on four, unique manned orbiters for sole access to space for all national security space systems. DoD studies and other independent evaluations have concluded that this does not represent an assured, flexible, and responsive access to space. While the DoD is fully committed to the STS, total reliance upon the STS for sole access to space in view of the technical and operational uncertainties represents an unacceptable national security risk. A complementary system is necessary to provide high confidence of access to space particularly since the Shuttle will be the only launch vehicle for all U.S. space users. In addition, the limited number of unique, manned Shuttle vehicles renders them ill-suited and inappropriate for use in a high risk environment. . . . The Air Force will take immediate action to acquire a commercial, unmanned, expendable launch vehicle capability to complement the STS with a first launch availability no later than FY 1988. These vehicles must provide a launch capability essentially equal to the original STS weight and volume specifications.[21]

Of course, the actions specified in the directive needed to be approved and funded by Congress before any CELVs were actually built.

Beggs and others in NASA's leadership were extremely upset and believed that the DoD was abandoning the Shuttle. He quickly wrote Weinberger asserting that the Shuttle could meet DoD requirements and that a CELV was not needed. If one were necessary, Beggs argued, it should be derived from Shuttle to meet DoD's requirements and to assist the Shuttle program from the greater production of components. Additionally, a Shuttle-derived vehicle would have the capability of meeting the nation's future space launch requirements.[22]

DoD officials began a long series of appearances before Congress and interagency meetings on the issue. NRO Director Pete Aldridge and others emphasized that the DoD was not giving up on the Shuttle but must have both it and the 10 CELVs, of which two a year would be launched beginning in 1988. They also argued that the mixed fleet would reduce DoD preemption of other U.S. government and commercial users of the Shuttle, permit the orbiters to be employed on more missions that required astronaut presence, and extend the life of the four orbiters. Congress approved $5 million in the FY 1985 budget for the DoD to take the initial steps in the development and acquisition of the CELV.[23]

NASA pressed for the right to submit a proposal for the new CELV, and the DoD reluctantly agreed. To avoid the prohibition of the government competing with industry, it decided that the winner of the bids submitted by industry would be compared with the NASA proposal. The DoD received four proposals in all—Martin Marietta's modified Titan III (designated Titan 34D7), General Dynamics' improved Atlas-Centaur, and two NASA Shuttle-derived vehicles. It initially selected the Titan 34D7 and then compared it with the NASA vehicles. On the grounds that the latter were much more costly and greatly exceeded the capacity needed to meet any planned DoD requirements, it chose the Titan 34D7 again.[24]

The Air Force needed an additional $30 million in FY 1985 to continue the development and acquisition of the Titan 34D7 and meet the initial launch date in 1988. NASA tried several tactics to delay development of the vehicle, such as calling for additional studies, but in the end the National Security Council intervened to resolve the dispute. On 14 February 1985, it met with Aldridge and Beggs and hammered out an agreement between them. The provisions directed that NASA and the DoD work together to make the Shuttle capable of making 24 flights per year, the DoD would commit to one-third of the Shuttle missions over the next 10 years, NASA would drop its opposition to the CELV, the DoD would procure 10 CELVs and would launch 2 a year beginning in 1988, and NASA and the DoD would develop a new pricing policy for the Shuttle to make its use more attractive to the DoD. These provisions were incorporated in National Security Decision Directive No. 164 entitled "National Security Launch Strategy," which Reagan issued later in February. With all the obstacles removed to building the new CELV, the DoD revealed the following month that it had also made the final decision to modify a limited number of Titan II intercontinental ballistic missiles being withdrawn from service to use as launch vehicles for small satellites from Vandenberg. This action had been planned for some time and NASA did not object.[25]

DoD's Participation in Further Shuttle Missions until the 1986 Challenger Accident

The Shuttle program continued to enjoy some success while the battles between NASA and the DoD over ELVs took place. *Discovery* was delivered to Kennedy in late 1983 and *Atlantis* in early 1985. Both were approximately 7,000 pounds lighter than *Columbia*. All the orbiters now incorporated lighter-weight solid rocket boosters and a lighter-weight external tank but not any thrust augmentation measures. NASA estimated that the steps taken still resulted in the payload capacity of even *Discovery* and *Atlantis* falling short of the original design goals by about 10,000 pounds at Kennedy and at least 3,000 pounds at Vandenberg.[26]

STS-41C in April 1984 deployed the Long Duration Exposure Facility, a 30-foot long cylindrical structure that held 57 unclassified experiments in four categories: power and propulsion; science; electronics and optics; and materials, coatings, and thermal systems. Among the DoD experiments were radar camouflage materials, laser optics, structural materials, solar power components, laser mirror coating, laser communication components, and advanced composite materials. Although scheduled to be retrieved after 11 months in orbit, due to Shuttle scheduling problems and the *Challenger* accident, it was not brought back to Earth until January 1990.[27] Astronauts deployed Navy Syncom communications satellites during *STS-41D* in early September 1984 and *STS-51A* in November 1984.[28]

STS-41G flew at an inclination of 57 degrees in October 1984 and carried another Office of Space and Terrestrial Applications payload, which included the second Shuttle Imaging Radar (SIR-B). Although having roughly the same maximum resolution as the SIR-A flown in 1981, it was more advanced with a tilting antenna that enabled imagery of a target from several angles and thus provided stereo imagery. When the Shuttle was in view of NASA's Tracking and Data Relay Satellite, the imagery was transmitted through it to a NASA ground station at White Sands. When out of view, the imagery was recorded on tape for later transmission to the ground station via the satellite.[29]

Over 40 domestic and foreign scientists participated in the SIR-B experiment to examine its use in archaeology, geology, cartography, oceanography, and agriculture. The sites to be imaged were small areas of the United States, Canada, South America, South and North Atlantic Oceans, Western Europe, Middle East, eastern and southern Africa, south Asia, Indonesia, Australia, New Zealand, and Japan. Due to hardware problems, less than half the imagery scheduled to be obtained was actually collected. For unknown reasons, the actual coverage included one unplanned long swath beginning in the Bay of Bengal and continuing through China into Mongolia. It is not known whether the national security agencies imposed restrictions on the imaging of certain areas and limited the dissemination of the product to scientific investigators as they had with *SEASAT-A* and evidently did with the SIR-A on *STS-2*. However, the press reported that the SIR-B data would be reviewed before public release and any that revealed ocean surveillance capabilities would be withheld.[30]

The *STS-41G* Office of Space and Terrestrial Applications payload also included the Large Format Camera, the most capable sensor NASA had flown in space to date for imaging the Earth and whose use the national security community had approved in 1976 on the condition that it review the proposed photographic plans. A 1984 General Accounting Office report indicated that now the DoD was very interested in the instrument, and but for this NASA would not have flown it, despite the considerable civilian scientific involvement in the experiment for purposes of oil and mineral exploration, mapping, and remote sensing.[31] However, there is no additional information on the DoD's interest in the camera and whether the imagery met any national security requirements.

Press reports stated that NASA's Defense Department Affairs Division requested that the ground tracks for the times in which the Large Format Camera was operating be withheld to avoid the impression that its mission was reconnaissance. The scientists involved in the experiment refused on the basis that the information was already available and that such secrecy was inappropriate for an acknowledged scientific project. The press also reported that along with the handheld photography and SIR-B imagery the Large Format Camera's roughly 2,300 images underwent a national security review before being released. It is not known whether any were withheld.[32]

STS-41G also carried a civilian Navy oceanographer who conducted a series of unknown experiments. Only some of the data he collected was made public. Adm. James Watkins, chief of naval operations, praised the individual's work and stated that it produced critical new information on the oceans.[33]

STS-51C in January 1985 was the third flight of Discovery and the first completely dedicated DoD mission. It was also the first Shuttle mission on which a DoD Manned Spaceflight Engineer payload specialist flew. Much more secrecy surrounded the mission than STS-4, which had carried three classified experiments in 1982.[34]

Neither NASA nor the DoD released any information on the STS-51C payload, other than the fact that it utilized an Inertial Upper Stage (its first flight since STS-6). However, the press widely reported that it was an advanced signals intelligence satellite to be placed in geosynchronous orbit. It is not known whether the spacecraft was dual-compatible or Shuttle-optimized. The spacecraft's launch had experienced several lengthy delays, illustrating some of the continuing problems in the Shuttle program. Technical issues with the Inertial Upper Stage originally delayed its launch scheduled for STS-10 in December 1983 to STS-41E in July 1984, and then again to STS-51C in December 1984 onboard Challenger. However, problems were found with many of Challenger's heat tiles after STS-51A, and this forced the mission to utilize Discovery the following month.[35]

There were no pre-mission press conferences or press kits distributed as there had been for every previous human spaceflight mission beginning with the first Mercury flight. Initially, the DoD wanted NASA to close the media facilities at Kennedy and keep out all unnecessary personnel in an attempt to launch the mission in complete secrecy. However, after NASA protested this far-reaching action it settled for NASA simply announcing that the liftoff would occur sometime within a three-hour period on a specific date. Limited information was released during the countdown to avoid disclosing the exact launch time. After ignition, the air-ground communications between the vehicle and mission control at Johnson were public. At the end of the orbital insertion phase of the flight (about 45 minutes after liftoff), however, the air-ground communications were encrypted and withheld. The only status reports given during the mission were brief updates on the orbiter every 8 hours. The landing time was announced 16 hours prior to landing. Air-ground communications were again made public during the reentry phase. All the press reports indicated that the mission was a success. These restrictions would remain in place for nearly all the subsequent dedicated DoD missions.[36]

It is not known what level of security the mission required or how it was achieved. The Shuttle undoubtedly had encryption devices, which presumably could provide protection for the various types of communications through the Sensitive Compartmented Information (codeword) level. Controlled Mode was apparently finally operational at the Johnson, Kennedy, and Goddard Centers, but it only afforded protection to preflight activities and mission control at the Secret level. The Air Force's Satellite Test Center controlled the Inertial Upper Stage and the payload, with Kennedy and Johnson stripping out the telemetry downlinked from the Inertial Upper Stage and payload and sending it to the Satellite Test Center. This facility's commands to the Inertial Upper Stage and payload were coordinated with and evidently sent through the Mission Control Center at Johnson. Once the payload deployed, the Satellite Test Center assumed full control of it and communicated directly with it.[37] Although its performance was degraded, the one Tracking and Data Relay Satellite in orbit presumably relayed classified and unclassified data to and from the Shuttle through the White Sands Ground Station for part of each orbit. The role of any other NASA Spaceflight Tracking and Data Relay Network or Air Force Satellite Control Facility ground stations when the Tracking and Data Relay Satellite was not used cannot be ascertained.

Astronauts deployed another Navy Syncom communications satellite during STS-51D in April 1985. However, the spacecraft's perigee kick motor failed to activate to send it into geosynchronous orbit after deployment from the *Discovery*'s cargo bay, rendering it useless. Another Syncom was placed in orbit during *STS-51I* in August 1985. During the mission, astronauts also retrieved, repaired, and redeployed the malfunctioning one from *STS-51D*.[38]

STS-51G in June 1985 carried the Strategic Defense Initiative's unclassified High Precision Tracking Experiment. To test the ability of a ground-based laser to stay pointed on a target in space, *Discovery* carried an 8-inch-diameter mirror that was to be hit and reflect back a low-power laser located at an Air Force facility in Hawaii. Although the first attempted test failed due to erroneous instructions sent by ground controllers, subsequent tests were successful.[39]

STS-51J in October 1985 was the first flight of *Atlantis* and the second dedicated DoD flight. Even though no information was released on the payload at that time, the media reported that it was two Defense Satellite Communications System satellites that were successfully launched on top of an Inertial Upper Stage. (Some observers in recent years have confirmed this with records the Air Force has subsequently released.) The strict publicity restrictions surrounding the entire mission remained in place. Once again, it is not known what level of security the flight required and, if above the Secret level, what procedures NASA and the DoD utilized to provide it.[40]

Astronauts deployed the small DoD Global Low Orbiting Message Relay Satellite during STS-61A in November 1985. It was designed to test a system to relay data from undersea sensors that tracked Soviet submarines.[41]

During *STS-61C* in early January 1986 the crew conducted two experiments related to the Cryogenic Infrared Radiance Instrument for the Shuttle to be flown on *STS-62A* later that year. One involved using an RCA infrared camera to observe areas of the Earth where Air Force planes were flying to determine whether their infrared signatures could be seen against the Earth's background radiation. The Air Force classified the type of aircraft involved in the test, their flight activities, and their locations. The other experiment employed an Air Force Particle Analysis Camera System to observe the amount and type of floating particulate contamination around the orbiter resulting from events such as water dumps and thruster firings. With this information, operations of the Cryogenic Infrared Radiance Instrument for the Shuttle could be planned around any periods of contamination.[42]

The Challenger Accident and Its Aftermath

NASA planned an ambitious schedule of 15 Shuttle missions in 1986 utilizing both the original and a new launch pad at Kennedy, the launch pad at Vandenberg scheduled to open later in the year, and all four orbiters. The DoD would fly four of the missions, two from each complex. In the first launch of an orbiter from Vandenberg, *STS-62A* would carry another Cryogenic Infrared Radiance Instrument for the Shuttle, deploy the Space Test Program *Teal Ruby* satellite with infrared sensors to demonstrate the ability to detect aircraft in flight from space, and conduct various other DoD experiments. The press reported that the second mission from Vandenberg would carry an advanced KH-11 imagery intelligence satellite. The payloads of the two DoD flights from Kennedy are not known. NASA had several high-visibility missions planned from Kennedy during 1986, including deploying the *Galileo* probe to Jupiter in the first use of the Centaur upper stage, the *Ulysses* spacecraft around the north and south poles of the Sun, and the *Hubble Space Telescope*. All three had to be placed in storage by NASA at great expense.[43]

The tragic loss of *Challenger* and all of her crew in the second mission of 1986 threw NASA and the DoD's launch schedules into chaos and left the nation in a precarious situation regarding the ability to place payloads in space. Shortly after the *Challenger* accident the Senior Interagency Group (Space) began a series of meetings on its effect on planned civilian and national security missions, whether a fifth orbiter should be built, and whether additional ELVs should be acquired. It reportedly submitted its recommendations to the National Security Council in May.[44]

NASA had immediately grounded the remaining three orbiters after the accident. There were only a limited number of ELVs available to both NASA and the DoD—7 Titan 34Ds (which only the DoD employed), 3 Deltas, 13 Atlases, and 10 Scouts. National security payloads, of course, had priority. The production lines for all four were closed, and it would be several years before they could deliver new vehicles if they were reopened. The first of the 10 CELVs the DoD was procuring would not be available until 1988, and their delivery could not be advanced. The DoD's conversion of the first of the limited number of Titan II intercontinental ballistic missiles into ELVs for small payloads would not be completed until 1987 at the earliest.[45]

Greatly compounding the problem of accessing space was the grounding of the six remaining Titan 34Ds in April 1986 after one exploded shortly after launch at Vandenberg, destroying the last HEXAGON scheduled for deployment and severely damaging the launch pad. This was the second consecutive Titan 34D failure after one had exploded soon after liftoff at Vandenberg in August of the previous year and destroyed what was reported to be a KH-11. The two accidents apparently left a single KH-11 in orbit to provide imagery intelligence, a situation that caused great concern but that could not be fixed until the Titan 34Ds resumed operations. Making matters worse, one of the three remaining Deltas carrying a National Atmospheric and Oceanic Administration weather satellite blew up shortly after launch at Cape Canaveral in early May 1986.[46]

Dr. William Graham, NASA's acting administrator from late 1985 to May 1986, and James Fletcher, who became administrator for the second time beginning on the latter date, made numerous public statements on the crisis. They noted that the country faced a shortfall in launch capacity for at least two to three years. Both acknowledged that priority must be given to national security payloads, followed by scientific payloads with specific launch windows, and then other U.S. government, foreign, and domestic commercial payloads. They strongly advocated that a fifth orbiter be built to replace *Challenger*, but stated that the country would require both the Shuttle and ELVs for the foreseeable future—a dramatic change in NASA's position from just a few months earlier.[47]

Pete Aldridge was the leading DoD official to publicly address the launch capacity problem, and he repeatedly made the following points. The DoD would need both the Shuttle and ELVs and strongly supported expanding production of ELVs and building a fifth orbiter. However, as had always been the case, the DoD opposed using any of its funds for a new orbiter. If the downtime of the Shuttles were less than a year there would be little negative effect on the DoD. However, longer delays in resuming flight coupled with the limited number of available ELVs would cause serious problems. A downtime of two years would result in more than 20 high-priority payloads waiting for launch. These were not only military and reconnaissance satellites, but also beginning in 1987 Strategic Defense Initiative research and development payloads. The DoD could exercise its priority rights and almost exclusively use the Shuttle when it resumed operations, but the DoD preferred to switch some payloads to ELVs when the cost or delay was not excessive in order to allow NASA to launch more civil, foreign, and commercial payloads on the Shuttle.[48]

NASA took several actions in the months after the accident that decreased the Shuttle's capabilities and accelerated the DoD's move away from it and to expand its fleet of ELVs. It decided that it would not use the Centaur upper stage on the Shuttle because of the unacceptable safety risks posed by its fuel's extreme volatility. This meant that NASA could not deploy *Ulysses* and *Galileo* from the Shuttle and that the DoD would not be able to place its payloads over 5,000 pounds (Milstar, Defense Support Program, and those from a still-classified program) into high-altitude orbits from the vehicle. NASA and the DoD together had spent almost $475 million on the Centaur. Additionally, NASA had purchased three sets of flight hardware for over $400 million. The DoD, however, planned to use the Centaur on the new CELV (now designated Titan IV) so it could place its heaviest payloads into high-altitude orbits from Kennedy. NASA also indefinitely postponed plans to employ a filament-wound case for the solid rocket boosters (which would have reduced the weight of an empty booster from 98,000 pounds to 65,000 pounds and increased the Shuttle's payload capacity by 4,600 pounds) and various thrust augmentation measures, including increasing the thrust level of the main engines to 109 percent. Aldridge stated that without these performance-enhancing measures, even the lighter-weight *Discovery* and *Atlantis* could only launch 54,000 pounds from Kennedy and 16,000 pounds from Vandenberg. This added to the number of national security payloads that could not be carried on the Shuttle from Kennedy and rendered the orbiters incapable of launching any reconnaissance satellites from Vandenberg.[49]

NASA's early estimates of a short grounding of the orbiters proved wildly inaccurate. It announced in July 1986 that the Shuttle would not fly again until early 1988 due to the time needed to develop and test improvements to the solid rocket booster insulation and joints. Over the strong opposition of his director of the Office of Management and Budget and chief of staff, President Reagan agreed with NASA and the DoD's recommendations and directed NASA to begin development of a replacement orbiter. Limited funding would be requested in the FY 1987 budget for the new vehicle, which was scheduled for delivery by 1992. At the same time, he barred NASA from soliciting any new contracts for the launch of commercial satellites in order to keep the Shuttle available for national security and civilian government payloads. The 44 launch contracts NASA then had for commercial payloads were not affected.

Because the Shuttle could not carry any reconnaissance spacecraft from Vandenberg, the DoD placed the Shuttle launch complex nearing completion there in a minimal caretaker status. It had spent over $3 billion on it. The annual cost of maintaining it in this status was $50 million. Primarily to launch Global Positioning System satellites, it also proposed expanding its ELV fleet by developing a new class of medium-lift vehicles more capable than the Titan IIs and Deltas. The DoD also received approval to acquire 13 more Titan IVs for a total of 23.[50]

The government resumed ELV launches in September 1986 when one of the two remaining Deltas launched a Strategic Defense Initiative payload from Cape Canaveral. Later that month, one of the remaining Atlases deployed a National Oceanic and Atmospheric Administration weather satellite from Vandenberg. An Atlas-Centaur at Cape Canaveral boosted a Navy Fleet Satellite Communications System satellite into geosynchronous orbit in December.[51]

NASA announced in October 1986 an ambitious Shuttle schedule of 46 missions for the first seven years following the planned resumption of flights in early 1988. The number of missions would grow gradually, from 5 the first year to 16 in 1994. National security payloads were 41 percent of the total (up from 33 percent in the pre-*Challenger* manifest).[52]

In November 1986 the Air Force cancelled all funding for the Shuttle mission control portion of the Consolidated Space Operations Center in Colorado for a variety of reasons. The DoD had spent almost $80 million to date on it and estimated that it would have cost another $103 million to develop an austere complex by 1992 or $383 million to build a complete one by the mid-1990s. However, the former would not have completely eliminated the dependence on Johnson's Mission Control Center and thus would not have afforded protection above the Secret level that Controlled Mode provided.[53]

President Reagan issued National Security Decision Directive 254 in late December 1986, which set forth a new U.S. space launch strategy basically incorporating the various policies adopted earlier in the year. It stated that the nation's launch capability would be based on both the Shuttle and ELVs. National security payloads would use both, and selected "critical" ones would be designed to be launched on either. The DoD would acquire additional ELVs and would retain launch complexes on both coasts. NASA would begin procurement of a new orbiter in FY 1987. The Shuttle would stop launching foreign and commercial payloads by 1995 that did not require a manned presence or the unique capabilities of the vehicle. NASA would not maintain a fleet of ELVs and would contract for ELV services if it needed additional launch capacity.[54]

The Air Force awarded McDonnell Douglas Astronautics a contract in January 1987 to build 20 of the new Delta II medium-lift launch vehicles to carry Global Positioning System spacecraft beginning in 1988. The Delta IIs would be launched from Cape Canaveral and be capable of boosting 11,400 pounds into low-Earth orbits and the 2,800-pound Global Positioning System into their orbits at 12,650 miles.[55]

NASA announced four months later that Shuttle launches would not be resumed until June 1988 (a further postponement of four months) and that the flight rate in the outyears would be reduced from 16 to 14. This projected maximum flight rate would not be reached until 1994.[56]

This action further accelerated DoD's move away from the Shuttle. It requested 25 more Titan IVs (7 to compensate for the reduced Shuttle availability, 2 for new missions, and 16 for FY 1994 and 1995 flights), 5 more Delta IIs, and 10 new medium-lift launch vehicles more capable than Deltas IIs to carry Defense Satellite Communications System satellites. The DoD sought approval to build a second Titan IV launch pad at both Cape Canaveral and Vandenberg to accommodate the increased flight rate. It also accelerated its study with NASA of several possible next generation space launchers that would be able to launch even heavier payloads than the Shuttle and Titan IV.[57]

At this point the DoD planned to use the Shuttle only for missions that required the presence of astronauts or for Shuttle-optimized payloads. It planned one mission in 1988, two in 1989, five in 1990, and two a year from 1991 through 1995. It had nine prepaid flights and was willing to pay the new increased price of $115 million a flight for the additional nine. Thereafter, it did not plan to utilize the orbiters to carry any payloads with the possible exception of a few small unclassified research and development ones. The residual Atlases, Titan 34Ds, and Scouts would be used up by 1991. Converted Titan IIs would enter service in 1988 and launch Defense Meteorological Support Program satellites, Space Test Program payloads, and still-classified signals intelligence satellites into low-Earth orbit. Titan IVs would now begin flying in 1989 (a delay of one year) and would carry Defense Support Program, Milstar, Defense Satellite Communications System, and still-classified spacecraft. Delta IIs would also enter service in 1989 and would launch Global Positioning

System satellites. The planned second new medium-lift launch vehicle would become operational in 1991.[58]

During 1987 the DoD continued to utilize the limited number of remaining ELVs. A signals intelligence satellite was reportedly successfully placed in orbit by a Titan IIIB in February, the first use of a Titan since the explosion of the Titan 34D the previous April. Several months later, lightning destroyed an Atlas-Centaur carrying a Fleet Satellite Communications System satellite shortly after liftoff, but another Atlas placed an apparent ocean surveillance spacecraft into orbit. The Titan 34D resumed service in April 1986 when it boosted a reported KH-11 into orbit in October. Another Titan 34D successfully launched a Defense Support Program satellite the following month.[59]

The DoD received approval for the second medium-lift launch vehicle and awarded General Dynamics in May 1988 the contract to build 11 of them (they were soon designated Atlas IIs). Designed to utilize the Centaur upper stage, it would be capable of placing 15,700 pounds in low-Earth orbit and 6,400 pounds in geosynchronous transfer orbit from Cape Canaveral. The DoD planned to use them primarily for the launch of Defense Satellite Communication System satellites, which it now wanted to move off the Titan IV. The acquisition of the Titan IVs, Titan IIs, Delta IIs, and Atlas IIs moved the DoD much closer to its goal of "assured access" to space. With the greatly decreased performance of the orbiters and the acquisition of the new ELVs, it decided to mothball the Shuttle launch complex at Vandenberg. The DoD reduced the number of planned Shuttle missions from 18 to the 9 prepaid flights and informed NASA that after *STS-46* in 1991 it would no longer require secure missions. The nine missions would only carry payloads that would be prohibitively expensive to move to ELVs. Most of the Global Positioning System, Defense Satellite Communications System, and Defense Support Program spacecraft manifested for the Shuttle would be moved to ELVs. This would not only result in slightly lower launch costs but permit these payloads to be launched sooner and ensure that NASA was able to launch its highest-priority science payloads with the Shuttle. However, realizing the goal of assured access to space came at a steep price. Pete Aldridge, now secretary of the Air Force, estimated in 1988 that the DoD had spent over $14 billion to date to purchase the new ELVs, build their launch pads, reconfigure and retest satellites, and conduct related activities.[60]

The Air Force successfully launched the first of the new ELVs in early September 1988 when a Titan II carried an apparent electronic intelligence satellite from Vandenberg. In February 1989, a Delta II deployed a Global Positioning System spacecraft in its initial launch. Employment of the Titan IV was delayed due to several factors, but it successfully boosted a reported Defense Support Program satellite in June 1989 during its first service.[61] The launches of these vehicles and the new Atlas II once it entered service increased dramatically as the DoD used up its small inventory of older ELVs and was completing its transition from the Shuttle.

DoD's Final Shuttle Missions

The Shuttle resumed flying when *Discovery* flew a four-day flight beginning in late September 1988. Its main payload was the second Tracking and Data Relay Satellite, which successfully reached geosynchronous orbit over the Pacific.[62]

The DoD flew some experiments on nondedicated missions in the following years. An unclassified one using the orbiter as a calibration target for the Ground-Based Electro-Optical Deep Space Surveillance sensors at the Air Force Maui Optical Site was conducted on a number of flights, beginning with *STS-29* in March 1989. The orbiters carried no special equipment for the tests and were only required to be in specified attitudes and lighting conditions. *STS-32* in January 1990 deployed another Navy Syncom communications satellite and retrieved the Long Duration Exposure Facility, which included DoD experiments.[63]

The DoD also flew its final eight dedicated missions from December 1988 until December 1992. The first five reportedly launched imagery or signals intelligence satellites. The crews also conducted Military Man-in-Space program land and ocean surveillance experiments. The extensive secrecy that surrounded the earlier dedicated missions remained in place for all five flights. Controlled Mode was still in place at the Johnson, Kennedy, and Goddard Centers and had to have been used for preflight activities and mission control. The Air Force Satellite Test Center must have continued to control the payloads and any Inertial Upper Stages in coordination with Johnson's Mission Control Center. Two Tracking and Data Relay Satellites were operational and were probably used in place of NASA's Spaceflight Tracking and Data Network or Air Force Satellite Control Facility ground stations to relay data to and from the orbiters.[64]

STS-39, *STS-44*, and *STS-53* were the last three dedicated DoD missions, and NASA and the DoD relaxed the secrecy surrounding them. Among other things, NASA issued press kits, held pre-mission news conferences, announced the launching and landing times, provided detailed status reports for most of the portions of the flights, and released many ground-air communications. Nevertheless, there was still some information withheld.[65]

STS-39 in April and May of 1991 carried a number of unclassified experiments related to missile defense and was the first Shuttle flight completely devoted to military research and development. One experiment was the AFP-675 payload with five different sensors, including the Cryogenic Infrared Radiation Instrumentation for Shuttle, which had flown as a classified payload on *STS-4* and had been scheduled to fly again in 1986. Another was the German-built Infrared Background Signature Survey Spas pallet, which was deployed and retrieved by the crew and whose ultraviolet/infrared devices imaged the orbiter's engine plumes, Earth/space backgrounds, and chemicals released by small canisters. Certain orbiter maneuvers were also imaged by the Naval Research Laboratory's Low Power Atmospheric Compensation Satellite about 250 miles away from the vehicle. The one known classified experiment was the Multi-Purpose Release Canister.[66]

STS-44 in November 1991 launched what was acknowledged to be a Defense Support Program satellite. The crew performed the most extensive Military Man-in-Space program land and ocean reconnaissance experiments to date.[67] *STS-53* in December 1992 reportedly deployed some sort of classified imagery intelligence satellite or perhaps a new combined imagery/signals intelligence satellite. Crew members conducted additional Military Man-in-Space program experiments. An experiment that involved releasing small metal balls to test the ability of U.S. ground-based space surveillance sensors to track them failed because of equipment malfunctions.[68]

After *STS-38* in November 1990, the DoD closed its secure work centers at the three NASA centers, and the Controlled Mode of operations ended.[69] It is not known how or where secure preflight activities and mission control for the orbiters and payloads during the last three dedicated flights took place. The two Tracking and Data Relay Satellites in orbit presumably handled most if not all of the communications between mission control and the orbiters.

Summary

STS-53 ended the longest and most complicated interaction between NASA and the national security agencies. The involvement of the defense and intelligence agencies resulted in a substantially more complex and expensive Shuttle program, and NASA bore the vast majority of the increased costs. Their participation also forced NASA to abandon its longstanding policies of openness and engaging in peaceful and scientific activities in space. Nevertheless, the benefits of the partnership to NASA were immense because the Space Transportation System would have very likely not been built without the participation and support of the national security community. The Shuttle was the last of the proposed post-Apollo, high-visibility programs that had any chance of development. Without it, NASA would have been greatly reduced in size and left only with space programs involving applications satellites, astronomy, and planetary probes. Additionally, the critical political support given to the program by the civilian leadership of the national security community saved it from drastic reduction or cancellation.

In contrast to NASA, the national security agencies received very few benefits from the partnership. The only apparent one is that the orbiters carried DoD experiments that could not have flown on ELVs. However, they were limited in number and were only geared to research and development. In the overall scheme of things these experiments proved to be of minimal value. The Shuttle did launch more than 10 DoD satellites, but ELVs could have carried them in the absence of the Shuttle. Due to a myriad of problems, the national security agencies never got the opportunity to determine the Shuttle's value in such areas as the on-orbit repair of satellites, the retrieval and return to Earth of spacecraft, building large structures in space, or the inspection of foreign satellites.

A huge but presently unknown negative resulting from the partnership was the likely degraded coverage in at least some national security space programs after the *Challenger* accident. The extended grounding of the orbiters, the small residual fleet of ELVs due to the ongoing transition to the Shuttle, and, to a lesser extent, the grounding of the Titan 34Ds after two launch failures all contributed to this situation. However, at this time it is impossible to determine the scope and duration of reduced coverage.

Another detrimental result of the partnership was the massive amount of money spent on the Shuttle program. The DoD spent around

$4 billion for the Vandenberg launch complex, the Shuttle mission control at the Consolidated Space Operations Center, and the Shuttle/Centaur upper stage, none of which ever became operational. It spent more than $1 billion on the Inertial Upper Stage, installation and operation of the Controlled Mode, and supplementary secure command and control measures to the Controlled Mode. A 1992 DoD study estimated that it cost over $20 billion just for the design and building of satellites for launch on the Shuttle instead of ELVs beginning in the late 1970s and their reconfiguration and retesting for launch on ELVs after the *Challenger* accident.[70] The extra costs (over and above the costs of continuing to procure and launch ELVs in the absence of the Shuttle program) incurred to rebuild and launch its fleet of ELVs beginning with the 10 CLVs in 1984 are unknown. In short, the Shuttle program fell far short in delivering the promised benefits to the national security community.

The Shuttle was not the only NASA spaceflight program in the 1970s and 1980s attempting to satisfy defense and intelligence requirements. The next chapter examines the increasing involvement of the national security agencies in the design and operation of NASA's applications satellites to meet these requirements and the conflicts and compromises reached over restricting the acquisition and dissemination of the scientific data collected.

NASA's Applications Satellites
and National Security Requirements

NASA's applications satellites during the 1970s and 1980s carried sophisticated sensors that collected higher-quality scientific data than previously. There was extensive interaction with the defense and intelligence agencies to ensure that the open use of technologies or the acquisition and dissemination of the data did not endanger national security and that the satellites also satisfied national security requirements when needed. As in the 1960s, conflicts occurred at times over proposed restrictions on the acquisition and dissemination of data.

NASA launched more capable polar-orbiting meteorological satellites during this period, including the third-generation Improved Tiros Operational Satellites beginning in 1970 and the fourth-generation TIROS-N starting in 1978. It also placed into orbit the last four Nimbus from 1970 to 1978 and the first geosynchronous meteorological satellite in 1974. There was close liaison with the national security community regarding these programs and TIROS-N utilized a spacecraft bus developed for the Defense Meteorological Satellite Program (DMSP). Although their orbits, sensors, and other factors made them incapable of contributing significantly to meeting strategic weather requirements, the civilian satellites helped satisfy some tactical weather requirements. Notwithstanding these limitations, the defense and intelligence agencies had to rely on them more when DMSP was only partially operational between 1975 and 1977 and completely inoperative from 1980 to 1982. Not surprisingly, the coverage of the civilian weather satellites was inadequate. Reflecting both the importance of the data collected to the United States and its potential value to adversaries, TIROS-N became only the second NASA satellite program to install equipment preventing unauthorized commands.

The national security community was deeply involved in the design and operation of NASA's *GEOS-3* geodetic satellite launched in 1975 and the subsequent *SEASAT-A* oceanographic satellite placed in orbit three years later. NASA agreed that both would carry specific radar altimeters and fly in certain orbits to acquire the best geodetic information to improve the accuracy of U.S. submarine–launched ballistic missiles, although in both cases civilian requirements could have been met with less capable instruments or different orbits. NASA successfully opposed encryption as a means of denying selected scientific data from these sensors to the Soviets, but for the first time ever in its applications satellite programs it consented to restricting the acquisition or dissemination of such data. Although the State Department and National Oceanographic and Atmospheric Administration (NOAA) strongly opposed this, it was only after the DoD decided in 1977 that the scientific data was no longer sensitive that President Carter ordered it be freely distributed. The national security community approved the use of the four other proposed sensors for *SEASAT-A*—including the Synthetic Aperture Radar—after determining that none contained sensitive technologies. However, because of the potential value of the Synthetic Aperture Radar to adversaries, it demanded that NASA encrypt the sensor's uplinks and downlinks. NASA successfully opposed this action but in the end agreed to limit the acquisition and dissemination of data from it and installed a device to prevent unauthorized commands.

NASA finally launched the first land remote sensing satellite, *Landsat 1*, in 1972. Because of the stringent resolution limits imposed by the defense and intelligence agencies in the 1960s, neither *Landsat 1* nor its successors were equipped with the high-quality sensors NASA and other federal civilian agencies originally had wanted but instead carried multispectral cameras. The intelligence agencies soon concluded that the data was valuable in improving agricultural forecasts, an issue with many domestic and foreign policy implications. In 1974, the White House approved NASA's proposal that *Landsat 2* conduct an extensive worldwide crop survey but directed that the coverage be scaled back to reduce the appearance of NASA as a collector of economic intelligence. Imagery from *Landsat 2* and its successors was regularly used by the CIA in estimating foreign agricultural production and by the DoD in making maps beginning in the 1980s.

Because of the growing overlap between the civilian and national security applications satellite programs, the White House periodically considered merging the separate efforts into a single program for reasons of economy. NASA, NOAA, and the defense and intelligence agencies opposed these actions for the most part. One exception was the joint National Oceanographic Satellite System developed in the late 1970s to provide both geodetic and oceanographic data for civilian and national security purposes, but in the end budget cutbacks forced the early termination of the project. The other was the brief NASA-DoD operation of the Landsat system during the early 1990s. There were also attempts by the White House to privatize some of the applications satellite programs for the same reason. NASA, NOAA, and the national security agencies opposed this with weather satellites, and congressional action stopped their impending privatization in 1984. The Landsat program was privatized the same year, but numerous problems terminated the effort in 1992.

Weather Satellites

NASA, the U.S. Weather Bureau, and the DoD agreed in the early 1960s to establish a National Operational Meteorological Satellite System that would meet all civilian and national security needs for weather data from space. They worked very closely on the project, but technical problems and schedule slippages quickly led to the withdrawal of the U.S. Weather Bureau and DoD. NASA continued it as a R&D project to develop and test sensors in space and began launching the Nimbus satellites in 1964. Because of the inability of NASA's first-generation Tiros satellites to satisfy strategic weather requirements, particularly collecting timely and accurate data over the Sino-Soviet Bloc for use in the targeting of imagery intelligence satellites, the NRO initiated the classified DMSP with the objective of having two spacecraft in near-polar orbits at all times. By the mid-1960s, it achieved this, and the program was providing information for strategic applications and was beginning to do the same for tactical applications.

NASA's nine second-generation Environmental Science Services Administration (ESSA) weather satellites launched into near-polar orbits from 1966 to 1969 provided improved civilian coverage and also helped satisfy some DoD tactical weather requirements. It developed and launched the spacecraft, while ESSA assumed their management after being checked out on orbit (this arrangement continued in all the subsequent programs with ESSA's successor, NOAA). The even-numbered ones carried an automatic picture transmission camera that, with a special ground terminal, enabled direct readout of local weather data when the satellite was within range. A number of DoD units (especially in Southeast Asia) had these terminals and used the downlinked information for operational purposes. The odd-numbered ones carried an advanced vidicon camera, whose global data was stored and then transmitted to the NOAA ground stations in Virginia or Alaska when within range. The ground stations relayed it to NOAA's weather center in Suitland, Maryland. NOAA shared it with the Air Force Global Weather Central and the Navy's Fleet Numerical Weather Center, as it would do in all of the following programs.[1]

NASA, NOAA, and the DoD closely coordinated their polar-orbiting weather satellite programs during the 1970s, although they had very different missions. The highest-priority objective of the civilian program was numerical weather analysis (taking current observations of weather and processing the data with computer models to forecast the future weather), followed by marine environment monitoring and prediction and other objectives. The most critical data needed to support these included atmospheric temperatures and humidity and water surface temperatures. Two spacecraft were to be operational at all times to provide global coverage four times a day. DMSP's highest-priority objectives were providing support to overhead imagery intelligence platforms and special military operations, followed by furnishing assistance to military operations worldwide, defense communications systems, and air defense and early warning radars. Among the most important data required was visible and infrared imagery at .35 mile constant resolution and the ability to acquire the former in low light. At least two spacecraft were to be operational at all times in sun-synchronous orbits at an inclination of 98 degrees and an altitude of approximately 850 miles to provide global coverage four times a day. To deny adversaries use of the data and to prevent unauthorized commands, both downlinks and uplinks were to be encrypted.[2]

NASA placed the third-generation *Improved Tiros Operational Satellite 1* and the follow-on *NOAA 1* through *NOAA 5* satellites into near-polar orbits between 1970 and 1975. In contrast to the second-generation ESSA spacecraft, the first two carried both automatic picture transmission and advanced vidicon cameras. These were the first civilian satellites to provide both direct readout of local weather data by special ground terminals and storage and downlinking of global weather data to the two U.S. ground stations. Beginning with *NOAA-2* in 1972, they carried a very high resolution radiometer that replaced both. This sensor provided daytime and nighttime cloud cover imagery at an improved maximum resolution of about one mile in the visible and infrared spectrums along the flight path but at a much-reduced resolution at the edges of the swath. They also flew two other radiometers to give temperature profiles from sea level to an altitude of 19 miles. Local data was directly read out by numerous civilian and U.S. military ground stations around the world. Global data was stored for downlinking to the two NOAA ground stations. The sensors enabled global cloud cover observations every 12 hours, compared with 24 hours of the ESSA satellites.[3]

NASA launched the final four Nimbus satellites into near-polar orbits between 1970 and 1978 to test advanced instruments for use in the operational weather satellite programs. Their primary objective was not mapping cloud cover but monitoring other conditions. The data was directly read out at numerous ground terminals worldwide and was stored for transmission to NOAA's Virginia ground station.[4]

The NRO successfully launched 10 Block 5A, 5B, and 5C DMSP satellites between 1970 and 1976. The primary sensor collected visible and infrared data at the improved resolution of .35 miles along the flight path, which was stored and downlinked for relay to the Air Force and Navy weather centers. Smoothed data with a constant resolution of 2.3 miles across the entire swath was directly read out by 12 tactical ground stations (including one on an aircraft carrier) around the world. These satellites also flew other sensors for vertical profiling of atmospheric temperatures and other purposes. In 1972, the DoD removed the Special Access Program restriction on DMSP and downgraded the program to the Secret level. The following year it declassified DMSP and began routinely providing the collected data to NOAA.[5]

The Air Force initiated development of the Block 5D spacecraft in 1972. It incorporated a completely new design and required much greater Earth-oriented pointing accuracy than its predecessors to support the new primary sensor, the Operational Linescan System. This sensor operated continuously and provided a resolution of .35 miles during the day and an important new capability to operate with a resolution of 2.3 miles at night with a one-quarter moon in the visible spectrum. The infrared subsystem provided .3 miles nadir resolution during both the day and night. In contrast to its predecessors, the resolution at the edges of the 1,840-mile-wide swath was only slightly degraded. This enabled faster computer processing for meeting strict forecasting deadlines for strategic applications and provided users at the direct readout sites to obtain much more accurate data on weather conditions in distant areas. The Block 5D had a much greater data storage capacity for downlinking to one of the two U.S. ground stations. It also carried additional sensors to acquire other types of data. RCA, which built all the earlier DMSP spacecraft and many of NASA's weather satellites as well, won the competition to build the Block 5D.[6]

The Office of Management and Budget proposed in late 1972 to save money by consolidating DMSP and the planned fourth-generation TIROS-N into one system or retaining two systems with more common characteristics. NASA, NOAA, and the DoD jointly examined the technical aspects the following year, as did the National Science Foundation in a separate effort. They did not find any major obstacles in this regard. Henry Kissinger, assistant to the president for national security affairs, directed the NSC Undersecretaries Committee (whose members were the undersecretary of state, deputy secretary of defense, the DCI, and the chairman of the Joint Chiefs of Staff) to examine the international and space policy implications of any merger. The DoD, Joint Chiefs of Staff, and CIA strongly opposed any civilian management of DoD meteorological satellites and any requirement that they accept civilian taskings. NOAA and NASA objected to a single system based on the separation of civilian and military activities set forth in the National Aeronautics and Space Act of 1958, the possible damage to the image of the United States as only engaging in peaceful activities in outer space, and a probable increase in the difficulties of NOAA and NASA meeting international obligations to acquire and freely exchange weather data. The NSC Undersecretaries Committee's report to President Nixon in December 1973 set forth three options: retention of civil and military systems with more common components, a single system under civilian management, or a single system under military management. The White House decided against consolidating the two systems under either civilian or military management and instead directed increased use of common components. Along these lines, NASA and NOAA agreed in 1974 to use the DMSP's Block 5D-2 satellite bus for the planned TIROS-N satellites.[7]

Following the success of the experimental spin scan cloud cameras on several Applications Technology Satellites in geosynchronous orbit during the late 1960s, NASA launched and initially operated two prototype Synchronous Meteorological Satellites in 1974 and 1975 and the first Geostationary Operational Environmental Satellites in 1975. Providing daytime and nighttime coverage of the entire Western Hemisphere from geostationary orbit, their primary mission was to furnish warnings of severe storms. Data was directly read out by numerous civilian and military ground terminals, including at the Air Force and Navy weather centers.[8]

NASA launched *TIROS-N* in 1978 and the first of the follow-on satellites (designated *NOAA 6*) in 1979. For the first time in the civilian weather satellite program, they were boosted into similar orbits to DMSP with an inclination of 98 degrees and an altitude of 850 miles. Each carried a variety of sensors that provided a wider range of more accurate data. These included the advanced very high resolution radiometer, which operated continuously and in both the visible light and infrared spectrums, had a maximum nadir resolution of .6 miles and conducted daytime and nighttime cloud mapping, detected snow and ice, and measured sea surface temperatures. Many hundreds of automatic picture transmission ground terminals around the world directly read out the data at a nadir resolution of 2.5 miles and a fewer number of high resolution picture transmission terminals at a nadir resolution of .6 miles. The U.S. military had 24 ground terminals which, for the first time, could directly read out data from both DMSP and the civilian satellites. The TIROS-N series had an increased storage capacity that enabled it to downlink more data to NOAA's two U.S. ground stations.[9]

TIROS-N was the first known NASA meteorological satellite program to be equipped with a device to protect against unauthorized commands. The national security agencies had first expressed concerns in the 1960s regarding civilian weather satellites being subjected to unauthorized commands in times of crisis and had wanted them to have emergency disabling or destruct devices. However, there is no evidence that they carried any. Sometime before November 1975, the classified "National Policy on the Security of Meteorological Satellite Information" was issued. Pursuant to it, both DMSP and TIROS-N employed a timer to lock out commands while not over U.S.-controlled ground stations. There was some unknown backup capability to override this lockout.[10]

The launch of the initial DMSP Block 5D-1 was delayed from 1974 to September 1976, and the second did not reach orbit until June 1977. This, along with decreasing performance of the Block 5C satellites on station and the loss of the last Block 5C during its early 1976 launch, led to the program's poor weather coverage from 1975 to 1977 and the downgrading of its status to only partially operational. The civilian polar-orbiting spacecraft in service for all or part of this period undoubtedly helped fill in the gaps in coverage for at least some tactical applications, although their resolution, pointing accuracy, nighttime capability in the visible light range, flexibility, and data delivery times were poorer. With respect to satisfying strategic requirements, their different crossing times over the Sino-Soviet Bloc and the delays in relaying their data to the Air Force Global Weather Central resulted in contributing far less. More specifically, the early morning DMSP satellite over the Sino-Soviet Bloc provided data for cloud-free estimates used to program the cameras in the imagery intelligence satellites that followed. The DMSP satellite that overflew the area later in the morning disclosed the accuracy of the estimates and furnished information for the building of a weather model by the Air Force Global Weather Central. To meet the different civilian weather modeling requirements, the first civilian polar-orbiting spacecraft crossed the USSR just after midnight and the second about eight hours later.[11]

The issue of merging the DMSP and civilian weather satellite programs was revisited after President Carter established the new Policy Review Committee (Space) in March 1977 to "formulate a statement of overall national goals in space, the principles which should guide U.S. government and private use of space and related activities, and a clearer definition of the roles and responsibilities of the federal government agencies involved." Chaired by the secretary of defense, its other members were the assistant to the president for national security affairs, the secretary of state, the DCI, the director of the Arms Control and Disarmament Agency, and NASA's administrator. The heads of the Office of Management and Budget, the Office of Science Technology Policy, and the Departments of Agriculture, Commerce, and Interior were to advise the group as needed.[12]

The committee's work directly led to the May 1978 Top Secret Presidential Directive/NSC-37, "National Space Policy." It covered all U.S. space activities—military, intelligence, and civilian—and tasked the Policy Review Committee (Space) "to review and advise on proposed changes to national space policy, to resolve issues referred to the committee, and to provide for orderly and rapid referral of open issues to the president for decision as necessary."[13]

NASA, NOAA, and other civilian agencies criticized Presidential Directive/NSC-37 for falling far short of their expectations. As a result, the president issued the Secret Presidential Directive/NSC-42, "Civil and Further National Space Policy," in October 1978. It created the Interagency Task Force on Integrated Remote Sensing Systems to assess the feasibility and policy implications of consolidation of the atmospheric, ocean, and land remote sensing programs. Chaired by NASA, the task force also had representatives from NOAA, CIA, Office of the Secretary of Defense, Office of Management and Budget, National Security Council, Office of Science and Technology, and the Departments of Agriculture, Interior, and State.[14]

Secretary of Defense Harold Brown and Secretary of Commerce Juanita Kreps set forth their views on convergence of the DMSP and civilian polar-orbiting programs to the director of the Office of Management and Budget in June 1979. Brown maintained that while most civil needs were met by a geosynchronous satellite system, defense needs were satisfied only by polar-orbiting satellites with stringent requirements for pointing accuracy, global coverage, local readout, assured availability, accommodation of special missions, and security. As a result, he supported a single system only if the DoD retained significant control over it. The CIA and NRO shared this position. Kreps opposed a single system predominantly managed by the DoD because civilian weather requirements and international obligations to collect and distribute meteorological data would not be fully met. Instead, she favored joint procurement of satellite buses and sensors where possible as the best means of achieving economies.[15]

The Interagency Task Force on Integrated Remote Sensing Systems submitted its classified report in August 1979. It contained three options with respect to weather satellites: separate but coordinated programs with increased technology sharing, a jointly managed single program to meet both civil and military needs, or a single program under either DoD or Department of Commerce management. The Policy Review Committee (Space) considered the issue in October 1979. Although Frank

Press, the president's science advisor and director of the Office of Science and Technology, strongly supported a single program, the group evidently recommended in the end retention of dual programs but with increased coordination through the existing Polar Orbiting Operational Meteorological Satellite Coordination Board. President Carter approved this course of action in his Secret Presidential Directive/NSC-54, "Civil Operational Remote Sensing," issued the following month.[16]

The DMSP coverage situation became much worse beginning in late 1979 and, ironically, the national security agencies ended up relying totally on civilian satellites for weather data from space during the next two years. The crisis began when the first Block 5D-1 spacecraft failed completely in September 1979. Three months later, the third Block 5D-1 also failed, and the fourth was placed in a backup mode due to electrical problems. In early 1980, the second Block 5D-1, which was in a drifting orbit and used for tactical support, ceased functioning. The last Block 5D-1, to be launched in July 1980, had to operate properly to prevent a complete gap in DMSP coverage. However, its booster malfunctioned and the spacecraft was lost. The gap did not begin to be closed until the successful launch of the first Block 5D-2 in December 1982. There were problems as well with civilian polar-orbiting satellite coverage during the approximately two years in which no DMSP spacecraft were operational. The useful life of *TIROS-N* ended in November 1980. NASA placed *NOAA-6* in orbit during June 1979, and it operated for over four years. However, NASA's next attempt to launch a spacecraft in this series failed in May 1980. *NOAA-7* reached orbit in June 1981, finally restoring the normal complement of two polar-orbiting operational civilian weather satellites. *Nimbus-5*, *Nimbus-6*, and *Nimbus-7* were also relaying data back to Earth during this period, which supplemented that of *TIROS-N* and its successors.[17]

The civilian satellites still had different missions than DMSP and poorer resolution, pointing accuracy, nighttime capability in the visible light range, flexibility, data delivery times, and different crossing times over the Eurasian landmass. Nevertheless, the national security community extensively used the Large Area Coverage mode of *TIROS-N*, *NOAA-6*, and *NOAA-7* during much of the period when no DMSP spacecraft were operational. An Air Force officer detailed to NOAA provided targets for it in terms of latitude and longitude and frequency of coverage for certain areas when the satellites were not within range of NOAA's two U.S. ground stations. The imagery, at a resolution of .6 miles, was stored for later downlink to one of the stations. However, the capacity of the recorders restricted use of this mode to about 10 percent of an orbit. It is not known whether the national security community made use of the alternative Global Area Coverage mode, which permitted more data to be stored and downlinked but at the decreased resolution of only 2.5 miles. Along these lines, an NRO history states that during this period data from the polar-orbiting civilian satellites and forecasts generated by the Air Force Global Weather Central model satisfied minimum strategic weather requirements.[18] It does not address tactical requirements, but the data from the civilian satellites received by the Air Force and Navy weather centers or directly by the 24 DoD ground terminals worldwide (including 8 on aircraft carriers) had to have satisfied them to an unknown extent. The 1980 edition of the *Aeronautics and Space Report of the President* only states that "DMSP support was being supplemented by NOAA weather systems, although the data was far from satisfactory for DoD requirements."[19] A more complete assessment of the contributions of the civilian weather satellites to meeting national security weather requirements and the impact of doing so on their civilian coverage during this period is not possible until further records are released.

The Reagan administration entered office believing that many government functions should be privatized, including the Landsat and civilian weather satellites. At the direction of the Office of Management and Budget in 1981, the Cabinet Council on Commerce and Trade examined whether there should be simultaneous transfer of both programs. The CIA and the Departments of Agriculture, Commerce, Defense, Interior, and State believed that there were major national security and international implications of commercializing civilian weather satellites and favored a long-term study on the matter. The former included concerns over dependence on a private sector system that would have to continue to provide critical backup data for military and intelligence missions and the difficulty in restricting this data going to other nations in times of crisis. With respect to the international implications, over 100 nations were then receiving data from the civilian weather satellites at little or no cost. If they were to be charged considerably more to obtain this information, the fear was that they would cut off the free flow to the United States of important weather observations they acquired from their ground stations, ships, and aircraft. The Office of Management and Budget, Office of Science Technology and Policy, and National Security

Council staff believed the concerns could be protected in any agreement for transfer and favored initiating the process soon.[20]

The Cabinet Council on Commerce and Trade decided in April 1982 to not privatize the civilian weather satellites. However, the Communications Satellite Corporation subsequently persuaded the Secretary of Commerce to change his position. Over the opposition of the CIA, DoD, and others, he sent a Decision Memorandum to the president in early 1983 recommending transfer of both the Landsat and civilian weather satellites as soon as possible. The document concluded that a U.S. firm could accommodate national security concerns and proposed an interagency board to oversee the process. President Reagan soon approved the Decision Memorandum and the establishment of the Interagency Board on Civil Operational Earth-Observing Satellite Systems to oversee the privatization efforts. The operational plan for weather satellites was that the U.S. government would purchase all their data for worldwide distribution and would also pay for its direct downlink to foreign ground stations with the appropriate receiving equipment. The Request for Proposals to be issued to industry addressed the foreign policy and national security concerns regarding transfer of the civilian satellites.[21]

Congress was overwhelmingly opposed to privatizing the civilian weather satellites, however. The Senate passed a resolution in the fall of 1983 against it and then voted to cut off all Department of Commerce spending to implement any transfer. The House soon approved a resolution opposing the commercialization effort as well. As a result of these actions, the Request for Proposals was never issued and the privatization effort died. The Land Remote Sensing Commercialization Act of 1984 expressly prohibited commercializing the civilian weather satellites.[22]

Geodetic Satellites

Satellites provided much more accurate geodetic measurements than traditional ground, ship, and airborne surveys and were the only way to establish a unified world reference system and to comprehensively map the Earth's gravity field. NASA and the DoD established the joint ANNA geodetic satellite program in the early 1960s to meet civilian and defense requirements, but NASA refused to participate until the DoD agreed that all the raw and processed data would be unclassified. Only the second and last ANNA satellite launched in 1962 reached orbit.

NASA and the DoD established the follow-on National Geodetic Satellite Program (NGSP) in 1964, and under it NASA launched a series of geodetic spacecraft often carrying sensors and flying in orbits the DoD requested. NASA's ground station instruments collected data from some sensors, while the DoD's acquired it from all of them. However, the DoD did not rely totally on the NGSP for satellite geodesy and placed some of these sensors on its satellites, which had other primary missions. Because NASA's objectives were scientific investigation and the DoD's were primarily to improve the accuracy of long-range missiles, they handled the raw data acquired and the finished product very differently. All information NASA received and produced was unclassified and disseminated freely, while much of what the DoD obtained and generated was classified. Although NASA objected to the classification and withholding, the DoD does not appear to have changed its policies.

Based in large part on the DoD's growing optical and electronic tracking of NGSP spacecraft and its own satellites carrying geomeasuring devices, the Defense Mapping Agency completed a new World Geodetic System in 1972 that superseded the one prepared six years earlier. It was classified, but eventually the U.S. Geological Survey released a degraded version.[23]

NASA and Air Force cameras continued to photograph the NGSP's *Passive Geodetic Earth Orbiting Satellite* balloon until it disintegrated in 1975, and NASA lasers continued to track the NGSP's *Geodetic Earth Orbiting Satellite II* until the mid-1970s. The DoD adopted the Navy's Doppler system as the primary geomeasuring device in this period. All Navy Transit navigation satellites carried the system, and the DoD also placed them on a limited number of CORONA and other unspecified satellites. It operated approximately 15 semipermanent and 50 mobile Doppler stations around the world.[24]

NASA received authorization from Congress in 1969 to proceed with *Geodetic Earth Orbiting Satellite III* as the final spacecraft in the NGSP. It would be boosted into a low-inclination orbit sometime between 1971 and 1973 to fill in gaps from the NGSP's previous satellites. The satellite would be the first to carry a radar altimeter in an Earth-orbital mission to acquire data for both geodetic and oceanographic applications. Due to the subsequent decision to end the NGSP without an additional satellite, NASA did not request any funds for the project in its FY 1972 budget. However, the DoD and NOAA soon contacted NASA and set forth specific requirements for the satellite and requested that the project continue.

As a result, NASA restarted it in July 1971 and gave it the new designation of *Geodynamics Experimental Ocean Satellite-3 (GEOS-3)*. The radar altimeter would be capable of accuracies between 20 and 50 centimeters and would generate much more accurate gravity models than from the NGSP. *GEOS-3* would also carry C- and S-band radar transponders, laser corner cube reflectors, and a Doppler transmitter for very precise tracking. (The DoD would expand its worldwide Doppler tracking network as part of this effort.) Additionally, NASA's *Applications Technology Satellite 6* would track the spacecraft and relay data to and from it in an experiment. NASA set the new launch date for July 1974. (As it turned out, this was postponed until April of the following year.) In October 1972, NASA solicited proposals for 13 separate scientific objectives and thereafter selected 48 domestic and foreign investigations.[25]

NASA flew its first radar altimeter during an Earth-orbital mission on Skylab in 1973. With an accuracy of between 1 and 2 meters, its primary objective was to obtain measurements needed for designing improved altimeters for *GEOS-3* and later satellites. The results were very encouraging.[26]

William Clements Jr., deputy secretary of defense, wrote James Fletcher, NASA's administrator, in early March 1975 stating that *GEOS-3*'s radar altimeter data was militarily important because it could be used to improve gravity models for the launch regions of U.S. and Soviet submarine–launched ballistic missiles (SLBMs). Why he sent this only a month before the scheduled launch is puzzling. The DoD had been deeply involved in the project and certainly knew of the potential value of the radar altimeter data to the USSR. In any event, Clements asked that the data be encrypted or, alternatively, that NASA limit its collection over specified ocean areas. He also requested that the launch scheduled for 12 April be delayed until resolution of the issue. Lastly, Clements asked that the data from the more capable radar altimeter planned for the follow-on *SEASAT-A* be encrypted. Fletcher rejected all the requests, but indicated he wanted to meet with the secretary of defense on *GEOS-3*.[27]

NASA and the director of defense research and engineering (DDR&E) staff reached an agreement in late March entitled "Operation, Data Acquisition, and Data Distribution Plan for the GEOS-C Program," which was quickly approved by Clements and Fletcher. The key points were (1) NASA would restrict the release, but not the collection, of data in areas between 2,000 and 5,000 nautical miles from U.S. missile sites (the regions in which Soviet ballistic missile subs operated), (2) with respect to satisfying French, Australian, and Canadian scientific investigators who wanted data from the restricted areas, NASA would make a "determined effort" to persuade them to accept data from outside them, (3) if they rejected this, they would only receive data along 5 degree ground track separation within the restricted areas, (4) it would give each domestic and foreign scientific investigator only the data required for their specific investigation, (5) it would maintain a log of all data distributed and not rerelease any data, and (6) it would transmit data in the mid-Atlantic only through an experimental satellite-to-satellite link (NASA's *Applied Technology Satellite 6*) to prevent French telemetry stations from acquiring data. With the agreement, the DoD dropped its request to delay the launch, and on 12 April NASA boosted *GEOS-3* into an orbit approximately 850 miles above the Earth and at an inclination of 115 degrees.[28]

President Gerald Ford directed the NSC Undersecretaries Committee to study the issues involved in the open dissemination of the radar altimeter data just a few days after the agreement was reached. The State Department and NOAA's opposition to its restrictions prompted this action. For the purposes of the study, the undersecretary of commerce and NASA's administrator were added to the Committee.[29]

The study was completed by the end of May, and in it both the DoD and NASA recommended that the existing agreement remain in force. The former argued that the radar altimeter data was needed to derive launch region gravity models that would enable the United States to develop in the future highly accurate SLBMs (0.1 nautical mile or better Circular Error Probable) capable of destroying hard targets. Similarly, the information from the restricted areas would assist the Soviets in improving the accuracy of their SLBMs and thereby acquire the same hard target kill capability in the future. While acknowledging that the agreement represented the first ever limitation on the dissemination of scientific data from one of its space programs, NASA supported it as the best compromise possible between national security and civilian scientific requirements. The State Department opposed the agreement, arguing that the DoD's concerns about Soviet utilization of the data were "unrealistic." It emphasized that no agency had estimated that the USSR wanted to develop a SLBM hard target kill capability. If the Soviets did so, they would have to undertake a massive effort and would likely acquire the necessary geodetic data on their own and not rely on that from *GEOS-3*. The State Department also believed that the agreement seriously harmed NASA's image of engaging only in open activities and would threaten future cooperation with civilian scientists. NOAA opposed the agreement solely on the last grounds. The CIA acknowledged that it would take an expensive program and probably 10 years for the USSR to develop SLBMs with a hard target kill capability. Nevertheless, the GEOS-3 gravimetric data would still be valid if it decided to pursue this course of action and that "from an intelligence point of view, the proposed NASA/Department of Defense agreement would seem to deny the Soviets the data they need to reduce gravimetric contributions to SLBM guidance errors to a level consistent with a hard target capability." President Ford received the report but made no decisions.[30]

The GEOS-3 radar altimeter was operating in the high-intensity mode (the most accurate and frequently used), but no data was being publicly released pursuant to the NASA-DoD agreement. The National Security Council's Senior Review Group examined the agreement in December 1975. Its members were the assistant to the president for national security affairs (who served as the chair), deputy secretary of state, deputy secretary of defense, deputy director of central intelligence, and the chairman of the Joint Chiefs of Staff. NASA's deputy administrator, NOAA's director, NRO's director, and the DDR&E participated in the meeting as well. The positions of the agencies had not changed since the NSC Undersecretaries Committee report in May. The fact that there had been no release of data thus far had not yet caused any complaints from the scientific community. However, in all likelihood it would the following year when the initial processing was expected to be completed and if the withholding continued. NASA had informed the staff of its House and Senate authorization committees of the issue, and for the time being they had not objected to the agreement. There were major concerns about the data from the more capable SEASAT-A radar altimeter scheduled to fly in 1978 and the need to quickly come to an agreement about its dissemination. All agreed that there would be no problem with withholding data if the DoD operated the satellites carrying the radar altimeters, but the deputy secretary of defense stated that it did not want to do so. In any event, there was no resolution of the differences among the agencies, and they decided to continue to meet to try to reach a settlement satisfactory to all concerned by April 1976.[31]

There is no indication from the available records that the issue was resolved by then. It appears that qualified scientists began receiving the GEOS-3 radar altimeter data at some point in 1976. French investigators rejected NASA's offer to provide data from only outside the restricted areas. They ended up receiving it from within them along 5 degree ground track separation only. It is not known whether Australian and Canadian investigators accepted NASA's offer to provide data from only outside the restricted areas. After President Carter directed in May 1977 that the more accurate *SEASAT-A* radar altimeter data should be freely distributed, all the *GEOS-3* data was released. The *GEOS-3* radar altimeter operated in the high-intensity mode until it began experiencing severe problems in October 1978. Ground controllers then commanded the instrument to function in the global mode until July 1979, primarily to satisfy NOAA and Navy requirements to monitor sea state and ice boundaries. Overall, the radar altimeter exceeded all expectations and provided the most complete set of geodetic and geophysical data ever collected over the oceans.[32]

Oceanographic Satellites

NASA began working in the early 1970s with universities, commercial interests, NOAA, the Naval Research Laboratory, the Navy's Fleet Numerical Weather Center, the Defense Mapping Agency, and other government agencies to develop *SEASAT-A*. It would be the first spacecraft to provide worldwide data for oceanographic applications continuously. Five instruments already tested on satellites or aircraft would acquire the various data—a radar altimeter, microwave wind scatterometer, Synthetic Aperture Radar (SAR), visible/infrared scanning radiometer, and scanning multichannel microwave radiometer. The radar altimeter was designed to have an accuracy of 10 centimeters or better, greater than those flown on Skylab and *GEOS-3*. The SAR would be the first imaging radar openly flown in space by the U.S. government. (The NRO flew a satellite with a SAR in December 1964 in Project QUILL. This is the only known time such a sensor was flown in space by the U.S. government prior to *SEASAT-A*.) *SEASAT-A* was to be a proof-of-concept mission in the development of a follow-on National Oceanographic Satellite System.[33]

As mentioned above, the deputy secretary of defense expressed serious concerns to Fletcher in early March 1975 about the plans for the collection and distribution of data from both the *GEOS-3* and *SEASAT-A* radar altimeters and requested that it be encrypted. Fletcher rejected this and, at his suggestion, an ad hoc NASA-DoD-CIA committee was set up to look at all of the issues concerning *SEASAT-A*. Jim Plummer, the

NRO director, was the lead DoD member. NASA agreed to temporarily delay procurement decisions and international requests for participation until resolution of the outstanding questions.[34]

NASA provided the committee with detailed descriptions of the sensors, the data processing and distribution procedures, and the operational plans. In 1978, NASA would boost *SEASAT-A* into a 500-mile high orbit at an inclination of 108 degrees. The satellite would provide repeat coverage every 36 hours. Its sensors had an expected operational life of one year. The SAR was to image a swath approximately 60 miles wide at a resolution as high as 25 meters. Because there was no recorder for the imagery, when the SAR was operating it could only be received in real time by one of three specially equipped NASA Spaceflight Tracking and Data Network ground stations in the United States or the one Canadian ground station when they were in contact with the satellite. All the data from the other four sensors would be both transmitted in real time and recorded for transmission to any of the network's ground stations around the world when the satellite was within range.[35]

With respect to the technologies of the five planned sensors, the DoD and CIA members concluded that no national security issues arose from their "open development, procurement, or integration into a space system." However, they had major concerns in four other areas. First, the radar altimeter data from certain ocean areas could not be publicly disseminated because of the potential to improve Soviet SLBM accuracy. Second, the SAR imagery needed protection because it could be used by a foreign power to detect U.S. naval vessels at sea or in port and could conduct ice surveillance for both surface and antisubmarine warfare operations. Third, the spacecraft needed uplink security measures to prevent foreign personnel from commanding the onboard computer to task the sensors for their own purposes and downlink security measures to prevent foreign ground stations from receiving data from any of them. Fourth, there was the potential that the data from two or more of the sensors might be combined to provide sensitive information such as the location of submerged submarines.[36]

Following the NASA-DoD-CIA committee's initial examination, the NRO director wrote NASA in June 1975 to confirm that the *SEASAT-A* technologies were not of concern under the current plans for the spacecraft's operation and that the DoD had no objections to NASA proceeding with the program. However, because of the military value of the data, it should be protected from "foreign military exploitation by encryption or other secure means." As with the radar altimeter data from *GEOS-3*, NASA was strongly opposed to encryption of any of the *SEASAT-A* data for several reasons. First, NASA's ground stations in foreign countries were not equipped to handle classified data and, in any event, were not permitted to do so pursuant to most of the agreements under which they operated. Second, NASA's relations with scientists around the world had always been on an open and unclassified basis and encryption would radically change this. Third, encryption would blur the separation between U.S. civilian and military space programs. NASA felt so strongly about the issue that it wanted the president to decide whether encryption must be used.[37]

There was no immediate resolution of the security questions, and the recently established Program Review Board continued examining them. The Board directed its Data and Information Release Committee (which had representatives from NASA, DDR&E, CIA, and the NRO) to prepare a report on *SEASAT-A*. The agencies had not changed their basic positions, but over several meetings NASA proposed to install a device to prevent unauthorized commands to the SAR, give the DoD advance notice of the areas that the SAR would image, not reorient the satellite so that the SAR could be pointed at a target, restrict the SAR's operation over sensitive sites, and permit any additional foreign ground stations beyond the one in Canada to receive SAR data only with the DoD's concurrence. NASA maintained its strong objections to the DoD's continued demand that the downlinks from all the sensors be encrypted. The Data and Information Dissemination Committee prepared a draft report in December. Because most of the questions concerning the SAR had been resolved, it stated that the key issue to resolve was one of national policy: whether NASA's long-standing practice of disseminating scientific data without restriction should be modified in the case of the radar altimeter data and, if so, whether encryption or some other method was best to achieve this.[38] It is not known what action the Program Review Board took on the report.

The Joint Chiefs of Staff wrote to the secretary of defense in March 1976 that while the above-mentioned NASA proposals eliminated many of the SAR's security concerns, encryption of the downlinks from it and the radar altimeter was still needed. Key staff members of the Senate Committee on Aeronautical and Space Sciences weighed in on the questions several months later in a meeting with representatives of DDR&E, the Joint Chiefs of Staff, NRO, and NASA. They counseled that it would be in the DoD's interest to resolve the security questions soon and that it should not ask NASA to encrypt any data, since that would cause huge political problems.[39]

NASA and DDR&E signed the apparently unclassified "Agreement for Real-Time User Data Demonstration for SEASAT" in the summer of 1976, but it did not address the security issues. Under the agreement, the Navy's Fleet Numerical Weather Center would receive the data from all the sensors except the SAR. After processing, it would release the information (except any agreed to be withheld) to other U.S. government agencies, approved domestic and foreign scientific investigators, and approved investigators in the fishing, shipping, and offshore drilling industries. Ultimately, all the information released would be archived at NOAA and be available to the public. The Fleet Numerical Weather Center, with assistance from the Naval Surface Weapons Center and the Defense Mapping Agency, would also determine the military value of all the non-SAR data. The Jet Propulsion Laboratory would receive and process the SAR imagery for civilian purposes, and the Naval Research Laboratory and Defense Mapping Agency would do the same for defense applications.[40]

Personnel from NASA, DDR&E, Navy, the National Security Agency, and other organizations held a series of meetings in late 1976 to try to finally resolve the remaining issue of encrypting the SAR's uplinks and downlinks. The DoD representatives maintained that this was the best course of action, but NASA continued to refuse to consider it. NASA again proposed a SAR Enable Disable Unit that, when commanded, would shut down the sensor for three days. The command could not be reversed during this period. At any time during the three days, a new command could be sent to continue the disablement for another three days. The unit would inform the NASA ground controllers when an unauthorized attempt to command the SAR was made. Although this fell far short of encryption, in the end the DoD agreed in principle to its use because it would deny SAR data to an adversary.[41]

Inclusion of the SAR Enable Disable Unit, of course, did not resolve the DoD's security concerns over the radar altimeter data. With respect to it, NASA in late 1976 proposed that the orbital inclination of SEASAT-A be changed from the planned 108 degrees to a polar or near-polar orbit and requested the DoD's views. NASA's stated reason was that the new orbit would provide more coverage of the ice in the polar regions. However, the DoD strongly suspected that NASA wanted to fly the new orbit because it would reduce the value of the radar altimeter data to the Navy and thus eliminate the need for any security protection for it. In late 1976, the DoD rejected the proposal since the orbital inclination of 108 degrees provided the most complete geodetic information for improving the accuracy of U.S. SLBMs.[42]

NASA then proposed in early 1977 that a less capable radar altimeter be flown. The DoD quickly rejected this, stating that only the current radar altimeter could provide the necessary geodetic data. It added that the costs of collecting the data from a DoD satellite to be developed would be as much as $40 million or, alternatively, collecting it by surface ships would take as many as 31 ship years at a cost of over $8 million annually per ship.[43]

Although it is not clear how the issue got elevated, the disagreement over security protection for the radar altimeter data was soon at the White House for resolution. In March 1977, President Carter directed the newly established Policy Review Committee (Space) to prepare recommendations on what, if any, restrictions should be placed on the release of the information. Harold Brown, the new secretary of defense, wrote the president that the DoD now had no objections to the release of any of the *SEASAT-A* radar altimeter data. The only explanation for this radical change in the DoD's position is contained in an internal White House memo from late April. It stated that although the Joint Chiefs of Staff continued to believe the data had the potential to give the Soviet SLBMs a hard target kill capability within 10 years, Brown, Frank Press, and the other principals did not agree. In this regard, either the intelligence agencies had concluded that the Soviets were not going to seek this capability or, if they were, they were acquiring the needed geodetic data from their own satellites and ship surveys. Undoubtedly based on the DoD withdrawing its opposition to the release of radar altimeter data, President Carter directed in early May 1977 to "collect the gravity data and distribute it openly." However, the secretary of defense quickly informed NASA's administrator that the president's decision only concerned *SEASAT-A* and that encryption must be an option for any follow-on system.[44]

NASA and the DoD finally reached a formal agreement covering all aspects of the SAR's operation in late 1977. It provided that *SEASAT-A* would have a SAR Enable Disable Unit and that NASA would give the DoD 24 hours of advance notice of its imaging plans. Only three U.S. ground stations (Alaska, California, and Florida), the Canadian ground station in Newfoundland, and a newly added ground station in England (operated by the European Space Agency) would be capable of receiving SAR data. However, NASA would control all SAR programming, and no commands would be sent from either of the two foreign stations. The SAR would not routinely and repetitively photograph any part of the United States except for some agricultural areas for research.[45]

NASA launched *SEASAT-A* in June 1978 into an orbit with an inclination of 108 degrees (the orbit the DoD had insisted on). The spacecraft operated until it lost power in early October. The problems could not be fixed, and NASA declared the satellite lost the following month. Just in this short period, the radar altimeter collected about 90 percent of the data acquired by the *GEOS-C* radar altimeter in over three years of operation. It contributed to meeting a number of the program objectives, including mapping the global ocean geoid.[46] It is not known whether the data satisfied all the Navy's requirements for improving the accuracy of its SLBMs.

The SAR proved useful for monitoring surface waves, polar sea ice, internal waves, and other ocean conditions and phenomena. Selected imagery revealed surface ships and their wakes but with insufficient resolution to determine their types. There is no evidence that submerged submarines could be detected. In conformance with the late 1977 agreement, NASA presumably gave the DoD 24 hours of advance notice of its plans to operate the SAR. Post-mission reports demonstrate that NASA abided by another provision of the agreement prohibiting the SAR from regularly imaging the United States.[47] It is not known whether the national security agencies reviewed any imagery before public release.

NASA's and NOAA's administrators, DDR&E, the director of the Office of Naval Research, and others agreed in August 1976 to set up an ad hoc group to begin planning for the follow-on National Oceanographic Satellite System, which would meet both civilian and military requirements. The group, chaired by NASA and with high-level representatives of all the interested agencies, would examine the sensors needed and whether there should be restrictions on the quality of any data collected and its dissemination. The fact of the existence of the group and much of its work would be classified, some at the codeword level. Malcolm Currie, DDR&E, quickly asked the Navy to do a comprehensive study on the military value from an operational follow-on system and the appropriate data dissemination policies to protect national security.[48]

The group submitted a classified concept paper in February 1978 on what was now designated the National Oceanic Satellite System. It called for the first satellite to be ready in 1983, but did not specify the sensors that should be carried. Both the uplinks and downlinks would be fully encrypted, but the downlinked data would be decrypted and distributed to U.S. government agencies and domestic and foreign scientists without restriction. The fact of the encryption would be publicly acknowledged. The concept paper recognized that this was a radical change in NASA and NOAA's customary practices and strongly recommended that the encryption be approved at the highest levels. Where NASA and NOAA disagreed with the DoD was on the restrictions on data dissemination in times of crisis or hostilities. NASA and NOAA believed that only the president should be able to stop the release of data and that this fact should be openly acknowledged. On the other hand, the DoD felt that this authority should rest with the secretary of defense with the concurrence of the secretary of state.[49]

The Interagency Task Force on Integrated Remote Sensing Systems, created in response to Presidential Directive/NSC-42, submitted its classified report on possible convergence of the separate weather, ocean, and land remote sensing programs in July 1979. It examined two main options. The first was to develop the National Oceanic Satellite System under either joint NASA-NOAA-DoD management or single agency management. The second was to place the oceanographic sensors on polar-orbiting meteorological spacecraft under an unspecified management structure.[50]

The Policy Review Committee (Space) reviewed the report in October 1979 and recommended that a decision on whether to have separate oceanographic satellites or combine them was a technical issue and left to the spacecraft developers. It also recommended that NASA, NOAA, and the DoD jointly manage at least the initial developmental stage of any oceanographic satellite, with NASA being the lead agency. Presidential Directive/NSC-54 from the following month mandated that if the decision were made to develop oceanographic satellites, the DoD, the Department of Commerce, and NASA would jointly develop, acquire, and manage them. It also directed that a committee with representatives from these organizations and the CIA, State Department, and National Science Foundation would make recommendations on policy issues to the Policy Review Committee (Space).[51]

The Carter administration approved inclusion of approximately $15 million in the FY 1981 budget request to Congress for beginning the National Oceanic Satellite System. It was to be managed jointly by NASA, NOAA, and the DoD, with NASA being the lead agency. NASA and NOAA would each contribute 25 percent to the total cost, while the DoD would contribute 50 percent. The planned instruments included a wide range of sensors, including an even more capable radar altimeter than that on *SEASAT-A*. Flying from Vandenberg, the Shuttle was to place the first satellite in orbit in 1985. It planned to use NASA's Tracking and Data Relay Satellite System for the transmission of all data. Congress, however, authorized and appropriated only a small amount of funds for the program in FY 1981. As an economy measure, the Reagan administration did not request any monies for it in the following year's budget. Although its development was supposedly just being deferred, funding for the program was never included in a subsequent budget, and this action terminated the National Oceanic Satellite System.[52]

The Navy still wanted even more accurate gravity models than provided by the *GEOS-3* and *SEASAT-A* radar altimeter data to improve the accuracy of the Trident II SLBMs, which were scheduled to enter service in approximately 1990. As a result, it launched the *Geodesy Satellite* into a near-polar orbit in March 1985 with an even more capable radar altimeter than the *SEASAT-A* instrument. During the first 18 months of operation, the data collected was classified and was not released to the public until 1995. Thereafter, the instrument operated in a degraded mode, and the information acquired was quickly released to the public until the satellite ceased functioning in early 1990.[53]

The Defense Mapping Agency prepared a new, classified World Geodetic System in 1984 to meet the demands of advanced weapons systems, support a wide range of navigational systems, and provide a more accurate geodetic model of the Earth. The radar altimeter data from *GEOS-3* and *SEASAT-A* was critical to its completion.[54]

Landsat Satellites

NASA finally overcame the opposition of the NRO and other national security agencies in the late 1960s to its robotic land remote sensing satellite, and the White House and Congress approved funding. However, the national security agencies had firmly rejected NASA's plans to employ sophisticated image-forming sensors and imposed strict limits on those that could be flown.

NASA launched *ERTS-A* (later designated *Landsat 1*) in July 1972. It and the follow-on satellites carried multispectral imaging devices, in contrast to those in the imagery intelligence satellite that operated in the visible light portion of the electromagnetic spectrum. NASA's aircraft sensor testing program and the Apollo 9 multispectral experiments had proven the potential value of multispectral sensors to a wide range of civilian disciplines. *Landsat 1* carried a Return Beam Vidicon camera that scanned in several spectral bands and produced a ground resolution as high as 80 meters. Its Multispectral Scanner also scanned in several spectral bands with the same maximum resolution. Operating in a near-polar orbit, the data acquired by the two sensors was either directly read out by one of the U.S. or Canadian ground stations when the satellite was within range or, when out of range, recorded on tape for later transmission to one of them. In keeping with President Nixon's 1969 pledge to freely share all the data and encourage international participation, over 300 domestic and foreign investigators became involved in the project, and the photography was available to any purchaser. However, the largest users were other federal agencies, including the Departments of Agriculture, Commerce, and the Interior. Scientists found the data very useful in a number of fields, including agriculture, forestry, mapping, geology, hydrology, and oceanography. *Landsat 1* exceeded its expected life by more than four years and returned imagery until early 1978.[55]

Shortly after *Landsat 1* reached orbit, the NRO director requested an evaluation of its intelligence value. The Defense Intelligence Agency replied that the imagery did not meet any current intelligence requirements but that it could in the future. The CIA's National Photographic Interpretation Center noted that it had only received second or third generation paper prints to date and was in the process of obtaining higher quality photographs from NASA to do a thorough analysis.[56] Undoubtedly it received them, but there is no available record on the results of the study.

One of the most important potential Landsat benefits was in improving estimates of agricultural production, an issue that had many domestic and foreign policy implications. The major challenge in this regard was developing and refining the techniques for analyzing the photography. At the time, the U.S. government produced two sets of agricultural estimates. The Department of Agriculture prepared unclassified ones on domestic and foreign crops for use by both the government and the private sector, while the CIA published classified reports almost exclusively focusing on communist countries. Both organizations used open sources such as press reports, meteorological data, and U.S. agricultural attaché reporting in preparing foreign production estimates, but the CIA additionally utilized classified data such as imagery from imagery intelligence satellites and clandestine agent reporting.[57]

The Department of Agriculture was criticized for failing to predict the extremely poor Soviet wheat crop in 1972 and the resulting massive Soviet purchases of U.S. grain during the summer months that year. Among other things, these led to greatly increased domestic prices for wheat, dairy, and meat products. In contrast, the CIA intelligence reports from the period were more accurate. However, Department of Agriculture officials who could have used them to better protect U.S. interests did not have access to them.[58]

The controversy over the grain sales resulted in steps to improve Department of Agriculture and CIA estimates and to increase the flow of information to policymakers. In early 1973, a new interagency group with representatives from the CIA, Department of Agriculture, and other organizations began monitoring the progress of the Soviet wheat crop and advising on U.S. export controls. High-level officials at the White House and Departments of Agriculture, Commerce, State, and Treasury received the group's reports. The CIA also started the in-house Project UPSTREET to improve its estimates of Soviet wheat production and to generally investigate the methodology and techniques of analyzing agricultural practices. The project concluded in 1974 that estimates would be much more accurate with better data collection (particularly weather information), faster information retrieval, and refinement of the computer wheat model.[59]

James Fletcher, NASA's administrator, wrote to George Shultz, assistant to the president, in September 1973 requesting that *Landsat 2* be authorized to conduct a pilot program to assess world production of one or more grains. It would utilize data from all sources, including *Landsat 2* and classified imaging satellites, and would employ analysts from both within and outside the federal government. Although some aspects would be classified, the program would be open for the most part. Fletcher asked that NASA be permitted to advance the spacecraft's launch from 1976 to 1974 so that forecasts could be made for 1975. He also requested that a national policy be adopted that recognized the need for an Earth resources survey program and that the data from it be treated "as national assets with appropriate domestic and international controls over their dissemination and use." Fletcher recommended that negotiations be held to address the international political concerns about remote sensing from space "based on the rights of the U.S. to global data in its own interest and to provide such earth resources survey services internationally as are appropriate." He offered to lead an interagency effort to draft the national policy.[60]

The White House distributed Fletcher's proposal to the national security community, and it generated controversy for two reasons. First, many feared that it would lead to further international efforts to limit the acquisition or dissemination of space-based imagery. Along these lines, there were already several countries at the United Nations questioning the right of *Landsat 1* to acquire or disseminate information about their territory without their consent. Second, many opposed NASA's involvement in obtaining economic intelligence. Fletcher corresponded extensively with the deputy secretary of defense, William Clements, on the pilot crop survey. Clements was deeply skeptical. It appears that DCI Colby met with Shultz on the issue, but there is no information on the meeting. The State Department forwarded to the National Security Council additional information provided by NASA for use by the Office of Management and Budget in reviewing the proposed pilot crop survey. In order to increase the chances for its approval, NASA was now stating that it would use no more than 12 percent of the imagery acquired by *Landsat 2* outside the United States for the survey; it would concentrate on wheat production in the United States, Argentina, Brazil, Canada, Australia, the USSR, and China; it would not interfere with the experiments of foreign scientists; and all of its data would be freely distributed.[61]

The Office of Management and Budget soon authorized NASA to advance the *Landsat 2* launch to early 1975 and use up to 12 percent of the photography for the pilot crop survey. No imagery would be withheld, although "appropriate levels of classification will be employed should these data be used for U.S. economic advantage purposes."[62] The restrictions did not completely eliminate the national security concerns as they would resurface shortly.

NASA launched *Landsat 2* in January 1975 with the same Return Beam Vidicon and Multispectral Scanner sensors, and it returned imagery until the early 1980s. International participation in the program continued to grow. Argentina, Australia, Brazil, the European Space Agency, India, Indonesia, Japan, and Thailand all acquired ground stations in the 1970s that enabled them to receive photography directly from the spacecraft.[63]

NASA, NOAA, and the Department of Agriculture initiated the Large Area Crop Inventory Experiment in 1974 after the Office of Management and Budget approved the pilot crop survey. The CIA provided some unknown technical assistance to the project. Its goals were to develop at-harvest wheat estimates that were within 10 percent of the true estimate at the national level 90 percent of the time, to establish the feasibility of acquiring and analyzing Landsat data within 15 days, and to determine how early in the crop year accurate wheat estimates could be produced. During Phase I from 1974 to 1975, the system was tested over the United States where ground truth was readily available. Phase II from 1975 and 1976 expanded the test areas to Canada and two regions of the USSR. During the last phase in 1976 and 1977, detailed study concentrated on the United States and the Soviet Union. The results were mixed, although the forecasts improved as the program progressed.[64]

Timely and accurate estimates of Soviet wheat production and early warnings of possible international purchases to make up for any shortfalls became even more critical intelligence goals in the mid-1970s. Indeed, the DCI listed this information as one of the six highest level intelligence objectives for FY 1976, and CIA reliance on Landsat photography increased.[65]

At the same time that the CIA was finding the imagery a useful source of intelligence, the national security agencies were becoming increasingly concerned that NASA's actions in this and other planned remote sensing programs jeopardized the National Reconnaissance Program. In early 1975, NASA's administrator and the NRO's director agreed to conduct a joint NASA-DoD study examining the continued viability of the July 1966 NSAM 156 Committee report (which, among other things, reaffirmed the limits on NASA's space-based image-forming sensors used to photograph the Earth), the evolution of open and classified Earth-sensing systems since the report, the current international political climate regarding remote sensing, and the possibilities of converging the two programs. The main DoD concern expressed in the April 1975 Top Secret/Codeword study was the growing convergence in technology and data quality between the two programs. Although NASA disputed some assertions the DoD made, they agreed that there were two key questions: What coordination should there be between the civil and military programs to avoid disclosure of classified capabilities or of military valuable information? What risks are there to the classified programs in the event of international opposition to space-based civil remote sensing?

The Space Policy Committee created by President Ford during the summer of 1975 examined these issues but did not resolve them by the time it was disestablished in January 1977.[66]

NASA launched *Landsat 3* in March 1978 with the same Return Beam Vidicon and Multispectral Scanner sensors as the first two. Although there were problems with the latter, the satellite continued operating until 1983.[67]

The Landsat program received extensive study during the Carter administration. Presidential Directive/NSC-42 from October 1978 directed that Landsat continue as a developmental program. The NASA-chaired Interagency Task Force on Integrated Remote Sensing Systems examined possibly integrating the current and future civilian and national security atmospheric, ocean, and land remote sensing systems and submitted its classified report in August 1979, which the Policy Review Committee (Space) soon reviewed. With respect to land remote sensing, the recommendations were to keep the two programs separate but to look at new management options for the civil program. Presidential Directive/NSC-54 issued in November 1979 incorporated these recommendations and assigned management responsibility for an operational civil land remote sensing system to NOAA and created the interagency Program Board for coordination and regulation. NASA would continue to acquire the spacecraft and launch them. The eventual goal was to have the private sector operate the satellites. The directive also increased the maximum permissible resolution for space-based civil remote sensing from 20 meters originally established by the 1965 NASA-NRO agreement to "at or better than ten meters . . . under controls and when such needs are justified and assessed in relation to civil benefits, national security, and foreign policy." Continued international participation in land remote sensing activities was strongly supported, as well as promotion of the development of complementary foreign systems to limit U.S. costs.[68]

The Reagan administration entered office strongly believing that many government functions could be privatized, including the Landsat and polar-orbiting civilian weather satellites. It supported the plans to develop, launch, and test the *Landsat 4* and *Landsat 5* satellites, but intended that any further satellites were contingent on commercializing the system. Chaired by the secretary of commerce, the Cabinet Council on Commerce and Trade examined the issue in depth beginning in 1981. The CIA and the Departments of Agriculture, Commerce, Defense, Interior, and State believed that since Landsat provided important economic, political, and national security benefits, any transfer should proceed slowly and must be accompanied by an increased federal commitment in the form of subsidies, tax incentives, or guaranteed loans to ensure that there would be no interruption in service. CIA officials repeatedly stressed its intelligence value in estimating foreign agricultural production and also raised the potential technology transfer issue if the private sector used improved sensors and sought to supply them to foreign nations. They also noted that the Defense Intelligence Agency had recently started using the imagery for preparing maps. In contrast, the Council of Economic Advisors, Office of Management and Budget, and National Security Council staff maintained that other sources of data were available to the various civilian and national security users and that market forces should determine whether the system continued after *Landsat 4* and *Landsat 5* reached the end of their expected operational lives later in the decade.[69]

NASA launched *Landsat 4* in July 1982 in the midst of the debates over privatization. The Thematic Mapper on *Landsat 4* provided improved spectral and spatial resolution. However, within a year of launch the satellite lost use of its direct downlink transmitters, and it could not transmit any imagery from the Thematic Mapper or Multispectral Scanner until it began utilizing the Tracking and Data Relay Satellite System in 1987.[70]

The Cabinet Council on Commerce and Trade recommended in December 1982 that Landsat be transferred to a U.S. firm as soon as possible in a competitive process, creation of an interagency body to oversee the process, and a near-term increase in federal funding for the system. It gave assurances that national security concerns could be accommodated. The White House approved the recommendations, and the Interagency Board on Civil Operational Earth-Observing Satellite Systems was established to oversee the privatization efforts. One of the key issues it debated was whether Landsat would be subsidized, since the total annual market for the product was much smaller than the cost of acquiring, launching, and operating a single satellite (the Office of Management and Budget was adamantly opposed to any subsidies). All the members, including NASA's, expressed concerns about how to restrict the dissemination of the product to adversaries in times of crisis.[71]

The Department of Commerce published the Request for Proposal for the purchase of the Landsat system in January 1984. Later that year,

Congress passed the Land Remote Sensing Commercialization Act approving the privatization effort. However, it also required the government to provide funding for the turnover.[72]

With *Landsat 4*'s downlink transmitter problems rendering it incapable of providing any imagery until the Tracking and Data Relay Satellite System became operational, NASA received extra monies and launched *Landsat 5* in March 1984 ahead of schedule. It carried the same two instruments as *Landsat 4*, but in 1987 the transmitter used with the now-operational Tracking and Data Relay Satellite System failed. Since there was no onboard recorder, the only imagery *Landsat 5* could downlink after this was that acquired in range of one of the U.S. ground stations.[73]

The Earth Observation Satellite Company (a joint venture between RCA and Hughes Aircraft) was awarded the contract to take over the Landsat system. After much wrangling, the Office of Management and Budget agreed in 1985 that the government would give the firm $295 million over several years to build and operate a U.S. ground station, take over operation of *Landsat 4* and 5, build and launch two new satellites, and market the data. NOAA remained responsible for system operations. After an initial payment of $125 million, the Reagan administration announced that the next payment would be substantially cut. Not surprisingly, the firm greatly raised the prices of the product it sold. This in turn depressed demand. Potential users did not invest in the equipment needed to process the data because it appeared that the system would stop operating soon. In 1989, NOAA announced that it did not have the funds for Landsat operations and that it would not request any additional monies. It also declared that it would soon direct the Earth Observation Satellite Company to turn off *Landsat 4* and 5.[74]

These actions resulted in the many and varied users of the data, as well as key members of Congress, pressing the new administration of President George H. W. Bush to reverse the privatization and save the program. Many of Landsat's supporters also argued that it had to be continued so other nations would not assume the leadership in the field, since the French and Soviets had launched their first remote sensing satellites in 1986 and 1987, respectively. Both provided considerably higher resolution imagery than Landsat—the French SPOT close to 10 meters and the Soviet spacecraft nearly 5 meters. The president approved the National Security Council's recommendations for additional funding to support *Landsat 4* and 5 and to build and launch *Landsat 6*.[75]

The national security agencies became even stronger supporters of the system after extensively using the multispectral photography during the Gulf War. The Earth Observation Satellite Company reportedly halted all other work at times to process it for the DoD within 24 hours of downlinking from the satellites. The imagery's chief use was for making image maps. At the outset of Desert Shield, there were few of the standard topographic line maps of the theater that were essential for combat operations. The Defense Mapping Agency combined Landsat multispectral imagery with the higher-resolution SPOT panchromatic imagery to make nearly 100 1:100,000 scale image maps that were utilized on an interim basis until they were replaced with 1:50,000 scale topographic line maps. Landsat photography was also used to identify oil fires and the location and movement of oil slicks in the Persian Gulf. One benefit of employing data from the civilian systems was that it was unclassified and could freely be shared with allies and, if needed, with the public.[76]

The Land Remote Sensing Policy Act of 1992 ended the privatization experiment for good. Reflecting the system's increased importance to the national security agencies, the legislation moved Landsat's operational control from NOAA and the private firm to DoD and NASA. While the DoD would fund development of the satellite and its instruments, NASA would fund construction of the ground-data processing and operations systems, launch and operate the satellite, and distribute the data. The law also mandated the selling of the data at cost and the provision of it to all purchasers on a nondiscriminatory basis. After the failed launch of *Landsat 6* in October 1993, the DoD decided for unknown reasons that the program was unnecessary to its needs and stopped funding *Landsat 7*. The following year, the White House directed that NASA acquire the satellite and reassigned operation of the spacecraft and ground stations from NASA to NOAA. NASA successfully placed *Landsat 7* in orbit in April 1999.[77]

Summary

NASA continued working closely with the national security agencies in the 1970s and 1980s to ensure that its increasingly sophisticated applications satellite programs met national security as well as civilian requirements when possible. Although having very different missions, its weather satellites continued to help satisfy some tactical requirements and, to a lesser extent, strategic requirements. In response to the growing concerns over the possibility of adversaries using the data from them in times of crisis, the fourth-generation TIROS-N were the first known NASA weather satellites to incorporate a device to prevent unauthorized commands. NASA agreed to carry advanced radar altimeters on two satellites and fly them in specific orbits to acquire geodetic data needed to improve the accuracy of U.S. SLBMs. In both cases, civilian requirements could have been met with less capable instruments and different orbits. Although successfully opposing the encryption of the downlinked scientific data to deny its use by the Soviets, for the first time ever in an applications satellite program NASA agreed to restrict the dissemination of some of it. The restrictions were short-lived, however, as the DoD finally concluded that it was not sensitive, and President Carter ordered that it be freely distributed.

The national security agencies approved the proposed sensors for NASA's oceanographic satellite, which was a demonstration for a planned follow-on joint NASA-DoD program. However, they had concerns over the value of the Synthetic Aperture Radar's imagery to adversaries. Once again, NASA successfully opposed encryption of it but did consent to limiting its acquisition and dissemination and to installing equipment to prevent unauthorized commands. Although development of the follow-on program started, budget cutbacks forced its early termination. Image-forming sensors for NASA's land remote sensing satellites continued to be subject to the restrictions originally imposed in the mid-1960s and, as a result, the spacecraft carried multispectral cameras. Although there was little intelligence interest in the photography initially, this changed quickly, and it began to be used for producing classified foreign agricultural estimates and making maps.

Possible convergence of the civilian and national security applications satellite programs was proposed periodically during the two decades. However, NASA, NOAA, and the defense and intelligence agencies successfully opposed such action in the weather and land remote sensing programs, except for a brief merger in the early 1990s in the latter to keep the program operational. The Reagan administration vigorously pursued privatization of the weather and land remote sensing programs during the 1980s. Agency and congressional opposition stopped the effort with the weather satellites. It occurred in the land remote sensing program, but only briefly.

Conclusion

The National Aeronautics and Space Act of 1958 mandated that NASA establish a space program separate and distinct from the national security community's and that it be peaceful, scientific, and open. Although these served as NASA's guiding principles, it could not and did not always follow them.

In addition to many NASA spaceflight programs receiving vital unclassified, overt support from the national security agencies during its first decades, several benefited greatly from their classified, covert assistance. The lunar photography programs during the Apollo era employed several classified cameras pursuant to agreements that hid their true origin and other information from the public. Beginning in the 1960s, these agencies closely monitored the planned employment of sensitive optical systems or pointing and stabilization systems in NASA's space-based astronomy programs. They apparently either provided such hardware or consented to their equivalent being used in several. NASA also obtained mirror blanks from the National Reconnaissance Office in the early 1970s for use in such programs. The NRO gave NASA access to its sophisticated optical manufacturing techniques for use in the Hubble project during this period. It subsequently permitted one of its contractors to use a highly classified facility to fabricate the *Hubble*'s mirrors with strict security controls, although it is unclear whether any classified manufacturing or testing techniques were employed. The NRO possibly allowed several other contractors to employ such facilities to build additional components. The NRO also furnished a shipping container used for classified satellites to transport the completed spacecraft from California to Florida. These different types of support to the lunar photography and astronomy programs, of course, did not change their peaceful and scientific nature.

The defense and intelligence agencies began in the 1960s to monitor and restrict NASA's Earth-imaging activities under great secrecy to eliminate any political and technological threats to the National Reconnaissance Program. The restrictions limited the types of image-forming sensors that could be used in the human spaceflight and robotic remote sensing programs. In the human spaceflight programs, the agencies also instituted a pre-flight review of all proposed photographic experiments and a post-flight screening of all imagery before public release. Waivers of the technical limitations were occasionally granted but only under strict conditions. No other NASA activity operated under such far-reaching constraints. To partially compensate for them, the defense and intelligence agencies established a procedure under which cleared personnel from NASA and other interested federal civilian agencies gained access to selected classified overhead imagery for use in civilian applications.

NASA's early spaceflight programs often served national security interests, but it was very sensitive about providing any classified support because of the need to maintain the image of engaging only in peaceful, scientific, and open activities. Mercury, Gemini, and Apollo missions flew many unclassified DoD experiments. NASA was concerned about the DoD's proposed classified experiments in Gemini, but ended up flying them under a compromise under which selected data was withheld. A joint National Geodetic Satellite Program was created in the mid-1960s to meet civilian and national security requirements, but NASA refused to participate until the DoD dropped its demand that much of the data from the first-generation satellites be classified. NASA managed the program thereafter, but conflicts continued over the DoD's classification of selected data it collected and analyzed from subsequent satellites under the program. This did not stop NASA's collaboration, however, and in several cases it flew instruments and orbits the DoD requested to improve the accuracy of long-range missiles. A joint National Operational Meteorological Satellite Program was also established in the early 1960s under NASA management to satisfy both civilian and national security objectives, but delays and technical problems caused the DoD to withdraw quickly, and NASA continued it as a R&D program. NASA continued collaborating on its other polar-orbiting weather satellites, and although they furnished little data for strategic applications, they provided it for some tactical applications.

NASA largely abandoned its guiding principles with the Shuttle, the first spaceflight program in which it would openly and repeatedly fly classified missions and withhold considerable information from the public. To ensure the critical political support of the defense and intelligence agencies, it formed an unprecedented partnership with them to build and operate the system to be a cheaper and more reliable launch vehicle for all U.S. government payloads. NASA readily accepted their requirements for a larger orbiter and more demanding performance specifications. It also agreed to pay the vast majority of the program's costs, give national security missions priority, and install the first ever secure command and control system. The national security community's civilian leaders during the Carter administration were very supportive of the Shuttle, and their intervention at the White House saved the program from drastic reduction or even cancellation when it was facing technical and financial problems. Shortly after the Shuttle became operational in the early 1980s, it flew a limited number of classified missions. However, its failure to meet the original performance specifications and flight schedule caused massive problems in the transition of national security payloads. Over NASA's strenuous opposition, these agencies obtained permission in 1984 to build a limited number of new expendable launch vehicles that satisfied these specifications. The *Challenger* accident two years later greatly accelerated their abandonment of the Shuttle and forced them to begin acquiring new types of expendable launch vehicles to place their payloads in space. This transition back to expendable launch vehicles was very expensive and may have caused a degradation of coverage in their space programs.

NASA also abandoned its guiding principles in several applications satellite programs during the Shuttle era. Collaboration with the national security agencies expanded as the more sophisticated sensors increasingly met national security as well as civilian requirements. Although the civilian weather satellites continued to have a different mission than the DoD's, they provided data for some tactical applications but were still limited in what they could furnish for strategic applications. This support became even more critical when no DoD weather satellites were operational at all from 1980 to 1982. Reflecting the importance of the data to both the United States and adversaries, NASA's fourth-generation weather satellites incorporated equipment to prevent unauthorized commands. NASA launched two satellites in the 1970s with specific geodetic instruments and orbits that the DoD wanted for improving the accuracy of U.S. submarine–launched ballistic missiles. Although it successfully opposed installing encryption devices to prevent foreign exploitation of the data, for the first time ever in an applications satellite program NASA agreed to limit its dissemination. The national security agencies approved the proposed sensors for NASA's oceanographic satellite launched in 1978. NASA again successfully resisted the demand to encrypt the data from one. However, NASA did agree to restrict the acquisition and distribution of selected data and to install a device to prevent unauthorized commands. A more capable joint NASA-NOAA-DoD follow-on oceanographic satellite that would have employed encryption and limited the distribution of data in times of crisis was cancelled because of budget cutbacks.

There was also collaboration and mutual assistance in areas not directly involving the operation of the civilian and national security space programs but related thereto, most of which was also hidden or classified. NASA's command and control networks furnished support to orbiting DoD satellites and missile tests during the 1960s and 1970s. From its earliest days, NASA regularly received the highest-level intelligence on the Soviet and other foreign space programs from the CIA and shared its expertise with the CIA and others in analyzing these programs. During the 1960s, NASA worked with the CIA and others in recovering space debris that had returned to Earth and establishing uniform policies to govern this process. Beginning early that decade, it collaborated with the DoD in acquiring space surveillance data and acted as a conduit for the release of unclassified information in this area. NASA also participated with the CIA and others in cover stories for the U-2, the follow-on A-12, and perhaps other reconnaissance vehicles.

NASA and the national security community have continued their extensive interaction, but there are very few details on the hidden or classified cooperation and support. For example, NASA's administrator and the NRO director formed the Space Partnership Council in 1997 to coordinate on a wide range of issues. The director of defense research and engineering, commander of the U.S. Strategic Command, commander of the Air Force Space Command, and the CIA's deputy director for science and technology subsequently became members.[1] Projects have included reducing the size of synthetic aperture radars used on reconnaissance satellites, improving automated data processing, developing sensors for the Lunar Reconnaissance Orbiter for possible later use on reconnaissance satellites, and using classified systems to monitor the Shuttle.[2] NASA, the Air Force, and Boeing agreed in 1998 to develop the X-37 unmanned reusable launch vehicle to test key airframe, propulsion, and operating technologies. The DoD assumed control of the project in 2004 and changed its focus to developing a reconnaissance platform. NASA continued to provide unspecified support, and the X-37 flew the first of several classified missions in 2010.[3] President Clinton directed in 1994 that a single weather satellite system be designed to meet both civilian and national security requirements—the objective of the earlier short-lived National Operational Meteorological Satellite Program. However, technical problems, cost growth, and schedule delays led to its cancellation in 2010.[4] Sensors on several NASA satellites provided important data to support tactical operations after 9/11.[5]

It is clear that from NASA's earliest days it did not always adhere to its guiding principles. The civilian and national security space programs had too many common interests and activities to remain truly apart. There was considerable overt and covert mutual support, and at times limitations were imposed on NASA's programs to ensure that they did not threaten national security. These truths do not in any way diminish NASA's remarkable accomplishments over the years, but they do provide a more complete and accurate history of it.

Notes

Introduction

1. Burrows, *This New Ocean*, 188–95, 214–18; McDougall, *Heavens and the Earth*, 141–56, 169–76.

2. Burrows, *This New Ocean*, 197–206; McDougall, *Heavens and the Earth*, 150–56.

3. Burrows, *This New Ocean*, 213–16; McDougall, *Heavens and the Earth*, 170–76; "Introduction to Outer Space" in Logsdon, ed., *Exploring the Unknown*, 1:332–34.

4. Burrows, *This New Ocean*, 213–16; McDougall, *Heavens and the Earth*, 170–76.

5. Burrows, *This New Ocean*, 213–16; McDougall, *Heavens and the Earth*, 170–76; *National Aeronautics and Space Act of 1958* in Logsdon, ed., *Exploring the Unknown*, 1:334–45.

6. The numerous works documenting the extensive assistance that NASA provided the defense and intelligence agencies in developing aircraft include Chambers, *Partners in Freedom: Contributions of the Langley Research Center to U.S. Military Aircraft of the 1990s*; Baals and Corliss, *Wind Tunnels of NASA*; Hartman, *Adventures in Research: A History of Ames Research Center, 1940–1965*; Muenger, *Searching the Horizon: A History of Ames Research Center, 1940–1976*; Crickmore, *Lockheed SR-71: Secret Missions Exposed*, 11; Rich and Janos, *Skunk Works: A Personal Memoir of My Years at Lockheed*, 210–11; Merlin, *Mach 3+: NASA/USAF YF-12 Flight Research, 1969–1979*, 1; Pedlow and Welzenbach, *The Central Intelligence Agency and Overhead Reconnaissance*, 278.

7. See, e.g., Launius and Jenkins, *To Reach the High Frontier*; Swenson, Grimwood, and Alexander, *This New Ocean*, 122–23, 154–56, 174–76, 185–90, 214–20, 644–48; Hacker and Grimwood, *On the Shoulders of Titans*, 57, 78–79, 104–5, 585–96; Brooks, Grimwood, and Swenson, *Chariots for Apollo*, 233, 249, 271, 283–84, 299, 213, 355–56.

8. See, e.g., Compton and Benson, *Living and Working in Space*, 15–18, 46–48; Day, "Invitation to Struggle"; Hays, "NASA and the Department of Defense," 215; Boone, *NASA Office of Defense Affairs*, 96–100; Erickson, *Into the Unknown Together*, 149–85, 477.

9. As readers will see, I rely heavily on declassified documents. Some have been released by the CIA under the automatic/systematic declassification review program established by Executive Order 12958 (as amended). These are available on the CREST (CIA Records Search Tool) electronic database in the library at the College Park National Archives. For the most part, however, I have had to file time-consuming declassification requests with the CIA, NASA, National Reconnaissance Office, Office of the Secretary of Defense, and others to obtain relevant documents. Many led to the eventual release of valuable material, but others did not or are still pending at the time of this writing.

Chapter 1. Forging Close Ties in NASA's Early Years

1. McDougall, *Heavens and the Earth*, 195–96; NASA, biography of T. Keith Glennan, http://history.nasa.gov/Biographies/glennan.html.

2. McDougall, *Heavens and the Earth*, 195–96; NASA, biography of Dr. Hugh L. Dryden, http://history.nasa.gov/Biographies/dryden.html; Pedlow and Welzenbach, *The Central Intelligence Agency and Overhead Reconnaissance*, 89–90; Hunley, *Birth of NASA*, 2–4.

3. McDougall, *Heavens and the Earth*, 197–200; Burrows, *This New Ocean*, 258–67; *Aeronautics and Space Report of the President, 1959 Activities*, vi.

4. Senate Committee, *NASA Authorization for Fiscal Year 1961*, 135–37.

5. Senate Committee, *NASA Authorization for Fiscal Year 1960*, 5–8.

6. Boone, *NASA Office of Defense Affairs*, 37–42.

7. Pedlow and Wetzenbach, *The Central Intelligence Agency and Overhead Reconnaissance*, 39–43, 51–54, 315–18.

8. Greer, "Corona"; Oder, Fitzpatrick, and Worthman, *GAMBIT Story*.

9. See., e.g., Henry G. Plaster, "Snooping on Space Pictures," *Studies in Intelligence*, entry #27, RG 263 (Records of the Central Intelligence Agency), National Archives and Records Administration, College Park, Md. (NARA); James Burke, "Missing Link," *Studies in Intelligence*; and Albert Wheelon and Sidney Graybeal, "Intelligence for the Space Race," *Studies in Intelligence*, CREST (CIA Records Search Tool database), NARA.

10. Turkey, for example, was the site of a series of radars beginning in 1955 that tracked Soviet rockets, missiles, and spacecraft during the early stages of their flights. Zabetakis and Peterson, "*Diyarbikar Radar*," CREST, NARA.

11. Wagoner, *United States Cryptologic History: Special Series Number 3*, 5, 19, 33.

12. Potts, *U.S. Navy/NRO Program C Electronic Intelligence Satellites*, 223–24.

13. Scientific and Technical Intelligence Research and Production Program, Fiscal Year 1959, 1 July 1958, Declassified Reference Collection, RG 263, NARA; Briefing Notes on the Intelligence Directorate and OSI/DD/S&T, January 1964, CREST, NARA.

14. David, "Two Steps Forward," 220–22.

15. *Critique of the Codeword Compartment in the CIA*, March 1977, 6, CREST, NARA; Potts, *U.S. Navy/NRO Program C Electronic Intelligence Satellites*, 34–35; memorandum from Dwight D. Eisenhower to the Secretary of State et al., 26 August 1960, CREST,

NARA; Johnson, *American Cryptology*, bk. 2, 405; Johnson, *American Cryptology*, bk. 3, 67; Richelson, *The U.S. Intelligence Community*, 510–14.

16. C. P. Cabell, General, USAF, Acting Director, to Dr. T. Keith Glennan, Director, NASA, 10 September 1958. The CIA provided this to the author pursuant to a Mandatory Declassification Review (MDR) request.

17. Memorandum from an unknown individual at NASA to an unknown individual at the CIA [date redacted], and attached Official Routing Slip and Action Sheet. The CIA provided these to the author in 2006 pursuant to an MDR request.

18. *National Aeronautics and Space Act of 1958*, in Logsdon, ed., *Exploring the Unknown*, 1:334–45.

19. The National Aeronautics and Space Council—Meeting of 20 October 1958, box #2, Minutes of NASC Meetings, RG 220 (Temporary Commissions, Committees, and Boards), NARA.

20. See, e.g., minutes of the 18 August 1961 meeting in box #2, Minutes of NASC Meetings, RG 220, NARA.

21. Chester L. Cooper, Deputy Assistant Director, National Estimates, to U.S. Intelligence Board, 20 November 1958; [Redacted], Executive Secretary to Dr. T. Keith Glennan, 9 December 1958. The CIA furnished these to the author pursuant to an MDR request. *National Intelligence Estimate 11-5-58: Soviet Capabilities in Guided Missiles and Space Vehicles*, 19 August 1958, 26–29, box #10, entry #29, RG 263, NARA. Restricted data is, among other things, information concerning the design, manufacture, or utilization of nuclear weapons. David, "Two Steps Forward," 221–22.

22. *National Intelligence Estimate 11-5-58: Soviet Capabilities in Guided Missiles and Space Vehicles*, 3 November 1959, 27–33, box #10, entry #29, RG 263, NARA.

23. Allen Dulles, Director, to Dr. Keith T. Glennan, Director, NASA, 21 March 1959, CREST, NARA.

24. T. Keith Glennan, Administrator, to Honorable Allen W. Dulles, Director of Central Intelligence, 27 March 1959, CREST, NARA.

25. [Redacted], Security Officer, DPD-DD/P, to Acting Chief, DPD, 24 April 1959, CREST, NARA.

26. James A. Cunningham Jr., Acting Chief, DPD-DD/P, to Deputy Director (Plans), 30 December 1959, CREST, NARA.

27. Hugh L. Dryden Appointment Calendar, 16 March 1959, Hugh L. Dryden Collection: NASA Headquarters History Office; Potts, *U.S. Navy/NRO Program C Electronic Intelligence Satellites*, 23.

28. Assistant Director, Current Intelligence, to Executive Secretary, USIB, 9 February 1960. The CIA provided this to the author pursuant to an MDR request.

29. *National Intelligence Estimate 11-5-59: Soviet Capabilities in Guided Missiles and Space Vehicles*.

30. Executive Secretary to Dr. Keith T. Glennan, Administrator, NASA, 19 October 1960, and [Redacted] to Executive Secretary, U.S. Intelligence Board, 21 October 1960, CREST, NARA.

31. *National Intelligence Estimate 11-5-60: Soviet Capabilities in Guided Missiles and Space Vehicles*, 3 May 1960, 8, box #11, entry #29, RG 263, NARA.

32. Summary of Committee Briefings, n.d., CREST, NARA.

33. The National Security Council created the U.S. Intelligence Board in 1958 to assist the DCI in establishing intelligence policy and program guidance, to report to the National Security Council and others on the foreign intelligence effort, and to produce national intelligence estimates and reports. Chaired by the DCI, its other members were from the Department of State, Defense Intelligence Agency (after 1961), National Security Agency, Atomic Energy Commission, and the intelligence organizations of the three services. The board established a number of committees, including the Guided Missiles and Astronautics Intelligence Committee. Among other things, its mandate was to develop overall objectives for intelligence collection in its field, evaluate the collection effort to meet these objectives, and produce reports on foreign missile and space programs for the board. Intelligence Community, 3 December 1971, CREST, NARA.

34. [Redacted] to Chief, Collection Staff, SI, 16 January 1959. The CIA provided this to the author pursuant to an MDR request. It has several large redactions in the body of the text.

35. [Redacted] to Chief [Redacted], 29 January 1959. The CIA provided this to the author pursuant to an MDR request.

36. Assistant Project Security Officer to Project Security Officer, 20 March 1956, NACA-AWS Upper Atmosphere Research Program; Chief, Operations Branch to Deputy Director (Plans), 7 April 1959, CREST, NARA.

37. Chronological Account of Handling of U-2 Incident, n.d., CREST, NARA.

38. Beschloss, *MAYDAY,* 46–52.

39. Ibid., 52–66, 243–304.

40. Staff Meeting Minutes, 7 October 1960; Memorandum for the Record, 22 September 1960; C. P. Cabell, General, USAF, Acting Director, to Walter T. Bonney, Director, Public Information, NASA, CREST, NARA.

41. National Aeronautics and Space Administration and Department of Defense: A National Program to Meet Satellite and Space Vehicle Tracking and Surveillance Requirements for FY 1959 and 1960, 8 January 1959; Space Council Jan.–June 1959, box #265, Records Relating to Atomic Energy Matters, 1944–1963, RG 59 (General Records of the Department of State), NARA; Sturdevant, "From Satellite Tracking to Space Situational Awareness," 5–23.

42. Sturdevant, "From Satellite Tracking to Space Situational Awareness," 5–23.

43. National Aeronautics and Space Administration and Department of Defense: A National Program to Meet Satellite and Space Vehicle Tracking and Surveillance Requirements for FY 1959 and 1960, 8 January 1959; W. E. Gathright to Mr. Hare et al., 16 August 1960, OCB Working Group on Outer Space Part 1, box #262, Records Relating to Atomic Energy Matters, 1944–1963, RG 59, NARA.

44. Ezell, *NASA Historical Data Book,* vol. 2, 521–96; Mudgway, *Uplink-Downlink;* Tsiao, *"Read You Loud and Clear!"*

45. Potts, *U.S. Navy/NRO Program C Electronic Intelligence Satellites,* 23, 50, 59; Paul T. Cooper, Brig. Gen., USAF, Assistant Director (Ranges and Space Ground Support), to Edmond Buckley, Office of Tracking and Data Acquisition, 15 March 1962, 111 NASA, box #6, accession #66-3170, RG 330 (Records of the Office of the Secretary of Defense), NARA.

46. Arnold, *Spying from Space;* Department of Defense Space and Space Related Program Data, Fiscal Years 1964–1969, n.d., 23–64 Satellite Program, box #15, SECAF Top

Secret 1956–1964, RG 340 (Office of the Secretary of the Air Force), NARA; Hawkes, "USAF's Satellite Test Center Grows."

47. Report of SFGEP Ad Hoc Study Team on Procedures for Reciprocal Support, 1 May 1964, 111 NASA, box #7, accession #68-5157, RG 330, NARA. The report also discusses smaller DoD tracking, data acquisition, and control capabilities at such facilities as the Air Proving Ground Center in Florida and the Churchill Research Range in Canada.

48. David, "Was It Really 'Space Junk'?" 47; Sturdevant, "From Satellite Tracking to Space Situational Awareness," 11.

49. Sturdevant, "From Satellite Tracking to Space Situational Awareness," 11.

50. Agreement between the Department of Defense and the National Aeronautics and Space Administration on Functions Involved in Space Surveillance of U.S. and Foreign Satellites and Space Vehicles, 16 January 1961, box #10, General Correspondence of NASC Executive Secretary Ed Welsh, 1961–1969, RG 220, NARA; James E. Webb, Administrator, to Dr. E. C. Welsh, Executive Secretary, National Aeronautics and Space Council, 4 November 1962, Webb-Cor. July–Dec. 1962 file, NASA Headquarters History Office.

51. Senate Committee, *NASA Authorization for Fiscal Year 1961*, 646, 664; Hall, *A History of the Military Polar Orbiting Meteorological Satellite Program*, 1.

52. Senate Committee, *Meteorological Satellites*, 98–101.

53. David, "Astronaut Photography," 186.

54. William Burke, Colonel, USAF, Acting Chief, DPD-DD/P, to Deputy Director (Plans), 17 July 1959, and Richard M. Bissell Jr. to Director of Central Intelligence, 20 July 1959, CREST, NARA.

55. Memorandum for the File, 19 November 1959, Space Council July–Dec. 1959, box #266, Records Relating to Atomic Energy Matters, 1944–63, RG 59, NARA.

56. T. Keith Glennan, Administrator, to Honorable Thomas S. Gates Jr., Deputy Secretary of Defense, 27 November 1959, *Tiros 1* file, NASA Headquarters History Office; Mr. Gathright to Mr. Farley, 30 December 1959, Satellites, Recon. 1959, box #279, Records Relating to Atomic Energy Matters, 1944–63, RG 59, NARA.

57. Herbert F. York to Dr. Keith T. Glennan, Administrator, NASA, 13 February 1960, 111 NASA-DOD, box #10, accession #66-3146, RG 330, NARA. The author has been unable to locate a copy of the Joint Chiefs of Staff document.

58. R. F. Courtney to Mr. Farley, 1 April 1960, Satellites, Navigation 1960, box #277, Records Relating to Atomic Energy, 1944–1962, RG 59, NARA; Hunley, *Birth of NASA*, 109–14.

59. Memorandum for the Record, 22 September 1960, and [Redacted], Chief, TISD to [Redacted] DPD/DD/P, 28 July 1960, CREST, NARA; *Proceedings of the International Meteorological Satellite Workshop, November 13–22, 1961*, 212–14.

60. John W. Finney, "Nations Offered Data from Tiros," *New York Times*, 26 September 1960, A5; Ezell, *NASA Historical Data Book*, vol. 2, 348–53.

Chapter 2. NASA, the CIA, and Foreign Intelligence during the Apollo Era

1. Senate Committee, *NASA Authorization for Fiscal Year 1966*, 7–8; Lambright, *Powering Apollo*, 82–88.

2. Lambright, *Powering Apollo*, 82–88; interview of Dr. Robert Seamans by Martin

Collins, 15 December 1988, http://www.nasm.si.edu.research/dsh/TRANSCRIPT/SEA-MAN10HTM (accessed 13 March 2012).

3. Lambright, *Powering Apollo*, 82–88; interview of Dr. Robert Seamans by Martin Collins, 15 December 1988.

4. NASA, biographies of aerospace officials and policymakers, http://www.history.nasa.gov/bioso-s.html (accessed 2 July 2010).

5. NASA, biography of Thomas A. Paine, . http://history.nasa.gov/Biographies/paine.html.

6. James E. Webb to Honorable Richard Helms, Director of Central Intelligence, 2 February 1968, CREST, NARA; NASA, biographies of aerospace officials and policymakers, http://www.history.nasa.gov/biosa-d.html (accessed 2 July 2010).

7. Boone, *NASA Office of Defense Affairs*, 50–60.

8. Memorandum for the Record, 28 February 1961, CREST, NARA.

9. [Redacted] Executive Secretary to James E. Webb, Administrator, NASA, 25 May 1961. The CIA provided this to the author pursuant to a Mandatory Declassification Review (MDR) request. [Redacted], Secretary, to [Redacted], Secretary, U.S. Intelligence Board, 31 May 1961 and [Redacted] to Executive Secretary, USIB, 19 June 1961, CREST, NARA.

10. *National Intelligence Estimate 11-5-61: Soviet Technical Capabilities in Guided Missiles and Space Vehicles*, 25 April 1961, 34–44, box #16, entry #29, RG 263, NARA.

11. Special Assistant for Planning to [Redacted], 12 September 1961; Memorandum for the United States Intelligence Board, 30 October 1961; [Redacted], Executive Secretary, to James E. Webb, Administrator, NASA, 1 February 1962. The CIA furnished these to the author pursuant to an MDR request.

12. John A. McCone, Director, to Dr. Hugh L. Dryden, Deputy Administrator, NASA, 10 November 1962, CREST, NARA.

13. Day and Siddiqi, "The Moon in the Crosshairs: Part 1," 468.

14. [Redacted], Executive Secretary, to James E. Webb, Administrator, NASA, 19 December 1962, CREST, NARA; *National Intelligence Estimate 11-5-62: The Soviet Space Program*, 5 December 1962, box #8, entry #29, RG 263, NARA.

15. *National Intelligence Estimate 11-5-62: The Soviet Space Program*, 5 December 1962, box #8, entry #29, RG 263, NARA.

16. Ibid.

17. Sherman Kent, Chairman, to the Director, 25 April 1963, CREST, NARA; James E. Webb, Administrator, to Dr. Simpson, AF, 11 October 1963, Intelligence and Correspondence file, Federal Agencies CIA NIEs box, NASA Headquarters History Office.

18. "ORR Position on Soviet Manned Lunar Landing Program," 19 February 1964, http://www/foia.cia.gov/browse_docs_full.asp?doc_no=0000969839&title=ORR+POSIT (accessed on 1 August 2007).

19. *TRW Space Log*, vol. 32.

20. Evert Clark, "NASA, with Budget Problem, Plans Stopgap Planetary Probes," *New York Times*, 23 December 1965, 8; Memorandum for the Record, 7 December 1965. The CIA furnished this to the author pursuant to an MDR request. If they still exist, the actual graphic materials remain classified.

21. Ezell, *NASA Historical Data Book*, vol. 1, 128.

22. John M. Clarke, Director of Budget, Program Analysis and Manpower, to Deputy

Director for Science and Technology, 27 October 1964; Memorandum for the Record, Morning Meeting of 27 February 1969; [Redacted] to Deputy Director for Intelligence, 28 December 1965, CREST, NARA.

23. As an example, Webb was asked to compare the U.S. and USSR's space programs at the March 1965 hearings of the Senate Committee on Aeronautical and Space Sciences on NASA's authorization for Fiscal Year 1966. He gave a general answer and then added, "I think if you wish details, these should be obtained perhaps in another forum, and maybe under executive hearings." Senate Committee, *NASA Authorization for Fiscal Year 1966, Part 1*, 26–27; Author's conversations with archivists in NARA's Center for Legislative Archives, October 2007.

24. Journal, Office of Legislative Counsel, 16 June 1964; Journal, Office of Legislative Counsel, 16 October 1964; Journal, Office of the Legislative Counsel, 14 June 1965, CREST, NARA.

25. Memorandum for the Record, 18 August 1964; Briefing Notice, 27 June 1968, CREST, NARA.

26. James E. Webb, Administrator, to Mr. Dembling, 15 August 1962, James E. Webb Papers, Truman Presidential Library; 90th Congress, 2nd Session, n.d., CREST, NARA.

27. U.S. Senate Committee on Aeronautical and Space Sciences Minutes, 13 January 1966, 31 August 1966, 2 February 1967, 21 March 1967, Committee Minutes file, 85th–94th Congress Decimal Files, box #66, Senate Aeronautical and Space Sciences Committee, RG 46 (U.S. Senate), NARA; Memorandum for the Record, 17 March 1967; Journal, Office of Legislative Counsel, 29 January 1969; Journal, Office of Legislative Counsel, 26 September 1969; Memorandum for the Record, 8 October 1970, CREST, NARA.

28. Memorandum for the Record, Morning Meeting of 10 May 1965, CREST, NARA.

29. W. F. Raborn to James E. Webb, Administrator, NASA, 25 June 1965. The CIA furnished this to the author pursuant to a MDR request. Webb's letter to Raborn has not been located. Memorandum for the Record, Morning Meeting of 20 July 1965, CREST, NARA.

30. Memorandum for the Record, Morning Meeting of 11 June 1965, CREST, NARA.

31. *National Intelligence Estimate 11-1-65: The Soviet Space Program*, 27 January 1965, box #16, entry #29, RG 263, NARA.

32. Ibid.

33. Albert D. Wheelon, Deputy Director for Science and Technology, to Deputy Director for Intelligence, 28 September 1965, CREST, NARA. As mentioned above, Webb's letter to Raborn has not been located.

34. [Redacted], Chief, Strategic Forces Division, OSR, to Director of Strategic Research, 14 December 1967, CREST, NARA.

35. Memorandum for the Record, Morning Meeting of 27 August 1965, CREST, NARA.

36. [Redacted] to Mr. Duckett, 11 October 1966, and Carl E. Duckett, Director, Foreign Missile and Space Activity Center, to Director of Central Intelligence, 6 November 1965, CREST, NARA.

37. Albert D. Wheelon, Deputy Director for Science and Technology, to Deputy Director for Intelligence, 28 September 1965.

38. Memorandum for the Record, Morning Meeting of 14 October 1965, CREST, NARA.

39. Carl E. Duckett, Director, Foreign Missile and Space Analysis Center, to Director of Central Intelligence, 6 November 1965.

40. James E. Webb, Administrator, to Admiral William F. Raborn, USN (Ret.), Director, Central Intelligence Agency, 10 December 1965, CREST, NARA.

41. Memorandum for the Record, Morning Meeting of 30 November 1965, CREST, NARA.

42. Albert D. Wheelon, Deputy Director for Science and Technology, to Director of Central Intelligence, 13 February 1966, CREST, NARA.

43. [Redacted], Executive Secretary, to Committee on Overhead Reconnaissance, 15 August 1966, CREST, NARA.

44. Memorandum for the Record, Morning Meeting of 4 January 1967; Memorandum for the Record, Morning Meeting of 5 January 1967; Memorandum for the Record, Morning Meeting of 3 February 1967; Memorandum for the Record, Morning Meeting of 7 July 1967; Memorandum for the Record, Morning Meeting of 12 July 1967, CREST, NARA.

45. John Kerry King, Director, Basic and Geographic Intelligence, to Assistant Deputy Director for Intelligence, 17 September 1971, CREST, NARA.

46. Brooks et al., *Chariots for Apollo*, 265–84.

47. Borman and Serling, *Countdown*, 189.

48. Shepard and Slayton, *Moon Shot*, 225–28.

49. Aldrin and McConnell, *Men from Earth*, 190–91.

50. Day, "Inconstant Moon."

51. *National Intelligence Estimate 11-1-67: The Soviet Space Program*, 2 March 1967, 4, 18–20, box #19, entry #29, RG 263, NARA.

52. Siddiqi, *Challenge to Apollo*, 609–11.

53. Brooks et al., *Chariots for Apollo*, 234–35.

54. *The President's Daily Brief*, 20 April 1967, http://www.archives.gov/declassification/iscap/decision-table.html (accessed 15 October 2012).

55. *The President's Daily Brief*, 24 April 1967, http://www.archives.gov/declassification/iscap/decision-table.html (accessed 15 October 2012).

56. Memorandum for the Record, Morning Meeting of 14 September 1967, CREST, NARA; Siddiqi, *Challenge to Apollo*, 611–14; *Memorandum to Holders of National Intelligence Estimate 11-1-67: The Soviet Space Program*, 4 April 1968, box #19, entry #29, RG 263, NARA; *Central Intelligence Bulletin*, 23 September 1968, 4; Outer Space, vol. II file, NSF Subject File, LBJ Presidential Library.

57. Brooks et al., *Chariots for Apollo*, 237–56; Memorandum for the Record, 28 October 1968, Apollo 8 file, box #3, accession #79-0687, RG 255 (National Aeronautics and Space Administration); Washington National Records Center, Suitland, Maryland (WNRC).

58. Memorandum for the Record, Morning Meeting of 11 March 1968; Memorandum for the Record, Morning Meeting of 18 March 1968, CREST, NARA; Siddiqi, *Challenge to Apollo*, 616–19.

59. *Memorandum to Holders of National Intelligence Estimate 11-1-67: The Soviet Space Program*, 2.

60. The one-page National Security Agency document is untitled and is dated 23 April 1968. The National Security Agency provided it to the author pursuant to a Freedom of Information Act (FOIA) request.

61. Siddiqi, *Challenge to Apollo*, 653–57; *The President's Daily Brief*, 20 July 1968, and *The President's Daily Brief*, 24 July 1968, http://www.archives.gov/declassification/iscap/decision-table.html (accessed 15 October 2012).

62. Brooks et al., *Chariots for Apollo*, 237–56; NASA, biography of Thomas A. Paine, http://history.nasa.gov/Biographies/paine.html ; Memorandum for the Record, 28 October 1968.

63. Brooks et al., *Chariots for Apollo*, 237–56.

64. The National Security Agency provided five heavily redacted intelligence reports to the author pursuant to an FOIA request. All were issued during the mission. Victor Cohn and George C. Wilson, "A Soviet Giant Step in Race to the Moon," *Washington Post*, 23 September 1968, A1.

65. Brooks et al., *Chariots for Apollo*, 265–84.

66. Memorandum for the Record, Morning Meeting of 10 October 1968, and Memorandum for the Record, Morning Meeting of 23 October 1968, CREST, NARA; Siddiqi, *Challenge to Apollo*, 658–62.

67. Siddiqi, *Challenge to Apollo*, 663–65; John Noble Wilford, "Soviet Manned Lunar Shot May Precede Apollo 8," *New York Times*, 20 November 1968, 31; "Soviet Press Agency Says Three Zond Flights Were Made as Tests for a Manned Shot to the Moon," *New York Times*, 24 November 1968, 29.

68. Memorandum for the Record, Morning Meeting of 12 November 1968; Memorandum for the Record, Morning Meeting of 15 November 1968; Memorandum for the Record, Morning Meeting of 18 November 1968; Memorandum for the Record, Morning Meeting of 20 November 1968, CREST, NARA.

69. Brooks et al., *Chariots for Apollo*, 265–84; NASA, biography of Thomas A. Paine; George E. Mueller, Associate Administrator for Manned Space Flight, to Dr. Thomas O. Paine, Acting Administrator, 11 November 1968, Apollo 8 file, box #3, accession #79-0687, RG 255, WNRC.

70. *The President's Daily Brief*, 26 November 1968, LBJ Presidential Library.

71. Inspector General's Survey of the Foreign Missile and Space Analysis Center, December 1968, http://www.gwu.edu/~nsarchiv/NSAEBB/NSAEBB54/docs/doc_39.PDF (accessed 6 August 2013).

72. *National Intelligence Estimate 11-1-67: The Soviet Space Program*, 2 March 1967, 4, box #19, entry #29, RG 263, NARA.

73. Day and Siddiqi, "The Moon in the Crosshairs: Part 2," 119.

74. *Memorandum to Holders of National Intelligence Estimate 11-1-67*, 4 April 1968, 1–2, box #19, entry #29, RG 263, NARA.

75. Day and Siddiqi, "The Moon in the Crosshairs: Part 2," 120; Memorandum for the Record, Morning Meeting of 10 September 1968, CREST, NARA.

76. T. O. Paine, Acting Administrator, to Honorable Richard M. Helms, Director, Central Intelligence Agency, 4 December 1968, and T. O. Paine, Administrator, to Dr. Mueller, Dr. Naugle, and Mr. Beggs, 31 January 1969; Intelligence Estimates and Correspondence file; Federal Agencies CIA NIEs, NASA Headquarters History Office; Day and Siddiqi, "The Moon in the Crosshairs: Part 2," 121. *The President's Daily Brief*, 3 January

1969, http://www.archives.gov/declassification/iscap/decision-table.html (accessed 15 October 2012).

77. *National Intelligence Estimate 11-1-69: The Soviet Space Program*, 19 June 1969, box #19, entry #29, RG 263, NARA.

78. Day and Siddiqi, "The Moon in the Crosshairs: Part 2," 123.

79. The five NIEs from 1970 to 1985 are in RG 263 at NARA or on the CIA website at http://www.foia.cia.gov.

80. *The Soviet Land-Based Ballistic Missile Program, 1945–1972: An Historical Overview*, http://www.archives.gov/declassification/iscap/decision-table.html (accessed 15 October 2012).

81. The Guided Missiles and Astronautics Intelligence Committee was established by the U.S. Intelligence Board's predecessor, the Intelligence Advisory Committee, in 1956. Its functions were to review, evaluate, coordinate, and produce intelligence on foreign missile and space programs. As of 1962, the members were from the CIA, National Security Agency, Atomic Energy Commission, State Department, Federal Bureau of Investigation, and Defense Intelligence Agency. Guided Missiles Intelligence Committee, 15 April 1957, CREST, NARA.

82. John H. Rubel to Mr. Gilpatric, 18 February 1963, and Albert D. Wheelon, Chairman, to Executive Secretary, USIB, 30 November 1962, CREST, NARA.

83. [Redacted], Executive Secretary, to the U.S. Intelligence Board, 7 March 1963 and Reorganization of the USIB Structure under Strengthened DCI Authority, n.d., CREST, NARA.

84. Henry G. Plaster, "Snooping on Space Pictures," *Studies in Intelligence*, entry #27, RG 263, CREST, NARA.

85. Robert C. Seamans Jr., Deputy Administrator, to Dr. Albert D. Wheelon, Deputy Director (Science and Technology), 27 May 1966, Seamans Chronological File 5/66, NASA Headquarters History Office; Carl E. Duckett, Director, Foreign Missiles and Space Analysis Center, to Deputy Director for Science and Technology, 19 November 1964, CREST, NARA.

86. Duckett to Deputy Director for Science and Technology, 19 November 1964.

87. Seamans to Wheelon, 27 May 1966; Inspector General's Survey of the Foreign Missile and Space Analysis Center, December 1968.

88. Seamans to Wheelon, 27 May 1966; NASA/FMSAC cooperation, 25 August 1965, CREST, NARA.

89. Memorandum for the Record by Albert Wheelon, Deputy Director for Science and Technology, 2 February 1965, CREST, NARA.

90. Albert D. Wheelon, Deputy Director, to Dr. George E. Mueller, Associate Administrator for Manned Space Flight, 16 February 1965, CREST, NARA.

91. Albert D. Wheelon, Deputy Director, to Dr. Robert C. Seamans Jr., Deputy Administrator, 14 April 1965, CREST, NARA.

92. Robert C. Seamans Jr., Associate Administrator, to Dr. Albert D. Wheelon, Deputy Director, 3 May 1965, CREST, NARA.

93. Guidelines Governing the Serving of Officials of the National Aeronautics and Space Administration (NASA) as Consultants on Advisory Panels of the Central Intelligence Agency (CIA), July 1965, CREST, NARA. The Manned Space Flight Panel had eight NASA members from the Manned Spacecraft Center (MSC); the Launch Vehicles

Panel had seven NASA members from the Lewis Research Center (LRC), Marshall Space Flight Center (MSFC), and Kennedy Space Center (KSC); the Launch and Test Facilities Panel had five NASA members from the MSC, MSFC, and KSC; the Scientific and Technical Satellites Panel had two NASA members from the Goddard Space Flight Center (GSFC); the Lunar and Planetary Probes Panel had six NASA members from headquarters and the Jet Propulsion Laboratory (JPL); the Aeronautics Panel had five NASA members from headquarters, the Dryden Flight Research Center, Ames Research Center, and LRC; the Advanced Research and Technology Panel had two NASA members from headquarters and LRC; the Tracking, Data Acquisition, and Reduction Panel had three members from headquarters, GSFC, and JPL.

94. Albert D. Wheelon, Deputy Director, to Dr. Robert C. Seamans Jr., Deputy Administrator, NASA, 8 March 1966. The CIA furnished this to the author pursuant to an MDR request.

95. Notes on First Meeting of Lunar and Planetary Panel, 20 June 1966. The CIA provided this to the author pursuant to an MDR request.

96. Albert D. Wheelon, Deputy Director for Science and Technology, to Director of Central Intelligence, 13 September 1965, CREST, NARA.

97. Space Intelligence Panel, n.d., and David S. Brandwein to [Redacted], 23 October 1968, CREST, NARA.

98. Space Intelligence Panel, n.d., and Carl E. Duckett to Dr. Joseph F. Shea, n.d., and Carl E. Duckett to Dr. Raymond L. Blispinghoff, n.d., CREST, NARA.

99. M. LeRoy Spearman, interview by author, Hampton, Va., 7 December 2006; M. Leroy Spearman, "Some NASA Aerodynamic Research Related to Foreign Systems" (paper presented at the 42nd AIAA Aerospace Sciences Meeting, 5–8 January 2004, Reno).

100. Spearman, interview; Special Briefings Listings (provided by Spearman to the author).

101. Richard Helms, Director, to the Honorable Clark M. Clifford, Secretary of Defense, 24 June 1968, CREST, NARA.

102. Special Briefings Listings; John S. Foster Jr. to Honorable Richard Helms, 12 August 1968, 111 NASA 1968, box #9, accession #73-0545, RG 330, NARA.

103. Memorandum for the Record, Morning Meeting of 13 September 1965, CREST, NARA; Harold B. Finger to Adm. W. F. Boone, 2 November 1965, Intelligence Estimates and Correspondence file, Federal Agencies CIA NIEs box, NASA Headquarters History Office.

104. Author's conversation with M. Leroy Spearman; GMAIC meeting, 5 May 1970. The CIA furnished this to the author pursuant to an MDR request. Minutes of Meeting, Weapons and Space System Intelligence Committee, 12 July 1983, CREST, NARA.

Chapter 3. Expanding Interaction in Old and New Areas

1. Memorandum for the Record, 22 December 1961, CREST, NARA.

2. Memorandum for the Record, 12 June 1962, CREST, NARA.

3. Pedlow and Welzenbach, *The Central Intelligence Agency and Overhead Reconnaissance*, 291–93.

4. Ibid.; [redacted], Special Assistant for Liaison, Office of Special Activities, to Dr. McMillan et al., 21 August 1963, CREST, NARA.

5. Lyman Kirkpatrick, Executive Director, to Deputy Director of Intelligence et al., 23 October 1963, CREST, NARA.

6. Ibid.

7. This section is based in large part on the author's "Was It Really 'Space Junk'?"

8. Max Frankel, "American Rebuts Soviet on Sputnik," *New York Times*, 5 August 1958, 12; Senate Committee, *Soviet Space Programs, 1962–1965*, 193–94.

9. Evert Clark, "Soviet Space Activity Reaches New Peak," *Aviation Week & Space Technology*, 22 October 1962, 27–28; "Reds Retrieve Bit of Sputnik," *Washington Post*, 8 January 1963.

10. Glenn Wolfe, Counselor of Embassy, to Robert Schneider, Office of Middle and Southern African Affairs, 29 September 1960; Records Relating to Atomic Energy Matters, 1944–1964, box #267, RG 59, NARA; R. Hart Phillips, "Pieces of Rocket Landed in Cuba," and John Finney, "Pentagon Embarrassed," *New York Times*, 1 December 1960, 10; L. X. Abernathy to Carl N. Jones, 24 June 1968, Space Fragment Files, NASA Headquarters History Office.

11. "Public Statement of Handling of Space Vehicle Fragments," General Correspondence of NASC Executive Secretary Edward Welsh, 1961–1969, box #17, RG 220 (Records of Temporary Committees, Commissions, and Boards), NARA.

12. Ibid.

13. R. J. Smith, Assistant Director, Current Intelligence, to Director of Central Intelligence, 18 January 1963, CREST, NARA; Memorandum for Record, 13 December 1962; E. C. Welsh to Honorable Robert C. McNamara, 7 January 1963; Memorandum for Record, 22 May 1963; General Correspondence of NASC Executive Secretary Edward Welsh, 1961–1969, box #17, RG 220, NARA.

14. Memorandum for Dr. Welsh, 29 May 1963, General Correspondence of NASC Executive Secretary Edward Welsh, box #18, RG 220, NARA.

15. L. X. Abernathy to Carl N. Jones, 24 June 1968.

16. Cheng, *Studies in International Space Law*, 129–32, 250–60.

17. Ibid.

18. Jeffrey C. Kitchen to Mr. Kohler, 30 June 1967, 1967–1968 Central Foreign Policy Files, box #3023, RG 59, NARA.

19. R. W. Hale, Memorandum for Record, 18 July 1968, General Correspondence of NASC Executive Secretary Edward Welsh, box #18, RG 220, NARA.

20. See, e.g., State to Amembassy Telegram No. 183370, 14 June 1968, 1967–1968 Central Foreign Policy Files, box #3029, RG 59, NARA.

21. State to Amembassy Moscow Telegram No. 070338, 6 May 1969 and Amembassy Stockholm to SecState Wash DC Telegram No. 735, boxes #3023 and #3029, 1067–1968 Central Foreign Policy Files, RG 59, NARA.

22. Paul T. Cooper, Brig. Gen., USAF, Assistant Director (Ranges and Space Ground Support) to Honorable John H. Rubel, Assistant Secretary of Defense (Deputy Director), 6 September 1961, 112.2 RSGS, box #11, accession #66-3146, RG 330, NARA.

23. Temple, *Shades of Gray*, 325.

24. Memorandum of Agreement, 7 November 1961, DoD-NASA Agreements, box #2, accession #70-5484, RG 330, NARA.

25. Temple, *Shades of Gray*, 375–78.

26. Harold Brown to Dr. Robert C. Seamans Jr., 15 October 1964, 111 NASA, box #8, accession #68-5157, RG 330, NARA; Robert C. Seamans Jr., Associate Administrator to Honorable Harold Brown, Director of Defense Research and Engineering, 10 December 1964, 111 NASA 1965, box #9, accession #69-3339, RG 330, NARA.

27. Ezell, *NASA Historical Data Book*, vol. 2, 521–96.

28. Report of SFGEP Ad Hoc Study Team on Procedures for Reciprocal Support, 1 May 1964, 111 NASA, box #7, accession #68-5157, RG 330, NARA. The report also discusses smaller DoD tracking, data acquisition, and command and control capabilities at such facilities as the Air Proving Ground Center in Florida and the Churchill Research Range in Canada.

29. Ibid.; Ezell, *NASA Historical Data Book*, vol. 2, 521–96; Memorandum for the Secretary of Defense re Department of Defense Support of National Aeronautics and Space Administration, 2 April 1962, 111 NASA, box #6, accession #66-3170, RG 330, NARA.

30. Minutes of the 18th Meeting, Space Flight Ground Environment Panel, Aeronautics and Astronautics Coordinating Board, 11 February 1964 and Report of the Ad Hoc Study Team on Procedures for Reciprocal Support, 1 May 1964; 111 NASA, box #7, accession 68-5157, RG 330, NARA. The DoD sites in foreign nations did not face the same obstacles in assisting NASA missions as did NASA's foreign facilities in supporting DoD missions. By 1964, the DoD had five agreements (Bahamas, Ascension Island, West Indies Federation, South Africa, Seychelles) under which tracking, data acquisition, and command and control facilities were operating without limitations. There were additional sites operating at overseas U.S. military bases under much broader agreements. Both access by the foreign governments to the data collected and access by their personnel to the facilities were very restricted. Report of the Ad Hoc Committee on International Agreements for Tracking and Data Acquisition Support of the Space Flight Ground Environment Panel, February 11, 1964, 111 NASA, box #7, accession #68-5157, RG 330, NARA.

31. Minutes of the 18th Meeting, Space Flight Ground Environment Panel, Aeronautics and Astronautics Coordinating Board, 11 February 1964 and Report of the Ad Hoc Study Team on Procedures for Reciprocal Support, 1 May 1964.

32. John H. Rubel to the Assistant Secretary of the Army (R&D) et al., 16 October 1961; 111 NASA-DOD, box #10, accession #66-3164, RG 330, NARA.

33. Paul T. Cooper, Brig. Gen., USAF, Assistant Director (Ranges and Space Ground Support), to Edmond Buckley, Office of Tracking and Data Acquisition, 15 March 1962, 111 NASA, box #6; accession #66-3170, RG 330, NARA; *TRW Space Log*, vol. 32.

34. Paul T. Cooper, Brig. Gen., USAF, Assistant Director (Ranges and Space Ground Support), to Edmond Buckley, Director, Office of Tracking and Data Acquisition, 8 November 1962, 111 NASA, box #6, accession #66-3170; J. P. Ruina, Director, to Assistant Director (Ranges and Space Ground Support), 20 April 1963; 36.1 Tetrahedral Research Satellite, box #1, accession #70-1552; Paul T. Cooper, Brig. Gen., USAF, Assistant Director (Ranges and Space Ground Support) to Edmond Buckley, Director, Office of Tracking and Data Acquisition, 26 April 1963; RSGS 1963, box #5, accession #81-0458, RG 330, NARA.

35. Minutes of the 24th Meeting, Space Flight Ground Environment Panel, Aero-

nautics and Astronautics Coordinating Board, 29 January 1965, 111 NASA 1965, box #9, accession #69-3339, RG 330, NARA.

36. Clifford J. Kronauer, Brig. General, USAF, Assistant Director (Ranges and Space Ground Support), to Edmond C. Buckley, Director, Office of Tracking and Data Acquisition, 16 November 1965, 111 NASA 1965, box #9, accession #69-3339, RG 330, NARA; Clifford J. Kronauer, Brig. Gen., USAF, Assistant Director (Ranges and Space Ground Support), to Edmond C. Buckley, 29 March 1967; M. W. Elliot, Brig. Gen., USAF, Assistant Director (Ranges and Space Ground Support), to G. M. Truszynski, Associate Administrator for Tracking and Data Acquisition, NASA, 15 April 1968; Victor W. Hammond, Assistant Director (Acting) (Ranges and Space Ground Support), 17 August 1970; Victor W. Hammond, Assistant Director (Acting) (Ranges and Space Ground Support) to Gerald M. Truszynski, Associate Administrator, Tracking and Data Acquisition, NASA, 16 October 1970. The Office of the Secretary of Defense provided the last four documents to the author pursuant to an MDR request.

37. W. M. Elliott, Col., USAF, Assistant Director (Ranges and Space Ground Support), to Edmond C. Buckley, Associate Administrator for Tracking and Data Acquisition, 24 July 1967. The Office of the Secretary of Defense provided this to the author pursuant to an MDR request. Amembassy London to SecState Washington, D.C., R 041445Z Apr 75 (available on the NARA website at www.nara.gov); House Committee, *Review of Tracking and Data Acquisition Program*, 282; Network Plans Sub-Panel Network Study Task Force (6 August to 14 October 1968), NASA 423–68 file, box #12, accession #72-A5501, RG 330, NARA; Tsiao, *"Read You Loud and Clear!"* 85–87; Cone, *United States Air Force Eastern Test Range*, 2-1.

38. M. W. Elliot, Brig. Gen., USAF, Assistant Director (Ranges and Space Ground Support), to Dr. Wilson, 18 June 1968, 112 R&SGS, box #19, accession #73-0545, RG 330, NARA.

39. Boone, *NASA Office of Defense Affairs*, 206–21.

40. John E. Foster Jr. to Dr. Homer E. Newell, Associate Administrator, 6 August 1968, Gen. Corr. of NASC Exec. Sec. Ed Welsh 1961–1969, box #10, RG 220, NARA.

41. AACB Annual Report, Space Flight Ground Environment Panel, Calendar Year 1968, 30 January 1969, Panel of Aeronautics Activity Reports, box #2, accession #73-2420, RG 330, NARA.

42. House Committee, *Review of Tracking and Data Acquisition Program*, 244, 272–73.

43. Memorandum to Co-Chairman, AACB, 1 May 1969, 112 R&SGS, box #11, accession #74-0017, RG 330, NARA.

44. Ibid.

45. Day, "Relay in the Sky," 56–62; Richelson, *Wizards of Langley*, 198–202; Ezell, *NASA Historical Data Book*, vol. 3, 326–27; NASA, *Tracking and Data Relay Satellite System*, http://msl.jpl.nasa.gov/Programs/tdrss.html (accessed 15 January 2012).

46. FY 1981 Director's Review, National Aeronautics and Space Administration, NLC-41-26-4-1-3, Carter Presidential Library; "TDRSS Program Security Classification Guide," 15 April 1982, TDRSS file, NASA Headquarters History Office; Charles, "Spy Satellites: Entering a New Era."

47. Ezell, *NASA Historical Data Book*, vol. 2, 348–55; *Compendium of Meteorological Satellites and Instrumentation*, 15–18.

48. Summary of Understandings, 8 June 1962, CREST, NARA; Tepper, *Meteorological Satellites*, 8–9; J. Wallace, Acting Director to Ambassador Bohlen, 4 October 1968, SP 1-1 USSR 1/1/67, box #3022, 1967–1969 Central Decimal Files, RG 59, NARA.

49. Plan for a National Operational Meteorological Satellite System, April 1961, 111 Relations w/NASA, box #10, accession #66-3146, RG 330, NARA; Senate Committee, *Meteorological Satellites*, 100–103.

50. Plan for a National Operational Meteorological Satellite System, April 1961, 111 Relations w/NASA, box #10, accession #66-3146, RG 330, NARA; Senate Committee, *Meteorological Satellites*, 100–103.

51. Military and Security Requirements and Capabilities of the National Operational Meteorological Satellite System, April 1961, 111 Relations w/NASA, box #10, accession #66-3146, RG 330, NARA.

52. Hall, *A History of the Military Polar Orbiting Meteorological Satellite Program*, 1–10.

53. Ibid.; Kaehn, "Military Applications Evolution and Future," 41–48.

54. Hall, *A History of the Military Polar Orbiting Meteorological Satellite Program*, 1–10; Perry, *A History of Satellite Reconnaissance*, vol. 2A, 203–44.

55. Hall, *A History of the Military Polar Orbiting Meteorological Satellite Program*, 1–10; John L. McLucas, Under Secretary of the Air Force, to the Director of Defense Research and Engineering, 4 January 1972, Sat. Program 366–71, box #7, accession #73-0011, RG 340, NARA.

56. Ezell, *NASA Historical Data Book*, vol. 2, 360; John Rubel to Hugh Dryden, Deputy Administrator, 22 January 1963, 209.9 Space, Satellites, and Guided Missiles, box # 16, accession #67-4860, RG 330, NARA.

57. Ezell, *NASA Historical Data Book*, vol. 2, 360–63. There are two declassified documents confirming that NASA weather satellites at some point had emergency disabling devices. One is a March 1978 NASA Concept Paper on oceanography in the 1980s which states that "United States meteorological satellites are required to incorporate at least a 'turn off' capability, but this fact is classified and the capability has never been exercised." The other is an April 1976 memorandum about communications security support for NASA's *SEASAT-A*. It states that there is a classified directive entitled "National Policy on the Security of Meteorological Satellite Information" which required that all such spacecraft contain equipment to stop by secure means the transmission of useful data during times of hostilities. The document also states that the TIROS-N series carried a timer to lock out commands when not over U.S.-controlled ground stations, with a backup capability to command the satellite during these periods. The Office of the Secretary of Defense provided these documents to the author pursuant to an MDR request.

58. J. W. Davis, Rear Admiral, USN, Deputy Director, Joint Staff to Secretary of Defense, 5 March 1964, 209.9 Space, box #16, accession #68-5157, RG 330, NARA; Hall, *A History of the Military Polar Orbiting Meteorological Satellite Program*, 12–16.

59. Eugene G. Fubini to Dr. Robert White, Chief of the Weather Bureau, Department of Commerce, 1 April 1964, 209.9 Space, box #16, accession #68-5157, RG 330, NARA.

60. Brockway McMillan to Chief, Joint Meteorological Satellite Program Office, 14 January 1965, 251–65 NASA, box #1039, entry A1 1-F, RG 340, NARA.

61. John S. Foster Jr. to Dr. Homer E. Newell, Associate Administrator for Space Sciences and Applications, 4 March 1966, 209.9 Space, box #16, accession #70-0057, RG 330, NARA; NASA, *Missions—ESSA—NASA Science, http://science.nasa.gov/missions/essa/* (accessed 10 January 2012); Special Assistant (National Intelligence) to Director of Defense Research and Engineering, 16 February 1967, 209 Guided Missiles and Satellites, box #17, accession #71-2285, RG 330, NARA.

62. John S. Foster Jr. to Dr. Homer E. Newell, Associate Administrator for Space Sciences and Applications, 4 March 1966; NASA, *Missions—ESSA—NASA Science*; Special Assistant (National Intelligence) to Director of Defense Research and Engineering, 16 February 1967.

63. Hall, *A History of the Military Polar Orbiting Meteorological Satellite Program*, 12–16; Senate Committee, *NASA Authorization for Fiscal Year 1982*, 483–84.

64. John S. Foster Jr. to Admiral W. F. Boone, USN, Ret., Asst. Administrator for Defense Affairs, NASA, 17 February 1967, 209 Guided Missiles and Satellites, box #17, accession #71-2285, RG 330, NARA.

65. John S. Foster Jr. to the Assistant Secretary of the Army (R&D) et al., 2 November 1967, 209 Guided Missiles and Satellites, box #16, accession #71-2285, RG 330, NARA.

66. NASA, *Missions—ESSA—NASA Science*.

67. David I. Liebman, Maj. Gen., USAF, Deputy Director, Joint Staff to Director of Defense Research and Engineering, 28 March 1970 and John W. Townsend, Deputy Administrator to I. Nevin Palley, Assistant Director (Space Technology), Office of the Director for Defense Research and Engineering, 13 January 1970, 209 Guided Missile and Satellites, box #1, accession #75-0047, RG 330, NARA.

68. Ezell, *NASA Historical Data Book*, vol. 3, 266–73; *TIROS M Spacecraft (ITOS 1) Final Engineering Report*, vol. 1, 1-I-1–1-I-2.

69. Hall, *A History of the Military Polar Orbiting Meteorological Satellite Program*, 26–34; Heacock, *Comparison of the Defense Meteorological Satellite Program (DMSP)*, II-7; Fuller, *Air Weather Service Support to the United States Army*, 114.

70. Frank B. Clay, Major General, USA, Deputy Director, Joint Staff to Secretary of Defense, 7 July 1970, 209 Guided Missiles and Satellites, box #1, accession #75-0047, RG 330, NARA.

71. Newell, *Beyond the Atmosphere*, 118; Supplement to Unmanned Spacecraft Panel Minutes, 21 September 1960 (Classified Portion), 111 NASA-DoD, box #10, accession #66-3146, RG 330, NARA.

72. *Selected Space Goals and Objectives and Their Relation to National Goals*, X-8.

73. House Committee, *United States Civilian Space Programs*, vol. 2, 313–16.

74. Summary Minutes, Meeting No. 4, Unmanned Spacecraft Panel, 24 August 1960, USP Minutes, box #1, accession #72-1288, RG 330, NARA; Supplement to Unmanned Spacecraft Panel Minutes, 21 September 1960 (Classified Portion).

75. Summary Minutes, Meeting No. 4, Unmanned Spacecraft Panel, 24 August 1960, USP Minutes, box #1, accession #72-1288, RG 330, NARA; Supplement to Unmanned Spacecraft Panel Minutes, 21 September 1960 (Classified Portion).

76. Summary Minutes, Meeting No. 4, Unmanned Spacecraft Panel, 24 August 1960, USP Minutes, box #1, accession #72-1288, RG 330, NARA; Supplement to Unmanned Spacecraft Panel Minutes, 21 September 1960 (Classified Portion).

77. House Committee, *United States Civilian Space Programs*, vol. 2, 317–21; Newell, *Beyond the Atmosphere*, 116–20.

78. House Committee, *United States Civilian Space Programs*, vol. 2, 317–21; Newell, *Beyond the Atmosphere*, 116–20; Harold Brown to Dr. Jerome B. Weisner, 23 December 1961, 209.9 Satellites, box #14, accession #66-3146, RG 330, NARA.

79. House Committee, *United States Civilian Space Programs*, vol. 2, 317–21; Newell, *Beyond the Atmosphere*, 116–20.

80. House Committee, *United States Civilian Space Programs*, vol. 2, 317–21; Newell, *Beyond the Atmosphere*, 116–20; Paul E. Nitze to the President, 5 November 1968, Outer Space, vol. II file, NSF Subject File, LBJ Presidential Library; Report of the Geodetic Satellite Sub-Panel of the Unmanned Spacecraft Panel, January 1969, 209 Missiles and Satellites, box #17; entry P57, RG 330, NARA.

81. B. W. Augenstein, Special Assistant (Intelligence and Reconnaissance) to Dr. Robert C. Seamans Jr., Associate Administrator, 23 December 1963, 111 NASA, box #6, accession #67-4860, RG 330, NARA.

82. House Committee, *United States Civilian Space Programs*, vol. 2, 317–24.

83. Ibid.

84. Pearlman, *A Study Program for Geodetic Satellite Applications*, 1–5.

85. Senate Committee, *NASA Authorization for Fiscal Year 1973*, 1006; Memorandum for Holders of USIB-D-46.4/33, 30 December 1969 and Memorandum for the United States Intelligence Board, 12 December 1969, CREST, NARA.

86. DoD Views of Classification Policy of MC&G Data, 9 January 1969, 123 Intell, box #13, accession #74-0017, RG 330, NARA. The actual manual has not been declassified, and this memo only summarizes it.

87. Paul E. Nitze to the President, 5 November 1968.

88. Charles L. Poor, Acting Assistant Secretary of the Army (R&D), to the Director of Defense Research and Engineering, January 1969, and Walter McCough Jr., Deputy Special Assistant (Threat Assessment), to Leonard Jaffe, Deputy Associate Administrator for Space Science and Applications, 1 December 1969, 209 Missiles and Satellites, box #17, entry P57, RG 330, NARA.

Chapter 4. Restrictions on Remote Sensing from Space

1. David, "What Should Nations Reveal about Their Spying from Space?" This section is based in part on this article.

2. Steinberg, *Satellite Reconnaissance*, 30–32, 40–42.

3. Memorandum for Secretary of State et al., 26 August 1960, CREST, NARA; Philip J. Farley to Mr. Merchant, 14 September 1960; Satellites, Reconnaissance file, box #279; Records re Atomic Energy Matters, 1944–1962, RG 59 (Records of the Department of State), NARA.

4. *Mercury Project Summary*, 215–24, 334–38.

5. Richelson, "Undercover in Outer Space," 308–20; Hall, "The NRO in the 21st Century," 5.

6. Chronology of Events Leading to UN Registration of Space Vehicles, n.d., CREST, NARA.

7. *National Security Action Memorandum 156*, 26 May 1962, CREST, NARA.

8. Memorandum for the Record, 2 June 1962, CREST, NARA.

9. Record of Action at the Five Hundred and Second Meeting of the National Security Council, 10 July 1962, http://state.gov/r/pa/ho/frus/kennedyjf/xxv/6022pf.htm (accessed 8 August 2008).

10. David, "The Intelligence Agencies Help Find Whales," 27–37. The author's Mandatory Declassification Review (MDR) request for the directive establishing the BYE-MAN special security control system was denied in its entirety. The denial was upheld on appeal.

11. Memorandum for the Record, 26 June 1963, CREST, NARA.

12. Memorandum from Deputy Under Secretary of State for Political Affairs to Secretary Rusk et al., 21 January 1964, http://www.state.gov/r/pa/ho/frus/johnsonlb/x/9015htm (accessed 8 August 2008).

13. Memorandum for the Record, 18 June 1964 and Memorandum for Assistant Deputy Director for Intelligence, 3 May 1966, CREST, NARA.

14. Ezell, *NASA Historical Data Book*, vol. 2, 335; NASA, *Nineteenth Semiannual Report to Congress*, 68–72.

15. Senate Committee, *NASA Authorization for Fiscal Year 1966*, 205–11.

16. *Gemini Summary Conference*, 292–93.

17. Memorandum for Deputy Administrator on NASA-DoD Interface on Lunar and Earth Sensors, June 1966. NASA provided this to the author pursuant to an MDR request. This chronology is the only reference to this McNamara-Webb correspondence that the author has been able to locate. The actual letters or memos in NASA and the Office of the Secretary of Defense files remain classified or have been lost or destroyed.

18. Brockway McMillan, Director, National Reconnaissance Office, to Dr. Robert C. Seamans Jr., Associate Administrator, NASA, 5 August 1965; Robert C. Seamans Jr. to Honorable Brockway McMillan, Director, National Reconnaissance Office, 24 August 1964. The NRO provided these documents to the author pursuant to an MDR request.

19. Memorandum for Deputy Administrator on NASA-DoD Interface on Lunar and Earth Sensors, June 1966.

20. Ibid.

21. Memorandum of Understanding between the Department of Defense and the National Aeronautics and Space Administration Concerning the Manned Space Flight Programs of the Two Agencies, 14 January 1966. The Office of the Secretary of Defense provided this to the author pursuant to an MDR request.

22. Memorandum for Deputy Administrator on NASA-DoD Interface on Lunar and Earth Sensors, June 1966.

23. Memorandum for the Manned Spaceflight Policy Committee, 11 April 1966. NASA provided this to the author pursuant to an MDR request.

24. Minutes of Second Meeting, Manned Space Flight Policy Committee, 14 April 1966. The NRO provided this to the author pursuant to an MDR request. All the MSFPC minutes the author has obtained are very summary in nature. Whether verbatim minutes were ever prepared and, if so, whether they still exist is unknown.

25. David, "The Intelligence Agencies Help Find Whales," 29–30. This section is based in part on this article.

26. NASA Earth Resources Survey Program, 13 May 1966, NSAM 156 Committee file, box #12, National Security File—Intelligence, LBJ Presidential Library. The infor-

mation on Project QUILL and the briefing NASA received on it comes from documents the NRO declassified in July 2012 and posted on its website at www.nro.mil.

27. Political Aspects of Disclosure of Space Reconnaissance Capabilities, 19 May 1966, NSAM 156 Committee file, box #12, National Security File—Intelligence, LBJ Presidential Library.

28. Evaluation of the Impact of the Proposed NASA Earth Resources Survey Program on the National Reconnaissance Program, 24 May 1966, NSAM 156 Committee file, box #12, National Security File—Intelligence, LBJ Presidential Library.

29. Spurgeon Keeney to Mr. Rostow, 10 June 1966, NSAM 156 Committee file, box #12, National Security File—Intelligence, LBJ Presidential Library.

30. Political and Security Aspects of Non-Military Applications of Satellite Earth-Sensing, 11 July 1966, NSAM 156 Committee file, box #12, National Security File—Intelligence; LBJ Presidential Library.

31. Memorandum for Holders of USIB-D-41.12/23, 29 August 1966, CREST, NARA.

32. DoD-NASA Coordination of the Earth Resources Survey Program, 23 September 1966. NASA provided this document to the author pursuant to an MDR request.

33. The NRO provided the SACC and MSFPC summary minutes the author has pursuant to an MDR request. Minutes of some of the meetings have apparently been lost or destroyed over the years and cannot be located now.

34. Minutes of the First Survey Applications Coordinating Committee, 29 September 1966. Minutes of the Second Survey Applications Coordinating Committee, 21 November 1966. The NRO provided these documents to the author pursuant to an MDR request.

35. United States Department of the Interior news release, 21 September 1966; Robert C. Seamans Jr., Deputy Administrator, to Walt W. Rostow, Special Assistant to the President, 22 September 1966; Charles E. Johnson to Dr. Robert Seamans Jr., 24 October 1966; Robert C. Seamans Jr., Deputy Administrator to Charles E. Johnson, National Security Council, 18 November 1966, Earth Resources Observation Satellite Interior Dept. file, box #14, National Security File—Files of Charles E. Johnson, LBJ Presidential Library.

36. [Redacted], Chief, DD/I Collection Guidance Staff to Mr. Bross, 3 January 1967 and [Redacted] to [Redacted], 8 February 1967, CREST, NARA.

37. Compton and Benson, *Living and Working in Space*, 82–90.

38. Leonard Jaffe to Dr. Seamans, 21 July 1967; Documentation Summary, n.d. NASA provided these to the author pursuant to an MDR request.

39. Leonard Jaffe to Dr. Seamans, 21 July 1967; Documentation Summary, n.d. NASA provided these to the author pursuant to an MDR request; Minutes of the Thirteenth Meeting of the Survey Applications Coordinating Committee, 12 May 1969. The NRO provided this to the author pursuant to an MDR request.

40. Minutes of the Third Survey Applications Coordinating Committee Meeting, 31 January 1967; Minutes of the Fifth Meeting Manned Space Flight Policy Committee, 20 April 1967; Minutes of the Seventh Survey Applications Coordinating Committee, 3 August 1967. The NRO provided all three documents to the author pursuant to an MDR request.

41. Richard Helms, Director, to Honorable Foy D. Kohler, Deputy Under Secretary

of State, 19 September 1967. NASA provided this to the author pursuant to an MDR request. The feasibility demonstration of the radar imager was the satellite launched by the NRO in December 1964 and codenamed QUILL. It only imaged parts of the United States and recorded the images on magnetic tape that was returned to Earth in a capsule similar to those carried by imagery intelligence satellites. The satellite only operated until mid-January 1965. Richelson, "Ups and Downs of Space Radars," 25.

42. Compton and Benson, *Living and Working in Space*, 99–104.

43. David Williamson Jr. NASA Executive Coordinator, Manned Space Flight Policy Committee to John Kirk, DOD Executive Coordinator, Manned Space Flight Policy Committee, 21 November 1968. NASA provided this to the author pursuant to an MDR request. Minutes of the Eleventh Survey Applications Coordinating Committee Meeting, 21 March 1968. Minutes of the Twelfth Survey Applications Coordinating Committee, 9 May 1968. The NRO provided these documents to the author pursuant to an MDR request. *Twentieth Semiannual Report to Congress*, 86–87.

44. Homer E. Newell, Associate Administrator, to Honorable Alexander H. Flax, Assistant Secretary of the Air Force (Research and Development), 19 April 1968. NASA provided this to the author pursuant to an MDR request.

45. Minutes of the Eleventh Survey Applications Coordinating Committee Meeting, 21 March 1968; Summary of Major SACC Coordination Activities, 9 May 1968 to 12 May 1969. The NRO provided these documents to the author pursuant to an MDR request.

46. Background Paper for Information of CIA Oversight Committees on the Partial Use of NRP Assets for Civil Applications, n.d., CREST, NARA.

47. [Redacted] to A/DD/S&T, 16 March 1973, CREST, NARA.

48. Ibid.

49. *Twenty-first Semiannual Report to Congress*, 81–82.

50. Minutes of the Thirteenth Survey Applications Coordinating Committee, 12 May 1969. The NRO provided this to the author pursuant to an MDR request.

51. Ezell, *NASA Historical Data Book*, vol. 3, 331–41.

52. Colwell, *Monitoring Earth Resources*, 1–15.

53. U. Alexis Johnson to the President, 11 December 1969, SP US 11/1/67 file, box #3009, 1967–1969 Central Decimal Files, RG 59, NARA.

54. Perry, *A History of Satellite Reconnaissance*, vol. 1, 205–7.

55. Lee A. DuBridge, Science Adviser, to David Packard and Richard Helms, 16 May 1969. NASA provided this to the author pursuant to an MDR request. A 1965 DoD-CIA agreement established the NRO's Executive Committee. Its responsibilities included approving the National Reconnaissance Program and its budget and assigning specific reconnaissance programs to the NRO's CIA or DoD components.

56. F. Robert Naka, Executive Secretary, to the NRP Executive Committee, 12 September 1969, CREST, NARA.

57. Richard Helms, Director, to Dr. John L. McLucas, 13 October 1969, and [Redacted] Chairman, "Open Skies" Task Force to Dr. Victor H. Reis, Assistant Director, Office of Science and Technology Policy, 1 April 1982, CREST, NARA; Chronology of NSAM 156 Actions, n.d. The NRO provided this last document to the author pursuant to an MDR request.

58. John R. McGuire, Chief, Forest Service, USDA et al., to Chairman, COMIREX, 16

September 1974, and Use of National Reconnaissance Program (NRP) Air and Satellite Photography for Civil Applications, 14 February 1975, CREST, NARA.

59. Use of National Reconnaissance Program (NRP) Air and Satellite Photography for Civil Applications, 14 February 1975.

60. Ibid.; *U.S. Reconnaissance Satellites: Domestic Targets,* http://www.gwu.edu/~nsarchiv/NSAEBB/NSAEBB229/index.htm (accessed 10 September 2012).

61. Compton and Benson, *Living and Working in Space*, 104–12, 182–90.

62. Ezell, *NASA Historical Data Book*, vol. 3, 331–41; Memorandum for the Record, 10 February 1971, CREST, NARA.

63. [Redacted] to [Redacted], National Security Council, 19 January 1971, CREST, NARA.

64. Memorandum for the Record, 10 February 1971.

65. Ezell, *NASA Historical Data Book*, vol. 3, 331–41.

66. Ibid.; Draft letter from Dr. McLucas, 16 March 1970, Document #1055, NRO Staff Records; [Redacted] to Dr. Proctor, 22 February 1971, and John L. McLucas to Director, CIA Reconnaissance Programs and Director, Program D, CREST, NARA.

67. [Redacted] to the 40 Committee, 21 January 1971; Edward E. David Jr. to Henry Kissinger, 11 December 1970; [Redacted] to the 40 Committee, 22 February 1971; Donald H. Steininger, Assistant Deputy Director for Science and Technology, to Director of Central Intelligence, 8 March 1971, CREST, NARA.

68. Myron W. Krueger, Technical Coordinator, Office of DoD and Interagency Affairs, to Dr. Donald Steininger, Assistant Deputy Director for Science and Technology, Central Intelligence Agency, 22 March 1971, CREST, NARA.

69. Wendell L. Bryant, Brig. Gen. USAF, Director of Special Activities, to Deputy Director for Science and Technology, 16 September 1971, CREST, NARA.

Chapter 5. Concerns over Human Spaceflight Program Experiments and Lunar and Astronomy Program Technologies

1. *Mercury Project Summary*, 213–28; Brugioni, *Eyes in the Sky*, 369; Author's conversation with Richard Underwood.

2. David, "Astronaut Photography," 187.

3. Boone, *NASA Office of Defense Affairs*, 83–96; Lambright, *Powering Apollo*, 108–22.

4. Boone, *NASA Office of Defense Affairs*, 83–96; Lambright, *Powering Apollo*, 108–22.

5. David, "Astronaut Photography," 187–88.

6. *Earth Photographs from Gemini*, 259–61; Author's conversation with Richard Underwood.

7. Brockway McMillan, Director, National Reconnaissance Office, to Dr. Robert C. Seamans Jr., Associate Administrator, 9 August 1965; Memorandum for Deputy Administrator on NASA-DoD Interface on Lunar and Earth Sensors, June 1966. NASA provided these to the author pursuant to an MDR request.

8. *Earth Photographs from Gemini*, 261–66.

9. David, "Astronaut Photography," 187–88; Brig. Gen. Andrew J. Evans, USAF, Director of Development, DCS/Research and Development, to Daniel J. Fink, Deputy Director, Strategic and Defensive Systems, DDR&E, 2 September 1965, 111 NASA 1965, box #9, accession #69-3339, RG 330, NARA; *NASA Program Gemini Working Paper No. 5040*.

10. Albert D. Wheelon, Deputy Director for Science and Technology, to Deputy Director for Intelligence, 21 September 1965, CREST, NARA.

11. W. F. Raborn to Administrator, NASA, 18 December 1965, CREST, NARA.

12. Author's conversation with Richard Underwood.

13. *Gemini Summary Conference*, 231–91.

14. Alexander H. Flax to Deputy Secretary of Defense, 12 March 1968. The author is grateful to Scott Wenger for providing this document.

15. David, "Astronaut Photography," 188–89.

16. Paul E. Worthman, Colonel, USAF, to Dr. Flax, 7 March 1968. The author is grateful to Scott Wenger for providing this document.

17. Flax to Deputy Secretary of Defense, 12 March 1968; Russell A. Berg, Brigadier General, USAF, Director, to General Smart, 1 April 1968. The author is grateful to Scott Wenger for providing these documents.

18. Minutes of the Survey Applications Coordinating Committee, Eleventh Meeting, 21 March 1968. The NRO provided this to the author pursuant to an MDR request.

19. Flax to Deputy Secretary of Defense, 12 March 1968; Berg to Smart, 1 April 1968.

20. NASA, *Apollo 6 Automated Earth Photographs, http://nssdc.gsfc.nasa.gov/database/MasterCatalog?sc=1968-025A&ds+* (accessed 27 December 2010); Minutes of Survey Applications Coordinating Committee, Twelfth Meeting, 9 May 1968. The NRO provided this to the author pursuant to an MDR request.

21. *Apollo Program Summary Report*, 3–90; Memorandum for the Record, 27 August 1968, CREST, NARA.

22. Memorandum for the Record, 30 October 1968, CREST, NARA; Summary of Major SACC Coordination Activities, 9 May 1968 to 12 May 1969. The NRO provided this to the author pursuant to an MDR request.

23. Donald H. Steininger to Dr. DuBridge, 28 April 1969. NASA provided this to the author pursuant to an MDR request.

24. David, "Astronaut Photography," 189; *Apollo Earth Photographs Index Maps*; Robert C. Seamans Jr., Deputy Administrator, to Honorable Alexander H. Flax, Assistant Secretary of the Air Force for Research and Development, 28 June 1966, Seamans Chron File, 6/66, NASA Headquarters History Office; Alexander H. Flax to Robert Seamans, 13 September 1966. NASA provided this last document to the author pursuant to an MDR request.

25. David, "Astronaut Photography," 189.

26. Ibid.

27. James C. Fletcher, Administrator, to Honorable Henry A. Kissinger, Assistant to the President for National Security Affairs, 20 May 1971. The author is grateful to Scott Wenger for providing this document.

28. Ibid.; James C. Fletcher, Administrator, to Honorable Henry A. Kissinger, Assistant to the President for National Security Affairs, 18 January 1972. The author is grateful to Scott Wenger for providing these documents.

29. Harold S. Coyle Jr., Major, USAF to Dr. McLucas, 9 November 1972, Document #58, NRO Staff Records; Carl E. Duckett, Deputy Director for Science and Technology, to the Director, 27 November 1972 and Memorandum for the Record, 4 December 1972, CREST, NARA.

30. Harold S. Coyle Jr., Major, USAF to Dr. McLucas, 9 November 1972, Document #58, NRO Staff Records; Carl E. Duckett, Deputy Director for Science and Technology, to the Director, 27 November 1972 and Memorandum for the Record, 4 December 1972, CREST, NARA.

31. *Apollo 14 Mission Report*, section 12.3; *Apollo 15 Mission Report*, 159; *Apollo 16 Mission Report*, 12-1; *Apollo 17 Mission Report*, 13-1.

32. Ed Lyon, interview by author, Arlington, Va., 15 June 2009.

33. George M. Low, Acting Administrator, to Honorable Henry A. Kissinger, Assistant to the President for National Security Affairs, 14 April 1971, CREST, NARA.

34. Donald H. Steininger, Assistant Deputy Director for Science and Technology, to Director of Central Intelligence, 3 March 1973, CREST, NARA.

35. Donald H. Steininger, Assistant Deputy Director for Science and Technology to Director of Central Intelligence, 9 April 1973, CREST, NARA.

36. *Skylab EREP Investigations Summary*, 1–5; *Skylab Earth Resources Data Catalog*, 8–12.

37. [Redacted] to Director of Central Intelligence, 19 April 1974, CREST, NARA; Day, "Astronauts and Area 51."

38. Ezell, *NASA Historical Data Book*, vol. 3, 108–13.

39. El-Baz, *Catalog of Earth Photographs*; Page and Williams, *Apollo-Soyuz Pamphlet No. 5*; Author's interview with Richard Underwood.

40. Ezell, *NASA Historical Data Book*, vol. 2, 300–331.

41. Perry, *A History of Satellite Reconnaissance*, vol. 2A, 164–74.

42. Ibid.

43. John H. Rubel to the Under Secretary of the Air Force, 21 February 1963; John H. Rubel to Dr. McMillan, 19 March 1963; Brockway McMillan, Assistant Secretary Research and Development to the Deputy Director of Defense Research and Engineering, 4 April 1963, 209.9 Space, Satellites, Guided Missiles, box #15, accession #67-4860, RG 330, NARA.

44. Robert C. Seamans Jr. to Honorable John H. Rubel, 5 April 1963; Assistant Secretary of Defense and Deputy DDR&E to the Secretary of Defense, 26 April 1963; Robert S. McNamara to Honorable James E. Webb, 27 April 1963; James E. Webb, Administrator to Honorable Robert S. McNamara, 31 May 1963; 209.9 Space, Satellites, Guided Missiles, box #15, accession #67-4860, RG 330, NARA.

45. Mitchell, "Showing the Way," 39–40.

46. Daily Log, Office of the Deputy Director, 9 July 1963, and Marshall S. Carter, Lieutenant General, USA, Deputy Director, to Deputy Director Defense (Research and Engineering), 24 July 1963, CREST, NARA; DoD/CIA-NRO Agreement on NASA Reconnaissance Programs, 28 August 1963. The NRO provided the last document pursuant to the author's MDR request.

47. Security Annex to DoD/CIA-NASA Agreement on NASA Reconnaissance Program, 28 August 1963. The NRO provided this to the author pursuant to an MDR request.

48. Mitchell, "Showing the Way," 40–41.

49. Byers, *Destination Moon*, 36–75; Hansen, *Spaceflight Revolution*, 318–35.

50. Byers, *Destination Moon*, 36–75; Hansen, *Spaceflight Revolution*, 318–35.

51. Mitchell, "Showing the Way," 40–41.

52. Ibid., 42; Memorandum for Deputy Administrator on NASA-DoD Interface on Lunar and Earth Sensors, June 1966. NASA provided this to the author pursuant to an MDR request. DoD/NASA Agreement on the NASA Manned Lunar Mapping and Survey Program, 20 April 1964, DoD-NASA Agreements, box #2, accession #70-5484, RG 330, NARA.

53. Mitchell, "Showing the Way," 42–43; Day, "Black Apollo."

54. Ezell, *NASA Historical Data Book*, vol. 2, 319–31; Byers, *Destination Moon*, 308–14.

55. Minutes of Survey Applications Coordinating Committee Meeting, 31 January 1967; Minutes of the Fifth Meeting Manned Space Flight Policy Committee, 20 April 1967; Minutes of Survey Applications Coordinating Committee, 3 August 1967. The NRO provided all three documents to the author pursuant to an MDR request.

56. David, "Astronaut Photography," 189–90.

57. NASA, *Apollo 15, 16, and 17 Mission Photography*, http://www.lpi.usra.edu/lunar/missions/apollo.

58. David, "Astronaut Photography," 189–90.

59. NASA, *Apollo 15, 16, and 17 Mission Photography*.

60. Ezell, *NASA Historical Data Book*, vol. 2, 232–63; Ezell, *NASA Historical Data Book*, vol. 3, 171–75; House Committee, *1970 NASA Authorization, Part 2*, 71–77.

61. Minutes of the Survey Applications Coordinating Committee, 21 November 1966. The NRO provided this document to the author pursuant to an MDR request. Dr. Hornig's letter to the NRO director on the Apollo Telescope Mount has not been located and reviewed for declassification. It is mentioned in the minutes to this SACC meeting.

62. Richard Helms, Director, to the Honorable Foy D. Kohler, Deputy Under Secretary of State, 19 September 1967. NASA provided this document to the author pursuant to an MDR request.

63. Minutes of the Sixth Meeting, Manned Space Flight Policy Committee, 28 November 1967; Minutes of the Survey Applications Coordinating Committee, 11 January 1968. The NRO provided these to the author pursuant to an MDR request. House Committee, *NASA Authorization for Fiscal Year 1968*, 210–11, 343–44; House Committee, *1970 NASA Authorization*, 78–80.

64. A Study—Recommendation G, Section XIV of a Report Forwarded by DCI to the NASM 156 Ad Hoc Committee, 19 September 1967, Examined. The NRO provided this to the author pursuant to an MDR request.

65. Ibid.

66. NASA, *Orbiting Astronomical Observatory*, http://science.nasa.gov/missions/oao/.

67. Richard Helms, Director, to Dr. John L. McLucas, 13 October 1969, CREST, NARA; Chronology of NSAM 156 Actions, n.d. The NRO provided this to the author pursuant to an MDR request.

68. NASA, *Orbiting Astronomical Observatory*, http://science.nasa.gov/missions/oao/; Watts, "An Astronomy Satellite Named Copernicus," 231–32.

69. John L. McLucas to Mr. Laird, 18 December 1972, http://www.nro.mil (accessed 20 January 2012); Day, "Mirrors in the Dark"; Day, "Out of the Black."

70. Smith, *Space Telescope*, 44–45.

71. Ibid.; Perry, *A History of Satellite Reconnaissance*, vol. 1, 168–70; Oder et al., *GAMBIT Story*, 60; Oder and Worthman, *HEXAGON Story*, 80, 81.

72. Smith, *Space Telescope*, 146; Pedlow and Welzenbach, *The Central Intelligence Agency and Overhead Reconnaissance*, 48–56; Perry, *A History of Satellite Reconnaissance*, vol. 1, 180–82; *A History of the HEXAGON Program*, 145–48; Oder and Worthman, *HEXAGON Story*, 86; Robarge, *Archangel*.

73. Source Evaluation Board for Optical Telescope Assembly, June 1977, 6-2-6-8, *Hubble Space Telescope* file, NASA Headquarters History Office; NASA, "Orbiting Astronomical Observatory-C," 8; Characteristics Idealist System, n.d., CREST, NARA.

74. Smith, *Space Telescope*, 110–14; *Exploring the Universe with the Hubble Space Telescope*, 48–50.

75. Smith, *Space Telescope*, 145–47. The available information indicates that the largest mirrors made up to that point for a spacecraft were those for the KH-11 electro-optical satellite, which was first launched in late 1976, On the first mission the primary mirror was 92 inches in diameter and slightly larger on later missions. The next largest were the 72-inch mirrors produced by Kodak for the KH-10 image-forming sensor to be flown on the Air Force's Manned Orbiting Laboratory, a program that was cancelled in 1969.

76. Smith, *Space Telescope*, 152–53. For example, Frank Lehan, an NRO consultant, submitted a Top Secret/BYEMAN report in 1972 which examined these possibilities. John S. Foster Jr. to Mr. Carl Duckett, Deputy Director, Science and Technology, Central Intelligence Agency, 9 August 1972, and Leslie C. Dirks, Acting Director of Special Projects, to Howard P. Barfield, Acting Assistant Director, Space Technology, O/DDR&E, 20 July 1972, CREST, NARA.

77. NRO, John L. McLucas to Mr. Laird, 18 December 18 1972.

78. Memorandum of Agreement for the Conduct of Intelligence and Civil Space Programs, 1 August 1975. The author thanks Jeffrey Richelson for providing a copy of this document.

79. [Redacted], Lt. Colonel, USAF, Deputy Director for Plans and Policy to Dr. Wade, OSD (ISA), 7 October 1975, Document #332, NRO Staff Records.

80. Smith, *Space Telescope*, 179–82.

81. Ibid.; Chaisson, *Hubble Wars*, 4–5; Burdick et al., "Clean Room," 203.

82. Chaisson, *Hubble Wars*, 4–5; Burdick et al., "Clean Room," 222–25; Senate Subcommittee, *Hubble Space Telescope and the Space Shuttle Problems,* 21; *A History of the HEXAGON Program*, 46–47, 69.

83. Vick, *KH-10 Dorian;* Byers, *Destination Moon,* 36–75; Hansen, *Spaceflight Revolution,* 318–35; NASA, *Observation of the Earth, Orbital and Suborbital Spaceflight Missions,* http://eol.jsc.nasa.gov/sseop/metadata/Apollo-Saturn_4-6_tables.htm; Smith, *Space Telescope,* 237; Oder et al., *GAMBIT Story,* 25, 61.

84. Smith, *Space Telescope*, 224, 237; NASA-Perkin-Elmer Contract, 19 October 1977, *Hubble Space Telescope* file, NASA Headquarters History Office.

85. Smith, *Space Telescope*, 228–30, 238–39, 290–91; Dunar and Waring, *Power to Explore,* 490, 502; *Hubble Space Telescope Optical Systems Failure Report,* 8-1–8-3; Rodney, *Hubble Space Telescope*, 13, 16–18. All the records concerning the limitations on NASA access remain inaccessible to the public.

86. Zimmerman, *Universe in a Mirror,* 106–11; Dunar and Waring, *Power to Explore,* 496–97; Smith, *Space Telescope*, 291.

87. Smith, *Space Telescope*, 360–61.

88. *DARPA Technology Transition*, 133; White Sands Missile Range, *Sensor Testing—Seismic*, http://www.asmr.army.mil/testcenter/services/st/Pages/Seismic.aspx (accessed 6 June 2013).

89. R. Jeffrey Smith and William Booth, "Pentagon Tests Could Have Found Hubble Defects before Launch, Officials Say," *Washington Post*, 30 June 1990, A16; Chaisson, *Hubble Wars*, 225; Senate Subcommittee, *Problems at National Aeronautics and Space Administration*, 53. On 10 July 1990, the Senate's Subcommittee on Science, Technology, and Space held a classified hearing on the *Hubble*'s problems. The transcript has not been released.

90. Chaisson, *Hubble Wars*, 4–5, 19–20.

91. *Technology Applications Report, 1993*, 41.

Chapter 6. The Shuttle: NASA's Radically New Partnership with the National Security Agencies

1. NASA, biographies of James Fletcher and George Low, http://www.hq.nasa.gov/office/pao/History/Biographies/fletcher.html and http://www.hq.nasa.gov/office/pao/History/Biographies/low.html.

2. NASA, biographies of Robert Frosch and Alan Lovelace, http://www.hq.nasa.gov/office/pao/History/Biographies/frosch.html and http://www.hq.nasa.gov/office/pao/History/Biographies/lovelace.html.

3. NASA, biographies of James Beggs, Hans Mark, William Graham, and Dale Myers, http://www.hq.nasa.gov/office/pao/History/Biographies/beggs.html and http://www.hq.nasa.gov/office/pao/History/Biographies/mark.html and http://www.hq.nasa.gov/office/pao/History/Biographies/graham.html and http://www.hq.nasa.gov/office/pao/History/Biographies/myers.html.

4. Heppenheimer, *Space Shuttle Decision*, 105–21.

5. Ibid., 125–36; Temple, *Shades of Gray*, 469; Hays, "NASA and the Department of Defense," 223–24.

6. Oder et al., *HEXAGON Story*, 81; *Joint DoD/NASA Study of Space Transportation Systems*, 16 June 1969, White House Documentation 1969–1973 file, NASA History Office.

7. Heppenheimer, *Space Shuttle Decision*, 166–70, 224–30; Hays, "NASA and the Department of Defense," 224; Memorandum for the Record, 28 January 1970 (Document II-30) and Agreement between the National Aeronautics and Space Administration and the Department of the Air Force Concerning the Space Transportation System (Document II-31), in Logsdon, ed., *Exploring the Unknown*, 2:366–68; Temple, *Shades of Gray*, 476–77.

8. Space Transportation System Committee: Summary of Activities for 1970 (Document II-32), in Logsdon, ed., *Exploring the Unknown*, 2:369–77; Heppenheimer, *Space Shuttle Decision*, 231–33; Senate Committee, *NASA Authorization for Fiscal Year 1973*, 1039.

9. Heppenheimer, *Space Shuttle Decision*, 231–33; Temple, *Shades of Gray*, 484; Hays, "NASA and the Department of Defense," 224–25.

10. Heppenheimer, *Space Shuttle Decision*, 386–87, 396–408; Temple, *Shades of Gray*, 487; Hays, "NASA and the Department of Defense," 224; NASA, biography of Fletcher.

11. Heppenheimer, *Space Shuttle Decision*, 231–33; Temple, *Shades of Gray*, 487–88;

Hays, "NASA and the Department of Defense," 225; Statement by President Nixon, 5 January 1972, http://history.nasa.gov/stsnixon.htm (accessed 20 October 2011).

12. "U.S. Launch Vehicles," *Aviation Week & Space Technology*, 13 March 1972, 122–23; Senate Committee, *NASA Authorization for Fiscal Year 1977*, 2094–95. It should be noted that the payload capacities of the ELVs often vary from one source to another. The size of the spacecraft that could be carried on the Titan IIID and Titan 34D is determined from the size of the HEXAGON imagery intelligence satellites (with the mapping camera module), which the former launched from 1971 to 1982 and the latter from 1983 to 1986.

13. Senate Committee, *NASA Authorization for Fiscal Year 1977*, 2094–95; "U.S. Launch Vehicles," *Aviation Week & Space Technology*, 3 March 1980, 124–26; *Assessment of Candidate Expendable Launch Vehicle for Large Payloads*, 6.

14. Heppenheimer, *History of the Space Shuttle*, 2:21–23; Ezell, *NASA Historical Data Book*, vol. 3, 113–16; Neal, "Space Policy and the Size of the Space Shuttle Fleet," 157–69.

15. Neal, "Space Policy and the Size of the Space Shuttle Fleet," 159; Senate Committee, *NASA Authorization for Fiscal Year 1973*, 1038–42.

16. Heppenheimer, *History of the Space Shuttle*, 2:80–83; House Subcommittee, *1974 NASA Authorization*, 475.

17. Jacob E. Smart, Assistant Administrator for DoD and Interagency Affairs, to Carl Duckett, 26 May 1972, and [Redacted], Assistant Deputy Director for Science and Technology, to Smart, n.d., CREST, NARA.

18. John S. Foster Jr. to Carl Duckett, Deputy Director, Science and Technology, Central Intelligence Agency, 9 August 1972, and Leslie C. Dirks, Acting Director of Special Projects to Howard P. Barfield, Acting Assistant Director, Space Technology, O/DDR&E, 20 July 1972, CREST, NARA.

19. James C. Fletcher, Administrator, to John S. Foster Jr., Director of Defense Research and Engineering, 11 October 1972, and Foster to Fletcher, 21 December 1972. The Office of the Secretary of Defense provided these to the author pursuant to an MDR request.

20. Oder et al., *HEXAGON Story*, 237–39; Memorandum for Director of Defense Research and Engineering, 30 October 1973, Space Shuttle 2964-73, box #11, RG 340, NARA.

21. Memorandum for Director of Defense Research and Engineering, 30 October 1973.

22. Heppenheimer, *History of the Space Shuttle*, 2:330–37; Temple, *Shades of Gray*, 495–96.

23. Heppenheimer, *History of the Space Shuttle*, 2:330–37; Temple, *Shades of Gray*, 496–98; Background Paper, 24 February 1976. The Office of the Secretary of Defense provided this to the author pursuant to an MDR request.

24. George M. Low, Deputy Administrator, and Malcolm Currie, Director of Defense Research and Engineering, to Secretary of Defense and Administrator, NASA, 28 May 1976. The Office of the Secretary of Defense provided this to the author pursuant to an MDR request. House Subcommittee, *United States Civilian Space Programs, 1958–1978*, 566–69; George Bush to the Honorable James T. Lynn, Director, Office of Management and Budget, 18 October 1976, http://history.state.gov/historicaldocuments/frus1969-76ve03/d135 (accessed 4 November 2011).

25. Bush to Lynn, 18 October 1976.

26. Malcolm R. Currie to the Honorable George Bush, Chairman, Committee on Foreign Intelligence, 21 October 1976, http://history.state.gov/historicaldocuments/frus1969-76ve03/d136 (accessed 4 November 2011).

27. Heppenheimer, *History of the Space Shuttle*, 2:384–85.

28. Memorandum from the President's Assistant for National Security Affairs to President Ford, 15 March 1976, Document #124; Memorandum from the President's Assistant for National Security Affairs to President Ford, 26 April 1976; National Security Decision Memorandum 333, 7 July 1976, Document #128, http://history.state.gov/historicaldocuments/frus1969-76ve03 (accessed 5 November 2011).

29. National Security Decision Memorandum 333, 7 July 1976, Document #128, http://history.state.gov/historicaldocuments/frus1969-76ve03 (accessed 5 November 2011). There were further studies on Shuttle survivability, including the Aerospace Corporation's "Space Shuttle Survivability Enhancement Study" and the Space and Missile Systems Organization/Defense Intelligence Agency's "Threat Assessment Report: Space Transportation System." However, they remain classified.

30. Hans Mark to Secretary of Defense, 1 August 1979, CREST, NARA; *Implications of Joint NASA/DoD Participation in Space Shuttle Operations*.

31. Memorandum for the Record, 23 August 1976, and Data and Information Release Committee Meeting Summary, 10 August 1977. The NRO provided these to the author pursuant to an MDR request.

32. *Technology Development*; *Science in Orbit*; *Orbiter Camera Payload System*.

33. Data and Information Release Committee Meeting Summary, 10 August 1977; *Space Transportation System Flight 2*.

34. Horton et al., *Shuttle/Spacelab MMAP Electromagnetic Environment Experiment*; Haber, *Space Shuttle Electromagnetic Environment Experiment*.

35. Data and Information Dissemination Committee Meeting Summary, 15 October and 30 December 1976. The NRO provided these to the author pursuant to an MDR request.

36. Data and Information Release Committee Meeting Summary, 30 December 1976 and 10 August 1977. The NRO provided these to the author pursuant to an MDR request.

37. Mark, *Space Station*, 71–73; Hays, "NASA and the Department of Defense," 226–27; Harold Brown to the President, 11 November 1977. The Office of the Secretary of Defense provided this to the author pursuant to an MDR request.

38. Memorandum for the Secretary of Defense, 25 November 1977. The Office of the Secretary of Defense provided this to the author pursuant to an MDR request.

39. Mark, *Space Station*, 71–73; Robert A. Frosch, Administrator to the President, 16 December 1977, White House Documents, NASA Headquarters History Office.

40. Presidential Directive/NSC-37, 11 May 1978, CREST, NARA.

41. Presidential Review Memorandum/NSC 23, 28 March 1977, http://www.jimmycarterlibrary.org/documents/prmemorandums/prm23.pdf (accessed 14 November 2011). Presidential Review Memorandum/NSC 23 from March 1977 established the NSC Space Policy Review Committee. Chaired by Frank Press, the President's science advisor, its other members included the secretaries of defense and state, assistant to

the president for national security affairs, NASA's administrator, and the director of the Arms Control and Disarmament Agency.

42. Harold Brown to the President, 26 June 1978. The Office of the Secretary of Defense provided this to the author pursuant to an MDR request.

43. Temple, *Shades of Gray,* 504–8.

44. Chairman, Policy Review Committee (Space), to the President, n.d., NLC-41-10-13-1-0, Jimmy Carter Presidential Library; Presidential Directive/NSC-42, 10 October 1978, CREST, NARA.

45. Robert A. Frosch, Administrator, to the Secretary of State et al., 20 November 1978, CREST, NARA; James T. McIntyre to Secretary of Defense et. al., 1 February 1979, Space Programs, box #279, Staff Offices—Domestic Policy Staff—Eizenstat, Jimmy Carter Presidential Library; "New Payload Could Boost Shuttle Cost," *Aviation Week & Space Technology,* 14 August 1978, 16–17; Kenneth L. Jordan Jr., Principal Deputy, Assistant Secretary Research and Development, to Under Secretary of Defense for Research and Engineering et. al., 24 April 1979, Space Program 2963-79, box #4, accession 81-0007, RG 340, NARA.

46. Jordan to Under Secretary of Defense for Research and Engineering et al., 24 April 1979; Temple, *Shades of Gray,* 506–7.

47. Temple, *Shades of Gray,* 504–8.

48. Hans Mark to Secretary of Defense and Director of Central Intelligence, 5 October 1979, CREST, NARA.

49. Heppenheimer, *History of the Space Shuttle,* 2:351–57; Robert A. Frosch and James T. McIntyre to the President, 11 November 1979, White House Documents, NASA Headquarters History Office; Jordan to Under Secretary of Defense for Research and Engineering et al., 24 April 1979.

50. Heppenheimer, *History of the Space Shuttle,* 2:351–57; Mark, *Space Station,* 101–4.

51. Issue Paper, Crosscutting Review of Space Programs, 1981 Budget, Issue #5: Space Shuttle Mission Control, n.d., NLC-41-27-4-2-1, Jimmy Carter Presidential Library; *DoD Participation in the Space Transportation System,* 20–21.

52. Sturdevant, "Two Steps Forward," 2867–69.

53. Mark, *Space Station,* 104–10; Hays, "NASA and the Department of Defense," 229–30.

54. Mark, *Space Station,* 104–10.

55. Hans M. Mark, Secretary of the Air Force and William J. Perry, Under Secretary of Defense for Research and Engineering, to Secretary of Defense, 29 October 1979, Space Program 2963-79, box #3, accession 81-0006, RG 340, NARA; Senate Subcommittee, *NASA Authorization for Fiscal Year 1983,* 163–66.

56. Mark and Perry to Secretary of Defense, 29 October 1979; Finke et al., *Continuing Issues (FY 1980) Concerning Military Use of the Space Transportation System,* 3–7.

57. Finke et al., *Continuing Issues,* 3–7; Dunar and Waring, *Power to Explore,* 320–24.

58. *DoD Participation in the Space Transportation System,* 6–15; House, *United States Civilian Space Programs,* vol. 2, 367–81.

59. *DoD Participation in the Space Transportation System,* 6–15.

Chapter 7. The National Security Agencies Abandon the Shuttle

1. Author's conversation with James Beggs; Mark, *Space Station*, 121–23.

2. John Noble Wilford, "Space Agency Sees Time of Sacrifice," *New York Times*, 20 September 1991, 38; Craig Covault, "NASA Curtails Shuttle Flights," *Aviation Week & Space Technology*, 21 June 1982, 16–18; *U.S. Human Spaceflight*, 35–40.

3. [Redacted] to Distribution, 7 December 1982, and Robert J. Hermann to Chief of Staff, Office of the Vice President, 11 March 1981, CREST, NARA. The latter sets forth the detailed transition plans for every NRO system, but this entire section is redacted.

4. Lowman, *Human Remote Sensor in Space*; Aber, *EB/ES 351 Manned Space Photography*.

5. Author's interview with Richard Underwood; Craig Covault, "Shuttle Earth Imagery Spurs Censorship Debate," *Aviation Week & Space Technology*, 22 October 1984, 18–21; James A. Williams, Lieutenant General, USA, Director, to John McMahon, Deputy Director for Central Intelligence, 2 February 1984, and J. M. Poindexter, Deputy Assistant to the President for National Security Affairs to James M. Beggs, Administrator, NASA, CREST, NARA.

6. *Space Transportation System Flight 2*; *Space Transportation System Flight 2, Addendum*; *Space Transportation System Flight 2, Management Plan*, 9-1–9-10; Craig Covault, "Shuttle Data Reveal New Earth Features," *Aviation Week & Space Technology*, 21 December 1981, 47–51.

7. National Security Decision Directive Number 8, 13 November 1981, CREST, NARA; Hays, "NASA and the Department of Defense," 231; Temple, *Shades of Gray*, 523.

8. Craig Covault, "Shuttle Flight Plan Paced by Payloads," *Aviation Week & Space Technology*, 7 June 1982, 82–86; "Cover Problem Spoils Test of Cirris," *Aviation Week & Space Technology*, 2 August 1982, 19–20.

9. Jonathan Harsch, "Secret Shuttle Cargo, Cryptic Messages: A More Military NASA?" *Christian Science Monitor*, 30 June 1982, 6; "Shuttle Flight Plan Paced by Payloads."

10. Dawson and Bowles, *Taming Liquid Hydrogen*, 169–77. There were also two smaller upper stages developed commercially for the Shuttle: the PAM-D (which could launch 900 pounds to geosynchronous orbit) and the PAM-A (which could place 3,000 pounds in that orbit). These were largely built for use with commercial payloads, although the Air Force ordered 28 PAM-Ds in 1983 to support the Global Positioning System program. Over 25 PAMs carried on ELVs and the Shuttle successfully launched a variety of NASA and commercial payloads beginning in 1980.

11. William P. Clark to the Vice President et al., 12 July 1982, CREST, NARA; Hays, "NASA and the Department of Defense," 231.

12. National Security Decision Directive Number 42, 4 July 1982.

13. Robert C. McFarlane, Deputy Assistant to the President for National Security Affairs, to William Schneider et al., 28 September 1982, and McFarlane to Members, Senior Interagency Group for Space, 26 November 1982, CREST, NARA.

14. James Beggs, Administrator, to Honorable Robert C. McFarlane, 8 December 1982, and Director, Intelligence Community Staff, to Deputy Director of Central Intelligence, 9 December 1982, CREST, NARA.

15. National Security Decision Directive Number 80, 3 February 1983, CREST, NARA.

16. Tsiao, *"Read You Loud and Clear!"* 269–73.

17. Hays, "NASA and the Department of Defense," 234–36; Thomas O'Toole, "Space Shuttle Friction Is Worsened," *Washington Post*, 16 March 1983, A21; John Noble Wilford, "Spy Satellite Reportedly Aided in Shuttle Flight," *New York Times*, 20 October 1981, C3.

18. Arlen J. Large, "U.S. Launches '84 Space Shuttle Schedule on Friday with the First of Nine Missions," *Wall Street Journal*, 30 January 1984, 10; *Military Space Operations, Shuttle and Satellite Computer Systems Do Not Meet Performance Objectives*, 8–12.

19. "U.S. Launches '84 Space Shuttle Schedule on Friday with the First of Nine Missions," "USAF Cancels July IUS Mission," *Aviation Week & Space Technology*, 13 February 1984, 22; House Subcommittee, *Department of Defense Appropriations for 1985, Part 6*, 690–98; Edward H. Kolcum, "Defense Dept. to Retain Expendable Launchers as Backup to Shuttle," *Aviation Week & Space Technology*, 18 March 1985, 115; Temple, *Shades of Gray*, 537.

20. Aldridge, *Assured Access*.

21. Spires, *Orbital Futures*, 894–97.

22. Ibid. The DoD was not the only agency beginning to move away from the Shuttle. In early 1985, the National Oceanic and Atmospheric Administration announced plans to build the next three Tiros weather satellites for launch only on Titan II intercontinental ballistic missiles the Air Force intended to convert into ELVs because configuring them for launch on the Shuttle was too expensive. Beggs publicly expressed his displeasure with NOAA's actions as well. Jay C. Lowndes, "NASA Challenges NOAA Plans for Three Non-Shuttle-Capable Tiros," *Aviation Week & Space Technology*, February 4, 1985, 24–25; Thomas O'Toole, "NASA Chief Hits Air Force Plan to Spurn Shuttle," *Washington Post*, February 25, 1985, A1.

23. Spires, *Orbital Futures*, 894–97; Aldridge, *Assured Access*.

24. Spires, *Orbital Futures*, 894–97; Aldridge, *Assured Access*.

25. Aldridge, *Assured Access*; National Security Decision Directive Number 164, February 25, 1985, http://www.fas.org/irp/offdocs/nsdd/nsdd-164.htm (accessed 21 December 2011).

26. NASA, NASA Orbiter Fleet, http://www.nasa.gov/centers/kennedy/shuttleoperations/orbiters (accessed 22 December 2011); Jenkins, *Space Shuttle*, 262.

27. *Long Duration Exposure Facility (LDEF), Mission 1 Experiments*; *Long Duration Exposure Facility (LDEF) Experiments M0003 Meteoroid and Debris Survey*.

28. NASA, NASA Orbiter Fleet.

29. Orloff, ed., *Space Shuttle Mission STS-41G Press Kit*.

30. Cimino et al., *Shuttle Imaging Radar B (SIR-B) Experiment Report*, 13–26, 137–44; Craig Covault, "Shuttle Earth Imagery Spurs Censorship Debate," *Aviation Week & Space Technology*, 22 October 1984, 18–21.

31. *Space Shuttle Mission STS-41G Press Kit*; *National Aeronautics and Space Administration's Fiscal Year 1985 Budget Requests That Support Department of Defense Programs*.

32. Covault, "Shuttle Earth Imagery Spurs Censorship Debate."

33. "Navy Chief Sees Gain in Space Shuttle Flight," *New York Times*, 22 March 1985, A19.

34. Craig Covault, "Military to Withhold Shuttle Liftoff Time," *Aviation Week & Space Technology*, 5 November 1984, 14–16; Pfannerstill, "Shuttle 51C Mission Report."

35. "Shuttle 51C Mission Report."

36. Ibid.

37. *STS and Defense Support Program Cargo Element;* NASA, *NSTS 1988 News Reference Manual,* http://science.ksc.nasa.gov/shuttle/technology/sts-newsref/stsref-toc.html.

38. Thomas O'Toole, "Spacewalkers Send Repaired Navy Satellite Spinning on Its Way," *Washington Post*, 2 September 1985, A3.

39. Boyce Rensberger, "Space Shuttle to Be Used in Star Wars Laser Test," *Washington Post*, 24 May 1985, A8; William J. Broad, "Laser Beam Hits 8-Inch Target in Space," *New York Times*, 22 June 1985, 11.

40. Day, "A Lighter Shade of Black: The (Non) Mystery of STS-51J."

41. Boyce Rensberger, "Challenger Crew Launches Military Research Satellite," *Washington Post*, 1 November 1985, A6.

42. "Mission 61C Payloads Test SDI Systems," *Aviation Week & Space Technology*, 20 January 1986, 20.

43. "USAF Declassifies Shuttle Mission Experiments," *Aviation Week & Space Technology*, 14 October 1984, 20; "Hectic Year to Begin with 2 January Missions," *Washington Post*, 30 December 1984, A15; Charles Mohr, "Pentagon Fears Future Delays on Future Spy Satellites," *New York Times*, 24 February 1986, B6; Thomas O'Toole, "Space Program Comes to Halt," *Washington Post*, 29 January 1986, A14. As an example of the storage costs, Robert Smith states that NASA paid $7 million a month just to store the *Hubble Space Telescope*.

44. Hays, "NASA and the Department of Defense," 236; Day, "Invitation to Struggle," 267; Walter Pincus, "NASA Officials Unsure of Need for New Shuttle," *Washington Post*, February 15, 1986, A9; David Isikoff and David Hoffman, "U.S. May Boost Shuttle's Military Role," *Washington Post*, 15 May 1986, A1.

45. Pincus, "NASA Officials Unsure of Need for New Shuttle"; Charles Fishman, "Security Experts See No Immediate Threat," *Washington Post*, 5 May 1986, A1.

46. Oder et al., *HEXAGON Story*, 118; "Titan Explosion Cripples U.S. Launch, Surveillance Capability," *Aviation Week & Space Technology*, 28 April 1986, 22; William E. Schmidt, "Third U.S. Rocket Fails, Disrupting Program in Space," *New York Times*, 4 May 1986, 1; Richelson, *America's Secret Eyes in Space,* 205–15.

47. "NASA Officials Unsure of Need for New Shuttle"; "Space Program Comes to Halt"; House, *Assured Access to Space: 1986,* 14–16.

48. House, *Assured Access to Space: 1986,* 21–28.

49. John Noble Wilford, "NASA Drops Plans to Launch Rocket from the Shuttle," *New York Times*, 20 June 1986, A1; "NASA Cancels Shuttle/Centaur Because of Safety Concerns," *Aviation Week & Space Technology*, 23 June 1986, 16; Dunar and Waring, *Power to Explore,* 321–23; Senate Subcommittee, *Air Force Space Launch Policy and Plans,* 4–22.

50. David E. Sanger, "Reagan Supports 4th Orbiter for Space Shuttle Program," *New York Times*, 12 June 1986, A1; "Air Force Presses to Mothball Vandenberg Shuttle Complex," *Aviation Week & Space Technology*, 28 July 1986, 16–17; David E. Sanger, "Air Force Plans Mid-Size Rocket for Space Fleet," *New York Times*, 24 June 1986, A1; Michael Isikoff, "President Gives NASA Green Light to Build Replacement for Challenger,"

Washington Post, 16 August 1986, A1; Molly Moore, "Air Force Chief Recounts Space Disaster's Lessons," *Washington Post*, 19 August 1986, A4.

51. William J. Broad, "Unmanned Rocket on Secret Mission Launched by U.S," *New York Times*, 6 September 1986, 1; Jay Matthews, "25-Year Old Atlas Rocket Orbits Weather Satellite," *Washington Post*, 18 September 1986, A3; "Military Satellite Lofted into Orbit," *New York Times*, 5 December 1986, A22.

52. Kathy Sawyer, "Military Payloads Dominate New Schedule for Shuttle," *Washington Post*, 4 October 1986, A1.

53. *Military Space Operations,* 13–16.

54. National Security Decision Directive 254, 27 December 1986. The Reagan Presidential Library provided this to the author.

55. Bruce A. Smith, "USAF Awards McDonnell Douglas Contract to Build, Operate MLVs," *Aviation Week & Space Technology*, 26 January 1986, 20.

56. Senate Subcommittee, *Air Force Space Launch Policy and Plans,* 1–50.

57. Ibid.

58. Ibid.

59. R. Jeffrey Smith, "Long-Grounded Titan Lofts Air Force Satellite," *Washington Post*, 13 February 1987, A6; John Noble Wilford, "$161 Million Space Mission Ruined as Rocket Goes Out of Control," *New York Times*, 27 March 1987, A15; "Air Force Rocket Launches Secret Satellite for Military," *Washington Post*, 16 May 1987, A16; William J. Broad, "2 Years of Failure End as U.S. Lofts Big Titan Rocket," *New York Times*, 27 October 1987, A1; Edward H. Kolcum, "USAF Titan 34D Launches Missile Warning Satellite," *Aviation Week & Space Technology*, 7 December 1987, 30.

60. "Vandenberg Shuttle Complex Will Go into Mothball Status," *Aviation Week & Space Technology*, 2 May 1988, 27–28; Edward H. Kolcum, "Mission to Mark Upsurge in Unmanned Space Mission Rate," *Aviation Week & Space Technology*, 23 May 1988, 83; Kathy Sawyer, "Air Force Braces as Pace or Unmanned Space Launches Soars," *Washington Post*, 28 November 1988, A4; *Compendium of Meteorological Satellites Assessment of Candidate Expendable Launch Vehicle for Large Payloads,* G-3–G-6.

61. Craig Covault, "New Missile-Warning Satellite to Be Launched on First Titan 4," *Aviation Week & Space Technology*, 20 February 1989, 34; *Aeronautics and Space Report of the President, 1988 Activities,* 71; *Aeronautics and Space Report of the President, 1989–1990 Activities,* 61.

62. Cass Peterson, "Shuttle Satellite Settles into Stationary Orbit," *Washington Post*, 1 October 1988, A7.

63. NASA, *NASA Space Shuttle Mission STS-29 Press Kit, http://science.ksc.nas.gov/shuttle/missions/sts-29/sts-29-press-kit.txt* (accessed 17 August 2011).; NASA, *Space Shuttle, Mission Archives, STS-32,* http://www.nasa.gov/mission_pages/shuttle/shuttlemissions/archives/sts-32.html (accessed 17 August 2011).; Kathy Sawyer, "Shuttle Orbits Navy Communications Satellite," *Washington Post*, 11 January 1990, A6.

64. Craig Covault, "Atlantis' Radar Satellite Payload Opens New Reconnaissance Era," *Aviation Week & Space Technology*, 12 December 1988, 26–27; NASA, *NASA Space Shuttle Mission STS-29 Press Kit*; Edward H. Kolcum, "Orbiting of Advanced Imaging Satellite Bolsters U.S. Intelligence Capabilities," *Aviation Week & Space Technology*, 14 August 1989, 30–31; NASA, *Space Shuttle, Mission Archives, STS-32,* http://www.nasa.gov/mission_pages/shuttle/shuttlemissions/archives/sts-32.html; Kathy Sawyer,

"Shuttle Orbits Navy Communications Satellite," *Washington Post*, 11 January 1990, A6; Edward H. Kolcum, "Atlantis Lofts AFP-731 Reconnaissance Satellite," *Aviation Week & Space Technology*, 5 March 1990, 22.

65. William J. Broad, "Secrecy Is Lifted from Military Shuttle Missions," *New York Times*, 22 April 1991, A12; William J. Broad, "Shuttle Is Off at Last, on 8-Day Military Mission without Secrecy," *New York Times*, 29 April 1991, B6; William J. Broad, "Shuttle Atlantis Is Launched with Military Satellite," *New York Times*, 25 November 1991, D8; "Shuttle Flight Set for Next Week to End 11-Year Partnership of Pentagon, NASA," *Washington Post*, 28 November 1992, A5.

66. Craig Covault, "Missile Surveillance Capability Advanced by Discovery's Tests," *Aviation Week & Space Technology*, 13 May 1991, 20–22.

67. Craig Covault, "Astronauts to Launch Warning Satellite, Assess Manned Reconnaissance from Space," *Aviation Week & Space Technology*, 18 November 1991, 65–69; Craig Covault, "Shuttle Deploys DSP Satellite, Crew Performs Reconnaissance Tests," *Aviation Week & Space Technology*, 2 December 1991, 23.

68. James R. Asker, "Shuttle Completes Military Mission," *Aviation Week & Space Technology*, 14 December 1992, 36.

69. "Secrecy Is Lifted from Military Shuttle Missions."

70. *Space Shuttle*, 8; *Military Space Operations*, 13; House Subcommittee, *1983 NASA Authorization*, 921; *DoD Participation in the Space Transportation System*, 13–14; Hall, *Military Space and National Policy*, 24.

Chapter 8. NASA's Applications Satellites and National Security Requirements

1. Allison, *NASA Technical Memorandum 80704*, 1–3.

2. *Polar Meteorological Satellite Program Options, July 23, 1979*, CREST, NARA. Much of the liaison at the working level appeared to have been through the Polar Orbiting Operational Meteorological Satellite Coordination Board. No declassified records of this group have been located.

3. Ibid, 3–6; Ezell, *NASA Historical Data Book*, vol. 3, 266–86; *TIROS M Spacecraft (ITOS 1) Final Engineering Report*, 1-I-1–1-I-2.

4. Allison, *NASA Technical Memorandum 80704*, 3–6; Ezell, *NASA Historical Data Book*, vol. 3, 266–86.

5. Allison, *NASA Technical Memorandum 80704*, 26–27; John L. McLucas, Under Secretary of the Air Force, to Director of Defense Research and Engineering, 4 January 1972, Satellite Program 366-71 file, box #7, accession #73-0011, RG 340, NARA.

6. Hall, *A History of the Military Polar Orbiting Meteorological Satellite Program*, 19, 37; *Status of Environmental Satellites and Availability of Their Data Products*, 3–51; Heacock, ed., *Comparison of the Defense Meteorological Satellite Program (DMSP)*, IV-1–IV-5.

7. Hall, *A History of the Military Polar Orbiting Meteorological Satellite Program*, 23–24; William A. Porter, Acting Chairman, to the President, 4 December 1973, CREST, NARA.

8. Ezell, *NASA Historical Data Book*, vol. 3, 271–79; Senate Subcommittee, *NASA Authorization for Fiscal Year 1982*, 483–84.

9. NASA, *NASA Science Missions—TIROS*, http://science.nasa.gov/missions/tiros/; *Advanced TIROS-N-NOAA G.*; *Comparison of the Defense Meteorological Satellite Program (DMSP)*, II-36, II-44, II-50, IV-23; National Oceanographic and Atmospheric Admin-

istration, NOAA Polar Orbiter Data User's Guide, http://www.ncdc.noaa.gov/oa/pod-guide/ncdc/docs/podug/ (accessed 12 July 2012).

10. Trip Report, 5 November 1975. The Office of the Secretary of Defense provided this to the author pursuant to an MDR request.

11. Hall, *A History of the Military Polar Orbiting Meteorological Satellite Program*, 26–34; *Comparison of the Defense Meteorological Satellite Program (DMSP)*, II-7.

12. Presidential Review Memorandum/NSC 23, 28 March 1977, http://www.jim-mycarterlibrary.org/documents/prmemorandums/prm23.pdf (accessed 5 July 2011).

13. Presidential Review Memorandum/NCS 23; *A Coherent U.S. Space Policy*, 28 March 1977, http://www.jimmycarterlibrary.org/documents/prmemorandums/prm23.pdf (accessed 7 May 2011); Presidential Directive/NSC-37, National Space Policy, 11 May 1978, CREST, NARA.

14. *Polar Meteorological Satellite Program Options, July 23, 1979,* and [Redacted] to the Director of Central Intelligence, 8 August 1979, CREST, NARA.

15. *Polar Meteorological Satellite Program Options, July 23, 1979,* and [Redacted] to the Director of Central Intelligence, 8 August 1979, CREST, NARA.

16. [Redacted] to Director of Central Intelligence, 27 November 1979, and [Redacted] to the Director of Central Intelligence, 4 October 1979, CREST, NARA. The section of the August 1979 report concerning weather satellites had a codeword annex, which remains classified.

17. Hall, *A History of the Military Polar Orbiting Meteorological Satellite Program*, 26–34; Davis, "History of the NOAA Satellite Program," 27; Allison, ed., *NASA Technical Memorandum 80704*, 5–7.

18. Hall, *A History of the Military Polar Orbiting Meteorological Satellite Program*, 26–34; National Security Support by the Civil Weather Satellites, 13 April 1988, CREST, CIA.

19. *Aeronautics and Space Report of the President, 1980 Activities*, 40.

20. Cabinet Council Decision Memorandum, 14 December 1981, CREST, NARA.

21. Eloise R. Page to Director of Central Intelligence, 4 March 1983, and [Redacted] to Director, Intelligence Community Staff, 18 July 1983, CREST, NARA.

22. "House Rejects Sale of Satellites to Industry," *Aviation Week & Space Technology*, 21 November 1983, 18; *Land Remote Sensing Commercialization Act of 1984 (Public Law 98–365)*.

23. Senate Committee, *NASA Authorization for Fiscal Year 1975*, 828; Reppy, *Secrecy and Knowledge Production*.

24. Senate Committee, *NASA Authorization for Fiscal Year 1973*, 1006; NASA. *Pageos and GEOS-2*, http://ilrs.gsfc.nasa.gov/satellite_missions/list_of_satellites/pag1_general.html and http://ilrs.gsfc.nasa.gov/satellite_missions/list_of_satellites/geo2_general.html (accessed 26 April 2011).

25. Pearlman, *A Study Program for Geodetic Satellite Applications*, 1–4; *NASA Geodynamics Program: Annual Report for 1979*, 8–11.

26. Ezell, *NASA Historical Data Book*, vol. 3, 342–43; Townsend, *An Initial Assessment of the Performance Achieved by the Seasat-1 Radar Altimeter*, 1–3.

27. Malcolm R. Currie to the Deputy Secretary of Defense, 2 April 1975. The Office of the Secretary of Defense provided this to the author pursuant to an MDR request.

28. Ibid.

29. GEOS-C Issue, Chronology and Key Papers, n.d. The Office of the Secretary of Defense provided this to the author pursuant to an MDR request. National Security Decision Memorandum 2, 20 January 1969, http://www.fas.org/irp/offdocs/nsdm-nixon/nsdm-2.pdf (accessed 1 May 2011).

30. GEOS-C Issue, Chronology and Key Papers; Memorandum for the President, 30 May 1975, http://history.state.gov/historicaldocuments/frus1969-76ve03/d110 (accessed 2 April 2012).

31. National Security Decision Memorandum 2; Jeanne Davis to Brent Scowcroft, 15 December 1975; Senior Review Group Meeting, 12/16/75-GEOS 3 Data file, box #15, U.S. National Security Council Institutional Records, Gerald Ford Presidential Library; Minutes of the Senior Review Group Meeting, 16 December 1975, http://history.state.gov/historicaldocuments/frus1969-76ve03/d120 (accessed 2 April 2012).

32. *NASA Geodynamics Program: Annual Report for 1979*, 11–12; William J. Perry to Dr. A. M. Lovelace, Acting Administrator, 7 June 1977; Deputy Director, Strategic and Space Systems to Assistant Director, Space and Advanced Systems, 3 October 1975. The Office of the Secretary of Defense provided these to the author pursuant to an MDR request. Regarding the use of the *GEOS-3* radar altimeter data by civilian scientists, the *Journal of Geophysical Research*, vol. 84, no. B8 (July 1979) contains 38 papers on the subject.

33. Dubach and Ng, *Compendium of Meteorological Space Programs*, SEA-3–SEA-4; House Subcommittee, *1976 NASA Authorization*, 18–21.

34. Malcolm R. Currie to the Deputy Secretary of Defense, 2 April 1975, and Adrian St. John, Major General, USA, Vice Director, Joint Staff, to the Secretary of Defense, 1 March 1976. The Office of the Secretary of Defense provided these to the author pursuant to an MDR request.

35. Robert S. Cooper, Assistant Director, Space and Advanced Systems to Mr. Mathews, NASA, 5 May 1975. The Office of the Secretary of Defense provided this to the author pursuant to an MDR request.

36. Ibid.; Memorandum for the Record, 19 May 1975. The Office of the Secretary of Defense provided this to the author pursuant to an MDR request.

37. J. W. Plummer to Dr. John E. Naugle, Deputy Associate Administrator, 11 June 1975, and Memorandum for the Record, 17 October 1975. The Office of the Secretary of Defense provided these to the author pursuant to an MDR request.

38. Data and Information Dissemination Committee Meeting Summary, 6 October 1975, 15 October 1975; John E. Naugle, Chairman, to Dr. Sayre Stevens, CIA et al., 22 December 1975. The NRO provided these to the author pursuant to an MDR request.

39. St. John to the Secretary of Defense, 1 March 1976; Memorandum for Record, 14 June 1976. The Office of the Secretary of Defense provided these to the author pursuant to an MDR request.

40. SEASAT-A Issue Chronology of Key Events, 29 September 1976, and Malcolm Currie to Assistant Secretary of the Navy (Research and Development), 29 September 1976. The Office of the Secretary of Defense provided these to the author pursuant to an MDR request. Senate Subcommittee, *NASA Authorization for Fiscal Year 1979*, 603–5.

41. [Redacted] to Military Assistant, Deputy Director for Strategic and Space Systems (ODDR&E), 4 October 1976; Malcolm Currie to A. M. Lovelace, Deputy Adminis-

trator, 1 November 1976. The Office of the Secretary of Defense provided these to the author pursuant to an MDR request.

42. Francis L. Williams, Director, Special Programs, Office of Applications to Adm. Ross N. Williams, ODDR&E (S&SS), 19 November 1976; Malcolm Currie to A. M. Lovelace, Deputy Administrator, 27 December 1976. The Office of the Secretary of Defense provided these to the author pursuant to an MDR request.

43. R. N. Parker, Acting Director, to A. M. Lovelace, Deputy Administrator, 11 February 1977. The Office of the Secretary of Defense provided this to the author pursuant to an MDR request.

44. PRM/NSC-23, National Space Policy, 28 March 1977, http://www.jimmycarter-library.org/documents/prmemorandums/prm23.pdf (accessed 7 May 2011). The Policy Review Committee (Space) was chaired by the Secretary of Defense and its other members were the assistant to the president for national security affairs, the secretary of state, the director of central intelligence, NASA's administrator, and the director of the Arms Control and Disarmament Agency. Michael Hornblow, Acting Staff Secretary to the Vice President et al., 2 May 1977. The Carter Presidential Library provided this to the author pursuant to an MDR request. William J. Perry to the Secretary of Defense, 14 April 1977, and Harold Brown to Dr. Robert A. Frosch, Administrator, 8 August 1977. The Office of the Secretary of Defense provided these to the author pursuant to an MDR request.

45. Ross N. Williams, Rear Admiral, U.S. Navy, Military Assistant, Strategic and Space Systems, to Director, Joint Staff, 15 December 1977. The Office of the Secretary of Defense provided this to the author pursuant to an MDR request.

46. Townsend, *An Initial Assessment of the Performance Achieved by the SEASAT-1 Radar Altimeter*, 1–3; Ezell, *NASA Historical Data Book*, vol. 3, 344.

47. Fu and Holt, *Seasat Views Oceans and Sea Ice with Synthetic Aperture Radar.*

48. Memorandum for Record, 27 August 1976; A. M. Lovelace, Deputy Administrator, to Malcolm Currie, Director of Defense Research and Engineering, 3 September 1976; Malcolm Currie to Assistant Secretary of the Navy (Research and Development), 29 September 1976. The Office of the Secretary of Defense provided these to the author pursuant to an MDR request.

49. Satellite Oceanography in the 1980s: A Concept Paper, 9 February 1978. The Office of the Secretary of Defense provided this to the author pursuant to an MDR request.

50. Christine Dodson, Staff Secretary, to the Secretary of State et al., 31 August 1979, CREST, NARA.

51. [Redacted] to the Director of Central Intelligence, 4 October 1979, CREST, NARA; *PD/NSC-54, Civil Operational Remote Sensing,* 16 November 1979, http://www.jimmycarterlibrary.org/documents/prmemorandums/pd54.pdf (accessed 7 May 2011).

52. Craig Covault, "Oceanic, Gamma Ray Plans Approved," *Aviation Week & Space Technology*, 21 January 1980, 51–54; "NOSS Design Study Contracts Awarded," *Aviation Week & Space Technology*, 7 July 1980, 25; House Subcommittee, *1980 NASA Authorization*, vol. 1, part 4, 2261–69.

53. "Geosat Data to Aid Trident 2 Accuracy," *Aviation Week & Space Technology*, 19 July 1982; NASA, NASA Science Missions, Geosat, http://science.nasa.gov/missions/geosat (accessed 13 May 2011).

54. Proposed Implementation of World Geodetic Reference System 1984, 15 July 1983, CREST, NARA.

55. House, *United States Civilian Space Programs*, vol. 2, 219–37; *Mission to Earth: Landsat Views the World*.

56. Keith S. Peyton, Captain, USAF, to Dr. McLucas, 28 August 1972, NRO Staff Records Collection.

57. Ibid.; USSR: Early August Prospects for Grain Production, 10 August 1977, CREST, NARA.

58. Marilyn Berger, "GAO Links Sales of Grain to Soviet, Food Price Rise," *Washington Post*, 10 July 1973, D9; GAO Report on Soviet Grain Purchases and the Role of Intelligence Past and Present, n.d., CREST, NARA.

59. GAO Report on Soviet Grain Purchases and the Role of Intelligence Past and Present, n.d., and George W. Allen, Director, Imagery Analysis Service, to Deputy Director, Office of Economic Research, 2 October 1974, CREST, NARA; Richelson, *Wizards of Langley*, 174–75.

60. James C. Fletcher, Administrator, to Honorable George P. Shultz, Assistant to the President, 5 September 1973, http://history.state.gov/historicaldocuments/frus1969-76ve03/d83 (accessed 24 April 2012).

61. Thomas R. Pickering, Executive Secretary, to Major General Brent Scowcroft, 26 October 1973, http://history.state.gov/historical documents/frus1969-76ve03/d87 (accessed 24 April 2012); Paul V. Walsh, ADDI to the Director, 19 October 1973, CREST, NARA.

62. Harold S. Coyle Jr., Lt. Colonel, USAF, to Dr. McLucas, 13 November 1973.

63. House, *United States Civilian Space Programs*, vol. 2, 219–37.

64. Ibid.; *Crop Forecasting by Satellite*; Draft Space Policy Alternatives Paper for Space Policy Review Committee, 21 August 1978, CREST, NARA.

65. W. E. Colby to the President, 10 October 1975, CREST, NARA.

66. [Redacted] to Mr. Williamson, NASA, 28 January 1975; Document #37, NRO Staff Records; Rutherford M. Poats, Acting Staff Director, to the Deputy Secretary of Defense et al., 29 September 1976, CREST, NARA; Leon Sloss to Mr. Sisco, 21 November 1975, http://history.state.gov/historical documents/frus1969-76ve03/d118 (accessed 25 April 2012); Donald G. Ogilvie to James E. Goodby, 6 October 1975; Document #989; Memorandum for the Record, 19 May 1976; Document #993; Leon Sloss, Chairman of the Working Group, to Space Policy Committee Working Group, 13 November 1975; Document #984, NRO Staff Records Collection.

67. House, *United States Civilian Space Programs*, vol. 2, 219–37.

68. Presidential Directive/NSC-42, 10 October 1978, CREST, NARA; Presidential Directive/NSC-54, 16 November 1979, http://www.jimmycarterlibrary.org/documents/pddirectives/pd54.pdf (accessed 7 July 2011); [Redacted], Chairman, "Open Skies" Task Force to Dr. Victor H. Reis, Assistant Director, Office of Science and Technology Policy, 1 April 1982, CREST, NARA.

69. Cabinet Council Decision Memoranda, 14 December 1981; Malcolm Baldridge to Members of the Cabinet Council on Commerce and Trade, 12 March 1982; R. E. Hineman, Acting Deputy Director for Intelligence, to Director of Central Intelligence and Deputy Director of Central Intelligence, 9 April 1982, and Eloise R. Page to Director of Central Intelligence, 4 March 1983, CREST, NARA.

70. NASA, Landsat 4, http://landsat.gsfc.nasa.gov/about/landsat4.html (accessed 8 July 2011).

71. Eloise R. Page to Director of Central Intelligence, 4 March 1983, and [Redacted] to Director, Intelligence Community Staff, 18 July 1983, CREST, NARA.

72. Allen and de Silva, "Landsat: An Integrated History," 6–22.

73. NASA, Landsat 5, http://landsat.gsfc.nasa.gov/about/landsat5.html (accessed 8 July 2011).

74. Thomas O'Toole, "U.S. to Pay for Landsat's Turnover to Private Industry," *Washington Post*, 18 May 1985, A2; Elizabeth Tucker, "Fund Battle Imperils U.S. Space Photos," *Washington Post*, 23 February 1987, WB1; Allen and de Silva, "Landsat: An Integrated History."

75. Allen and de Silva, "Landsat: An Integrated History"; James R. Asker, "Congress, White House Weigh Overhaul of Landsat Program," *Aviation Week & Space Technology*, 28 October 1991, 23; *Headquarters, United States Space Command, Command History, January 1993 to December 1994*, 218.

76. House, *Scientific, Military, and Commercial Applications of the Landsat Program*.

77. Allen and de Silva, "Landsat: An Integrated History."

Conclusion

1. House Subcommittee, *NASA–Department of Defense Cooperation in Space Transportation*, 20.

2. Jeremy Singer, "NASA and NRO to Team Up on Lunar Spacecraft," http://www.space.com/751-nasa-nro-team-lunar-spacecraft.html (accessed 17 September 2013).

3. See, e.g., Alicia Chang, "Air Force Brings Unmanned Space Plane Home," *Washington Post*, June 18, 2012, A7.

4. *Polar Satellites*, 1–4.

5. Frank Morring Jr., "Mikulski Cautions NASA on Defense," *Aviation Week & Space Technology*, August 28, 2002, 39–40; Craig Covault, "Navy Enlists NASA in the War on Terror," *Aviation Week & Space Technology*, April 8, 2002, 30–31.

Selected Bibliography

Aber, James S., *EB/ES 351 Manned Space Photography*. http://academic.emporia.edu/
aberjame/geospat/space/spae.htm (accessed 26 July 2011).

Advanced TIROS-N-NOAA G. Washington, D.C.: National Aeronautics and Space Administration, 1985.

Aeronautics and Space Report of the President, 1959 Activities. Washington, D.C.: National Aeronautics and Space Administration, 1960.

Aeronautics and Space Report of the President, 1980 Activities. Washington, D.C.: National Aeronautics and Space Administration, 1980.

Aeronautics and Space Report of the President, 1988 Activities. Washington, D.C.: National Aeronautics and Space Administration, 1990.

Aeronautics and Space Report of the President, 1989–1990 Activities. Washington, D.C.: National Aeronautics and Space Administration, 1991.

Aldridge, E. C. Pete. *Assured Access: "The Bureaucratic Space War."* http://ocw.mit.edu/
courses/aeronautics-and-astronautics/16-885j-aircraft-systems-engineering (accessed 21 December 2011).

Aldrin, Buzz, and Malcolm McConnell. *Men from Earth*. New York: Bantam Books, 1989.

Allen, James M., and Shanaka de Silva. "Landsat: An Integrated History." *Quest: the History of Spaceflight Quarterly* 12, no. 1 (2005), 6–22.

Allison, L. J., ed. *NASA Technical Memorandum 80704, Meteorological Satellites*. Greenbelt: National Aeronautics and Space Administration, 1980.

Apollo 14 Mission Report. Houston: National Aeronautics and Space Administration, 1971.

Apollo 15 Mission Report. Houston: National Aeronautics and Space Administration, 1971.

Apollo 16 Mission Report. Houston: National Aeronautics and Space Administration, 1972.

Apollo 17 Mission Report. Houston: National Aeronautics and Space Administration, 1973.

Apollo Earth Photographs Index Maps, Apollo Missions 6, 7, and 9. Houston: Manned Spacecraft Center, 1970.

Apollo Program Summary Report. Houston: National Aeronautics and Space Administration, 1975.

Arnold, David Christopher. *Spying from Space—Constructing America's Satellite Command and Control Systems.* College Station: Texas A&M University Press, 2005.

Assessment of Candidate Expendable Launch Vehicle for Large Payloads. Washington, D.C.: National Academy Press, 1984.

Baals, Donald D., and William R. Corliss. *Wind Tunnels of NASA.* Washington, D.C.: National Aeronautics and Space Administration, 1981.

Beschloss, Michael. *MAYDAY: Eisenhower, Khrushchev, and the U-2 Affair.* New York: Harper & Row, 1986.

Boone, Adm. W. Fred. *NASA Office of Defense Affairs—The First Five Years.* Washington, D.C.: National Aeronautics and Space Administration, 1970.

Borman, Frank, and Robert J. Serling. *Countdown: An Autobiography.* New York: Silver Arrow Books, 1988.

Brooks, Courtney G., James M. Grimwood, and Loyd S. Swenson Jr., *Chariots for Apollo: A History of Manned Lunar Spacecraft.* Washington, D.C.: National Aeronautics and Space Administration, 1979.

Brugioni, Dino. *Eyes in the Sky: Eisenhower, the CIA, and Cold War Aerial Espionage.* Annapolis: Naval Institute Press, 2010.

Burdick, L. A., A. E. Hultquist, and K. D. Mason. "Clean Room for Hubble Space Telescope." *SPIE Vol. 777 Optical System Contamination: Effects, Measurement, Control* (1987): 203–10.

Burrows, William E. *This New Ocean: The Story of the First Space Age.* New York: Random House, 1998.

Byers, Bruce K. *Destination Moon: A History of the Lunar Orbiter Program.* Washington, D.C.: NASA Headquarters, 1977.

Chaisson, Eric. *The Hubble Wars: Astrophysics Meets Astropolitics in the Two-Billion-Dollar Struggle over the Hubble Space Telescope.* New York: Harper Collins, 1994.

Chambers, Joseph R. *Partners in Freedom: Contributions of the Langley Research Center to U.S. Military Aircraft of the 1990s.* Washington, D.C.: National Aeronautics and Space Administration, 2000.

Charles, Daniel. "Spy Satellites: Entering a New Era." *Science* 243, no. 4898 (24 March 1989): 1541–43.

Cheng, Bin. *Studies in International Space Law.* Oxford: Clarendon Press, 1997.

Cimino, Jo Bea, Benjamin Holt, and Annie Holmes Richardson. *The Shuttle Imaging Radar B (SIR-B) Experiment Report.* Washington, D.C.: National Aeronautics and Space Administration, 1988.

Colwell, Robert N. *Monitoring Earth Resources from Aircraft and Spacecraft*. Washington, D.C.: National Aeronautics and Space Administration, 1971.

Compendium of Meteorological Satellites and Instrumentation. Greenbelt: National Aeronautics and Space Administration, 1973.

Compton, W. David. *Where No Man Has Gone Before: A History of Apollo Lunar Exploration Missions*. Washington, D.C.: National Aeronautics and Space Administration, 1989.

Compton, W. David, and Charles D. Benson. *Living and Working in Space: A History of Skylab*. Washington, D.C.: National Aeronautics and Space Administration, 1983.

Cone, Bruce. *The United States Air Force Eastern Test Range—Range Instrumentation Handbook*. Patrick Air Force Base: United States Air Force Eastern Test Range, 1976.

Crickmore, Paul F. *Lockheed SR-71: Secret Missions Exposed*. Oxford: Osprey, 1993.

Crop Forecasting by Satellite: Progress and Problems. Washington, D.C.: Government Printing Office, 1978.

DARPA Technology Transition. Arlington: Defense Advanced Research Projects Agency, 1997.

David, James. "Astronaut Photography and the Intelligence Community: Who Saw What?" *Space Policy* 22 (2006): 185–93.

———. "The Intelligence Agencies Help Find Whales: Civilian Use of Classified Overhead Photography under Project Argo." *Quest: The History of Spaceflight Quarterly* 16, no. 4 (2009): 27–36.

———. "Two Steps Forward, One Step Back: Mixed Progress under the Automatic/Systematic Declassification Review Program." *American Archivist* 70, no. 2 (Fall/Winter 2007): 220–38.

———. "Was It Really 'Space Junk'? U.S. Intelligence Interest in Space Debris that Returned to Earth." *Astropolitics* 3 (2005): 43–65.

———. "What Should Nations Reveal about Their Spying from Space? An Examination of the U.S. Experience." *Space Policy* 25, no. 2 (2009): 117–27.

Davis, Gary. "History of the NOAA Satellite Program." *Journal of Applied Remote Sensing* 1 (2007): 1–28.

Dawson, Virginia P., and Mark D. Bowles. *Taming Liquid Hydrogen: The Centaur Upper Stage Rocket, 1958–2002*. Washington, D.C.: National Aeronautics and Space Administration, 2004.

Day, Dwayne A. "Astronauts and Area 51: The Skylab Incident." *Space Review*, 9 January 2006. http://www.thespacereview.com/article/531/1.

———. "Black Apollo." *Space Review*, 29 November 2010. http://thespacereview.com/article/1743/1 (accessed 17 December 2010).
http://www.thespacereview.com/article/1734/1.

———. "Inconstant Moon: CIA Monitoring of the Soviet Manned Lunar Program," *Space Review*. http://thespacereview.com/article/878/1.

———. "Invitation to Struggle: The History of Civilian-Military Relations in Space." In *Exploring the Unknown: Selected Documents in the History of the U.S. Civil Space Program*, edited by John Logsdon, 223–71. Washington, D.C.: National Aeronautics and Space Administration, 1996.

———. "A Lighter Shade of Black: The (Non) Mystery of STS-51J." *Space Review*, 4 January 2010. http://thespacereview.com/article/1536/1.

———. "Mapping the Dark Side of the World, Part 2." *Spaceflight* 40 (1998): 303–10.

———. "Mirrors in the Dark." *Space Review*. http://thespacereview.com/article/1371/1.

———. "Out of the Black." *Space Review*. http://thespacereview.com/article/2100/1.

———. "Relay in the Sky: The Satellite Data System." *Space Chronicle: Journal of the British Interplanetary Society* 59, Suppl. 1 (2006): 56–63.

Day, Dwayne A., and Asif Siddiqi. "The Moon in the Crosshairs: CIA Intelligence on the Soviet Manned Lunar Programme, Part 1—Launch Complex." *Spaceflight* 45 (November 2003): 466–75.

———. "The Moon in the Crosshairs: CIA Intelligence on the Soviet Manned Lunar Programme, Part 2—The J Vehicle." *Spaceflight* 46 (March 2004): 113–25.

DoD Participation in the Space Transportation System: Status and Issues, MASAD-81-6. Washington, D.C.: General Accounting Office, 1981.

Dubach, Leland L., and Carolyn Ng. *Compendium of Meteorological Space Programs, Satellites, and Experiments.* Washington, D.C.: National Aeronautics and Space Administration, 1988.

Dunar, Andrew J., and Stephen P. Waring. *Power to Explore: A History of Marshall Space Flight Center, 1960–1990.* Washington, D.C.: National Aeronautics and Space Administration, 1999.

Earth Photographs from Gemini III, IV, and V, SP-129. Washington, D.C.: National Aeronautics and Space Administration, 1967.

El-Baz, Farouk, ed. *Catalog of Earth Photographs from the Apollo-Soyuz Test Project.* Washington, D.C.: National Aeronautics and Space Administration, 1979.

Engineering Report 8374—Briefing Charts for the NASA Meeting at Perkin-Elmer on the Project: Study for an Optical Technology Apollo Extension System (OTES). Perkin-Elmer Electro-Optical Division, 1966.

Erickson, Mark. *Into the Unknown Together: The DOD, NASA, and Early Spaceflight.* Maxwell Air Force Base, AL: Air University Press, 2005.

Exploring the Universe with the Hubble Space Telescope. Washington, D.C.: U.S. Government Printing Office, 1989.

Ezell, Linda Neumann. *NASA Historical Data Book.* Vol. 1, *NASA Resources 1958–1968.* Washington, D.C.: National Aeronautics and Space Administration, 1988.

———. *NASA Historical Data Book.* Vol. 2, *Programs and Projects, 1958–1968.* Washington, D.C.: National Aeronautics and Space Administration, 1988.

———. *NASA Historical Data Book.* Vol. 3, *Programs and Projects 1969–1978.* Washington, D.C.: National Aeronautics and Space Administration, 1988.

Federici, Gary. *From the Sea to the Stars: A History of U.S. Navy Space and Space-Related Activities.* Washington, D.C.: Naval Historical Center, 1997.

Finke, Reinald G., Charles J. Donlan, and George W. Brady. *Continuing Issues (FY 1980) Concerning Military Use of the Space Transportation System, IDA Paper P-1531.* Arlington, VA: Institute of Defense Analysis, 1980.

Fu, Lee-Lueng, and Benjamin Holt. *Seasat Views Oceans and Sea Ice with Synthetic Aperture Radar.* Washington, D.C.: National Aeronautics and Space Administration, 1982.

Fuller, John F. *Air Weather Service Support to the United States Army—Tet and the Decade After.* Scott Air Force Base, IL: Military Airlift Command History Office, 1979.

Gemini Summary Conference. Houston: National Aeronautics and Space Administration, 1967.

Greer, Kenneth E. "Corona." In *Corona: America's First Satellite Program,* edited by Kevin Ruffner, 3–39. Washington, D.C.: Central Intelligence Agency, 1995.

Haber, Fred. *Space Shuttle Electromagnetic Environment Experiment—Phase A Definition Study.* Washington, D.C.: National Aeronautics and Space Administration, 1976.

Hacker, Barton C., and James A. Grimwood. *On the Shoulders of Titans: A History of Project Gemini.* Washington, D.C.: National Aeronautics and Space Administration, 1977.

Hall, R. Cargill. *A History of the Military Polar Orbiting Meteorological Satellite Program.* Chantilly, VA: Office of the Historian, National Reconnaissance Office, 2001.

———. *Military Space and National Policy: Record and Interpretation.* Arlington, VA: George C. Marshall Institute, 2006.

———. "The NRO in the 21st Century—Ensuring Global Information Supremacy." *Quest: The History of Spaceflight Quarterly* 11, no. 3 (2004): 6–12.

———. *SAMOS to the Moon: The Clandestine Transfer of Reconnaissance Technology between Federal Agencies.* Chantilly, VA: Office of the Historian, National Reconnaissance Office, 2001.

Hansen, James R. *Spaceflight Revolution, NASA Langley Research Center from Sputnik to Apollo.* Washington, D.C.: National Aeronautics and Space Administration, 1995.

Hartman, Edwin P. *Adventures in Research: A History of Ames Research Center, 1940–1965.* Washington, D.C.: National Aeronautics and Space Administration, 1970.

Hawkes, John. "USAF's Satellite Test Center Grows." *Aviation Week* (30 May 1960): 57–63.

Hays, Peter. "NASA and the Department of Defense: Enduring Themes in Three Key Areas." In *Critical Issues in the History of Spaceflight,* edited by Steven J. Dick and Roger D. Launius, 199–238. Washington, D.C.: National Aeronautics and Space Administration, 2006.

Heacock, E. Larry, ed. *Comparison of the Defense Meteorological Satellite Program (DMSP) and the NOAA Polar-orbiting Operational Environmental Satellite (POES) Program.* Washington, D.C.: National Oceanic and Atmospheric Administration, 1985.

Headquarters, United States Space Command, Command History, January 1993 to December 1994. U.S. Space Command History Office, 1994.

Heppenheimer, T. A. *History of the Space Shuttle.* Vol. 2, *Development of the Shuttle, 1972–1981.* Washington, D.C.: Smithsonian Institution Press, 2002.

———. *The Space Shuttle Decision, NASA's Search for a Reusable Space Vehicle.* Washington, D.C.: National Aeronautics and Space Administration, 1999.

A History of the HEXAGON Program. Chantilly: Center for the Study of National Reconnaissance, 2012.

Horton, J. B. *Shuttle/Spacelab MMAP Electromagnetic Environment Experiment—Phase B Definition Study Preliminary Report.* Philadelphia: General Electric, 1975.

The Hubble Space Telescope Optical Systems Failure Report. Washington, D.C.: National Aeronautics and Space Administration, 1990.

Hunley, J. D., ed. *The Birth of NASA: The Diary of T. Keith Glennan.* Washington, D.C.: National Aeronautics and Space Administration, 1993.

Implications of Joint NASA/DoD Participation in Space Shuttle Operations, GAO/NSI-AD-84-13. Washington, D.C.: General Accounting Office, 1983.

Jenkins, Dennis. *Space Shuttle: The History of the National Space Transportation System—The Beginning through STS-75*. Marceline, MO: Walsworth, 1996.

Johnson, Thomas R. *American Cryptology during the Cold War, 1945–1989*. Bk. 2, *Centralization Wins, 1960–1972*. Ft. Meade, MD: Center for Cryptologic History, National Security Agency, 1995.

———. *American Cryptology during the Cold War, 1945–1989*. Bk. 3, *Retrenchment and Reform, 1972–1980*. Ft. Meade, MD: Center for Cryptologic History, National Security Agency, 1995.

Kaehn Jr., Brig. Gen. Albert J. "Military Applications Evolution and Future." In *The Conception, Growth, Accomplishments, and Future of Meteorological Satellites*, 41–48. Washington, D.C.: National Aeronautics and Space Administration, 1982.

Lambright, W. Henry. *Powering Apollo: James E. Webb of NASA*. Baltimore, MD: Johns Hopkins University Press, 1995.

Launius, Roger D., and Steven J. Dick, eds. *Critical Issues in the History of Spaceflight*. Washington, D.C.: National Aeronautics and Space Administration, 2006.

Launius, Roger D., and Dennis R. Jenkins, eds. *To Reach the High Frontier: A History of U.S. Launch Vehicles*. Lexington: University Press of Kentucky, 2002.

Logsdon, John M., ed. *Exploring the Unknown: Selected Documents in the History of the U.S. Civil Space Program*. Vol. 1, *Organizing for Exploration*. Washington, D.C.: National Aeronautics and Space Administration, 1995.

———. *Exploring the Unknown: Selected Documents in the History of the U.S. Civil Space Program*. Vol. 2, *External Relationships*. Washington, D.C.: National Aeronautics and Space Administration, 1996.

The Long Duration Exposure Facility (LDEF), Mission 1 Experiments, SP-473. Washington, D.C.: National Aeronautics and Space Administration, 1984.

Long Duration Exposure Facility (LDEF) Experiments M0003 Meteoroid and Debris Survey. El Segundo: Aerospace Corporation, 1984.

Lowman Jr., Dr. Paul D., *The Human Remote Sensor in Space—Astronaut Photography*. http://rst.gfsc.nasa.gov/Sect12/Sect12_4html (accessed 26 July 2011).

Mack, Pamela Mack. *Viewing the Earth: The Social Construction of the Landsat Satellite System*. Cambridge, MA: MIT Press, 1990.

Mark, Hans. *The Space Station: A Personal Journey*. Durham, NC: Duke University Press, 1987.

McDougall, Walter A. *The Heavens and the Earth: A Political History of the Space Age*. New York: Basic Books, 1985.

Mercury Project Summary, Including Results of the Fourth Manned Orbital Flight, SP-45. Washington, D.C.: National Aeronautics and Space Administration, 1963.

Merlin, Peter W. *Mach 3+: NASA/USAF YF-12 Flight Research, 1969–1979*. Washington, D.C.: National Aeronautics and Space Administration, 2002.

Military Space Operations, Shuttle and Satellite Computer Systems Do Not Meet Performance Objectives, GAO/IMTEC-88-7. Washington, D.C.: General Accounting Office, 1988.

Mission to Earth: Landsat Views the World. Washington, D.C.: National Aeronautics and Space Administration, 1976.

Mitchell, Vance. "Showing the Way: NASA, the NRO, and the Apollo Lunar Reconnaissance Program, 1963–1967." *Quest: The History of Spaceflight Quarterly* 17, no. 4 (2010): 38–45.

Mudgway, Douglas J. *Uplink-Downlink: A History of the Deep Space Network, 1957–1997.* Washington, D.C.: National Aeronautics and Space Administration, 2001.

Muenger, Elizabeth A. *Searching the Horizon: A History of Ames Research Center, 1940–1976.* Washington, D.C.: National Aeronautics and Space Administration, 1985.

NASA Geodynamics Program: Annual Report for 1979. Washington, D.C.: National Aeronautics and Space Administration, 1980.

NASA Program Gemini Working Paper no. 5040: Gemini V Air-to-Ground Transcription. Houston, TX: National Aeronautics and Space Administration, 1965.

National Aeronautics and Space Administration (NASA). *Apollo 6 Automated Earth Photographs.* http://nssdc.gsfc.nasa.gov/database/MasterCatalog?sc=1968-025A&ds+* (accessed 27 December 2010).

——. *Apollo 15, 16, and 17 Mission Photography.* http://www.lpi.usra.edu/lunar/missions/apollo (accessed 28 December 2010).

——. *Biographies of Aerospace Officials and Policymakers.* http://www.history.nasa.gov/bio-html (accessed 2 July 2010).

——. *Biographies of T. Keith Glennan, Dr. Hugh L. Dryden, Thomas A. Paine, Robert Frosch, Alan Lovelace, James Fletcher, George Low, James Beggs, Hans Mark, William Graham, and Dale Myers.* http://www.hq.nasa.gov/office/pao/History/ prsnnl.html (accessed 18 May 2011).

——. *Landsat 4.* http://landsat.gsfc.nasa.gov/about/landsat4.html (accessed 8 July 2011).

——. *Landsat 5.* http://landsat.gsfc.nasa.gov/about/landsat5.html (accessed 8 July 2011).

——. *Missions—ESSA—NASA Science.* http://science.nasa.gov/missions/essa/ (accessed 10 January 2012).

——. *NASA Orbiter Fleet.* http://www.nasa.gov/centers/kennedy/shuttleoperations/orbiters (accessed 22 December 2011).

——. *NASA Science Missions—TIROS.* http://science.nasa.gov/missions/tiros/ (accessed 18 April 2011).

——. *NASA Space Shuttle Mission STS-29 Press Kit.* http://science.ksc.nas.gov/shuttle/missions/sts-29/sts-29-press-kit.txt (accessed 17 August 2011).

——. *Nineteenth Semiannual Report to Congress, January 1–June 30, 1968.* Washington, D.C.: Government Printing Office, 1968.

——. *NSTS 1988 News Reference Manual.* http://science.ksc.nasa.gov/shuttle/technology/sts-newsref/stsref-toc.html (accessed 7 May 2012).

——. *Observation of the Earth, Orbital and Suborbital Spaceflight Missions.* http://eol.jsc.nasa.gov/sseop/metadata/Apollo-Saturn_4-6_tables.htm (accessed 5 October 2011).

——. *Orbiting Astronomical Observatory.* http://science.nasa.gov/missions/oao/ (accessed 2 April 2012).

——. "Orbiting Astronomical Observatory-C." News Release 72-156/156A. 17 August 1972.

————. *Pageos and GEOS-2.* http://ilrs.gsfc.nasa.gov/satellite_missions/list_of_satel-lites/pag1_general.html and http://ilrs.gsfc.nasa.gov/satellite_missions/list_of_satellites/geo2_general.html (accessed 26 April 2011).

————. *Space Shuttle, Mission Archives, STS-32.* http://www.nasa.gov/mission_pages/shuttle/shuttlemissions/archives/sts-32.html (accessed 17 August 2011).

————. *Tracking and Data Relay Satellite System.* http://msl.jpl.nasa.gov/Programs/tdrss.html (accessed 15 January 2012).

The National Aeronautics and Space Administration's Fiscal Year 1985 Budget Requests That Support Department of Defense Programs-GAO/NSIAD-84-120. Washington, D.C.: General Accounting Office, 1984.

National Oceanographic and Atmospheric Administration. *NOAA Polar Orbiter Data User's Guide.* http://www.ncdc.noaa.gov/oa/pod-guide/ncdc/docs/podug/ (accessed 12 July 2012).

Neal, Valerie. "Space Policy and the Size of the Space Shuttle Fleet." *Space Policy* 20 (2004): 157–69.

Newell, Homer E. *Beyond the Atmosphere: Early Years of Space Science.* Washington, D.C.: National Aeronautics and Space Administration, 1980.

Oder, Frederic C., James C. Fitzpatrick, and Paul E. Worthman. *The GAMBIT Story.* Chantilly, VA: National Reconnaissance Office, 1988.

Oder, Frederic C., Paul E. Worthman, and [redacted]. *The HEXAGON Story.* Chantilly, VA: National Reconnaissance Office, 1988.

Orbiter Camera Payload System, Final Report 19 December 1980. Itek Optical Systems, 1980.

Orloff, Richard, ed. *Space Shuttle Mission STS-41G Press Kit.* Washington, D.C.: National Aeronautics and Space Administration, 1984.

Page, Thornton, and Lou Williams. *Apollo-Soyuz Pamphlet No. 5: The Earth from Orbit.* Washington, D.C.: National Aeronautics and Space Administration, 1977.

Pearlman, Dr. Michael R. *A Study Program for Geodetic Satellite Applications.* Washington, D.C.: National Aeronautics and Space Administration, 1972.

Pedlow, Gregory W., and Donald E. Welzenbach. *The Central Intelligence Agency and Overhead Reconnaissance: The U-2 and OXCART Programs, 1954–1974.* Washington, D.C.: Central Intelligence Agency, 1992.

Perry, Robert. *A History of Satellite Reconnaissance.* Vol. 1, *CORONA.* Chantilly, VA: National Reconnaissance Office, 1973.

————. *A History of Satellite Reconnaissance.* Vol. 2A, *SAMOS.* Chantilly, VA: National Reconnaissance Office, 1973.

Pfannerstill, John A. "Shuttle 51C Mission Report." *Spaceflight* 27 (September/October 1985): 364–66.

Polar Satellites—Agencies Need to Address Potential Gaps in Weather and Climate Data Coverage (GAO-11-945T). Washington, D.C.: U.S. Government Accountability Office, 2011.

Potts, Ronald. *U.S. Navy/NRO Program C Electronic Intelligence Satellites, 1958–1977.* Chantilly, VA: National Reconnaissance Office, 1998.

Proceedings of the International Meteorological Satellite Workshop, November 13–22, 1961. Washington, D.C.: National Aeronautics and Space Administration, 1961.

Reppy, Judith. *Secrecy and Knowledge Production*. http://www.fas.org/sgp/library/index.html (accessed 1 November 2012).

Rich, Ben E., and Leo Janos. *Skunk Works: A Personal Memoir of My Years at Lockheed*. Boston: Little, Brown, 1994.

Richelson, Jeffrey T. *The U.S. Intelligence Community*. 5th ed. Boulder: Westview Press, 2008.

———. *The Wizards of Langley: Inside the CIA's Directorate of Science and Technology*. Boulder: Westview Press, 2001.

———. *America's Secret Eyes in Space: The U.S. Keyhole Spy Satellite Program*. New York: Harper & Row, 1990.

———. "Undercover in Outer Space: The Creation and Evolution of the NRO, 1960–1963." *International Journal of Intelligence and Counterintelligence* 13 (2000): 308–20.

———. "Ups and Downs of Space Radars." *Air Force Magazine* 92, no. 1 (January 2009): 24–27.

Robarge, David. *Archangel: CIA's Supersonic A-12 Reconnaissance Aircraft*. Washington, D.C.: Central Intelligence Agency, 2012.

Rodney, George A. *Hubble Space Telescope, SRM&QA Observations and Lessons Learned*. Washington, D.C.: National Aeronautics and Space Administration, 1990.

Science in Orbit: The Shuttle & Spacelab Experience—1981–1986. Washington, D.C.: National Aeronautics and Space Administration, 1988.

Selected Space Goals and Objectives and Their Relation to National Goals. Columbus: Batelle Memorial Institute, 1969.

Shepard, Alan, and Deke Slayton. *Moon Shot: The Inside Story of America's Race to the Moon*. Atlanta: Turner, 1994.

Shuttle and Satellite Computer Systems Do Not Meet Performance Objectives GAO/IMTEC-88-7. Washington, D.C.: General Accounting Office, 1988.

Siddiqi, Asif A. *Challenge to Apollo: The Soviet Union and the Space Race, 1945–1974*. Washington, D.C.: National Aeronautics and Space Administration, 2000.

Skylab Earth Resources Data Catalog. Houston, National Aeronautics and Space Administration, 1974.

Skylab EREP Investigations Summary. Washington, D.C.: National Aeronautics and Space Administration, 1978.

Smith, Robert. *The Space Telescope: A Study of NASA, Science, Technology, and Politics*. Cambridge: Cambridge University Press, 1989.

Space Shuttle, the Future of the Vandenberg Launch Site Needs to Be Determined, GAO/NSIAD-88-158. Washington, D.C.: General Accounting Office, 1988.

Space Transportation System Flight 2, OSTA Scientific Payload, Data Management Plan. Washington, D.C.: National Aeronautics and Space Administration, 1981.

Space Transportation System Flight 2, OSTA Scientific Payload, Data Management Plan, Addendum. Washington, D.C.: National Aeronautics and Space Administration, 1982.

Space Transportation System Flight 2, OSTA-1 Scientific Payload Data Management Plan. Washington, D.C.: National Aeronautics and Space Administration, 1981.

Spires, David N. *Orbital Futures, Selected Documents in Air Force History*. Vol. 2. Peterson Air Force Base, Colorado: Air Force Space Command, 2004.

The Status of Environmental Satellites and Availability of Their Data Products. Houston: National Aeronautics and Space Administration, 1977.

Steinberg, Gerald M. *Satellite Reconnaissance: The Role of Informal Bargaining*. New York: Praeger, 1983.

STS and Defense Support Program Cargo Element, Payload Integration Plan, February 19, 1980. Houston: National Aeronautics and Space Administration, 1980.

Sturdevant, Rick W. "From Satellite Tracking to Space Situational Awareness: The USAF and Space Surveillance, 1957–2007." *Air Power History* (Winter 2008): 5–23.

———. "Two Steps Forward, One Step Back: U.S. Military Human Spaceflight, 1979–1999." AIAA SPACE Conference & Exposition 2012: 2865–76.

Swenson Jr., Loyd S., James M. Grimwood, and Charles C. Alexander. *This New Ocean: A History of Project Mercury*. Washington, D.C.: National Aeronautics and Space Administration, 1998.

Technology Applications Report, 1993. Washington, D.C.: Ballistic Missile Defense Organization, 1994.

Technology Development: Future Use of NASA's Large Format Camera Is Uncertain—GAO/NSIAD-90-142. Washington, D.C.: General Accounting Office, 1990.

Temple III, L. Parker. *Shades of Gray: National Security and the Evolution of Reconnaissance*. Reston: American Institute of Aeronautics and Astronautics, 2005.

Tepper, Morris. *Meteorological Satellites*. Washington, D.C.: National Aeronautics and Space Administration, 1963.

TIROS M Spacecraft (ITOS 1) Final Engineering Report. Vol. 1. Washington D.C.: National Aeronautics and Space Administration, 1970.

Townsend, William F. *An Initial Assessment of the Performance Achieved by the Seasat-1 Radar Altimeter*. Washington, D.C.: National Aeronautics and Space Administration, 1980.

TRW Space Log. Vol. 32. Redondo Beach: TRW Space and Electronics Group, 1997.

Tsiao, Sunny. *"Read You Loud and Clear!" The Story of NASA's Spaceflight Tracking and Data Network*. Washington, D.C.: National Aeronautics and Space Administration, 2008.

Twentieth Semiannual Report to Congress, July 1–December 31, 1968. Washington, D.C.: National Aeronautics and Space Administration, 1969.

Twenty-first Semiannual Report to Congress, January 1–June 30, 1969. Washington, D.C.: National Aeronautics and Space Administration, 1969.

U.S. Aeronautical and Space Activities, January 1 to December 31, 1959. Washington, D.C.: National Aeronautics and Space Administration, 1960.

U.S. Human Spaceflight: A Record of Achievement, 1961–1998. Washington, D.C.: National Aeronautics and Space Administration, 1998.

Vick, Charles P. *KH-10 Dorian*. http://www.globalsecurity.org/space/systems/kh-10.htm (accessed 29 September 2011).

Wagoner, H. D. *United States Cryptologic History. Special Series Number 3, Space Surveillance Sigint Program*. Ft. Meade: National Security Agency, 1980.

Watts, Jr., Raymond N. "An Astronomy Satellite Named Copernicus." *Sky and Telescope* 44, no. 4 (October 1972): 231–33.

Zabetakis, Stanley G., and John F. Peterson. "The *Diyarbakir Radar*." *Studies in Intelligence* (Fall 1964): 41–47.

Zimmerman, Robert. *The Universe in a Mirror: The Saga of the Hubble Telescope and the Visionaries Who Built It.* Princeton: Princeton University Press, 2008.

Congressional Publications

U.S. Congress. House. Committee on Sciences, *NASA Authorization for Fiscal Year 1968.* 90th Cong., 1st sess., 3 March 1967.

U.S. Congress. House. Subcommittee on Space Science and Applications, Committee on Science and Astronautics. *1970 NASA Authorization.* 93rd Cong., 2nd sess., 11 March 1969.

U.S. Congress. House. Subcommittee on Manned Space Flight, Committee on Science and Astronautics, *1970 NASA Authorization.* Part 2. 93rd Cong., 2nd sess., 2 April 1969.

U.S. Congress. House. Subcommittee on Aeronautics and Space Technology, Committee on Science and Astronautics. *Review of Tracking and Data Acquisition Program.* 93rd Cong., 1st sess., 24 October 1974.

U.S. Congress. House. Subcommittee on Space Science and Applications, Committee on Science and Technology. *1983 NASA Authorization.* Vol. 2. 97th Cong., 2nd sess., 24 February 1982.

U.S. Congress. House. *United States Civilian Space Programs*, Vol. 2, *Applications Satellites.* 98th Cong., 1st sess., 1983.

U.S. Congress. House. Subcommittee on Space Science and Applications, Committee on Science and Technology. *Assured Access to Space: 1986.* 99th Cong., 2nd sess., 26 February 1986.

U.S. Congress. House. Subcommittee on Space and Aeronautics, Committee on Science and Technology. *NASA-Department of Defense Cooperation in Space Transportation*, 108th Cong., 2nd sess., 18 March 2004.

U.S. Congress. Senate. Committee on Aeronautical and Space Sciences, *NASA Authorization for Fiscal Year 1960.* 86th Cong., 1st sess., 22 March 1959.

U.S. Congress. Senate. Committee on Aeronautical and Space Sciences, *NASA Authorization for Fiscal Year 1961.* 86th Cong., 2nd sess., 28 March 1960.

U.S. Congress. Senate. *Meteorological Satellites.* 87th Cong., 2nd sess., 1962.

U.S. Congress. Senate. Committee on Aeronautical and Space Sciences, *NASA Authorization for Fiscal Year 1965.* Part 2. 88th Cong., 2nd sess., 4 March 1964.

U.S. Congress. Senate. Committee on Aeronautical and Space Sciences, *NASA Authorization for Fiscal Year 1966.* Part 1. 89th Cong., 1st sess., 8 March 1965.

U.S. Congress. Senate. *Soviet Space Programs, 1962–1965.* 89th Cong., 2nd sess., 1966.

U.S. Congress. Senate. Committee on Aeronautical and Space Sciences. *NASA Authorization for Fiscal Year 1973.* 92nd Cong., 2nd sess., 22 March 1972.

U.S. Congress. Senate. Committee on Aeronautical and Space Sciences, *NASA Authorization for Fiscal Year 1975.* 93rd Cong., 2nd sess., 13 March 1974.

U.S. Congress. Senate. Committee on Aeronautical and Space Sciences. *NASA Authorization for Fiscal Year 1977.* 94th Cong., 2nd sess., 17 February 1976.

U.S. Congress. Senate. Subcommittee on Science, Technology, and Space, Committee on Commerce, Science, and Transportation. *NASA Authorization for Fiscal Year 1979.* 95th Cong., 2nd sess., 8 March 1978.

U.S. Congress. Senate. Subcommittee on Science, Technology, and Space, Committee on Commerce, Science, and Transportation, *NASA Authorization for Fiscal Year 1980*. Vol. 1, part 4. 96th Cong., 1st sess., 21 Feb. 1979.

U.S. Congress. Senate. Subcommittee on Science, Technology, and Space, Committee on Commerce, Science, and Transportation, *NASA Authorization for Fiscal Year 1982*. 97th Cong., 1st sess., 7 April 1981.

U.S. Congress. Senate. Subcommittee on Science, Technology, and Space, Committee on Commerce, Science, and Transportation. *NASA Authorization for Fiscal Year 1983*. 97th Cong., 2nd sess., 23 February 1982.

U.S. Congress. Senate. Subcommittee on Strategic Forces and Nuclear Deterrence, Committee on Armed Services. *Air Force Space Launch Policy and Plans*. 100th Cong., 1st sess., 6 October 1987.

U.S. Congress. Senate. Subcommittee of the Committee on Appropriations, *Problems at National Aeronautics and Space Administration*. 101st Cong., 2nd sess., 17 May 1990.

U.S. Congress. Senate. Subcommittee on Science, Technology, and Space, Committee on Commerce, Science, and Transportation, *Hubble Space Telescope and the Space Shuttle Problems*. 101st Cong., 2nd sess., 10 July 1990.

Index

Duncan, Charles, 206
Dyna-Soar, 5

Earth Observation Satellite Company, 274–75
Earth Resources Experiment Package, 130–31, 165–67
Earth Resources Observation Satellite, 119
Earth Resources Survey Aircraft Program, 109, 131–32
Earth Terrain cameras, 165–67
Eastern Test Range, 79, 82–83
Eastman Kodak (Kodak), 153
  contract for *Hubble* backup mirror, 182–84
  contracts with NASA and NRO, 183
  Lunar Orbiter camera and, 169, 171–72
Echo 1 satellite, 96
Eisenhower, Dwight, 1–3, 13–14, 24–26, 31–32
Electromagnetic Emanation Experiment, 20
ELVs. *See* Expendable launch vehicles
Environmental Science Services Administration (ESSA), 126, 129–30
  weather satellites of, 86, 91–93, 247
ERTS-A. See Landsat
ERTS-B. See Landsat
ESSA. *See* Environmental Science Services Administration
Evans, Harry, 45
Expendable launch vehicles (ELVs), 195, 215–16, 220, 226–29, 243
  backup, 189, 208, 210–11, 215–16
  inventory and production of, 189, 200, 219, 226, 234–40
Explorer satellites, 27

Fairchild mapping and stellar cameras, 174
Finger, Harold, 45
Fink, Daniel, 112
Finland, 76
Fisher, Adrian, 107

Flax, Alexander, 111–14, 118
Fletcher, James, *142*, 163, 166, 190, 197–98, 235, 257, 260–61, 269–70
FMSAC. *See* Foreign Missile and Space Analysis Center
Ford, Gerald, 7, 199–201, 258, 272
Foreign Missile and Space Analysis Center (FMSAC), 36, 40–42
  briefings and reports of, 40–42, 46, 56–57, 60–61
  NASA personnel detailed to, 60–61
40 Committee, 132–33, 163–64, 166–67
Foster, John, 66, 83–84, 112, 117–18, 121–22, 197
Frosch, Robert, 124, 190, 202, 206, 211–13

Gagarin, Yuri, 37
GALAXY, 60
Galileo, 224, 234, 236
GAMBIT 1 satellites, 6, 17, 43, 120, 125, 129, 151, 173
GAMBIT 3 satellites, 17, 125, 129
Garbarini, Robert, 111
Gemini, 5, 35, 128, 278
  V, 155–56
  DoD and, 150–51, 154–58
  experiments for, 155–58
  missions of, 156–58
  objectives of, 154
  review of photography from, 157–58
Gemini Program Planning Board (GPPB), 154–55
General Dynamics, 191, 228, 239
General Electric, 84, 127
Geodesy, 95–96, 255
Geodetic Earth Orbiting Satellite II, 99–101, 256
Geodetic Earth Orbiting Satellite III, 99–101, 256–57
Geodetic Objective Plan, 95
Geodetic Satellite Policy Board, 98
Geodetic satellites, 4–5, 12, 98–100
  ANNA, 81, 96–98, 154, 255
  Beacon 1 and 2, 96
  Beacon Explorer, 98–99

GAMBIT 1, 6, 7, 43, 120, 125
GAMBIT 3, 17, 125, 129
HEXAGON, 127–28, 179, 182, 192, 210
KH-11, 11, 84
Landsat, 204, 268–71, 273, 275
radar, 115
TALENT-KEYHOLE, 117, 120–21, 129
U-2, 17
use for civilian applications, 120–21,
124–26, 129–30
Improved Tiros Operational System
(ITOS) satellites, 93–94
and military requirements, 247–48
NOAA 1 through 5, 94, 248
Inertial Upper Stage, 210, 218, 226, 231,
243
Injun 3 satellite, 81
Intelligence, 17–19, 20–21, 36, 40–43
clearances needed to access, 18–19
influence on Apollo, 8, 47–57
on Soviet Union's space program, 3,
13, 19–22, 36, 37–58, 61–67, 69
on Soviet Union's wheat production,
268–73
"The Intelligence Agencies Help Find
Whales: Civilian Use of Classified
Overhead Photography under Project
Argo," 6
Intelligence Board, U.S., 17–18, 20–21,
30, 59–60, 67, 99, 117, 129, 131
Committee on Imagery Requirements
and Exploitation of, 126, 159, 164
Committee on Overhead Reconnais-
sance of, 117–18, 158
Guided Missiles and Astronautics
Intelligence Committee of, 23,
59–62, 67
Joint Atomic Energy Intelligence
Committee of, 23
Scientific Intelligence Committee of,
23
Interagency Board on Civil Operational
Earth-Observing Satellite Systems,
255
Interagency Review Team, 161, 164
Interagency Task Force on Integrated

Remote Sensing Systems, 252–53,
266, 272
Interdepartmental Contingency Plan-
ning Committee, 71
Interim Upper Stage, 189, 206
International Geophysical Year (1957-
1958), 96
International Solar Polar. See Ulysses
"Invitation to Struggle: The History of
Civilian-Military Relations in Space," 8
Itek Corporation, 163, 174, 203
bids on Hubble of, 182–83, 185
contracts with NASA and NRO of,
179–80, 203
ITOS. See Improved Tiros Operational
System

Jaffe, Leonard, 119
Jet Propulsion Laboratory (JPL), 14, 27,
263
assistance to CIA of, 59–60
lunar probe programs of, 168–69
SEASAT-A and, 263
Johnson, Charles, 114
Johnson, Lyndon B.
A-12 cover story and, 71
administration of, 40, 101, 104, 121,
126
on outer-space treaties, 74–75
Johnson, U. Alexis, 107, 116
Johnson Space Center (JSC), 223, 232
Joint Chiefs of Staff, 30, 59, 71, 87, 90,
93–95, 227
establishing geodetic requirements, 95
establishing weather satellite require-
ments, 31, 87–88, 93, 95
security devices for NASA satellites
and, 262
Joint Meteorological Group, 30
Joint Meteorological Satellite Advisory
Committee, 30, 90
Joint Meteorological Satellite Program
Office, 90, 93
"Joint State-Defense NASA Message," 75
JPL. See Jet Propulsion Laboratory
JSC. See Johnson Space Center

National Academy of Sciences, 109, 128
National Advisory Committee on Aeronautics (NACA), 2–3, 13
  participation in U-2 cover story, 23–24
National Aeronautics and Space Act of 1958, 2, 15, 277–78
National Aeronautics and Space Administration (NASA)
  activities of, 9, 120–23, 152
  budget of, 40–41, 121, 123, 126, 180, 211–12
  clearances of, 21, 37, 47, 109, 117
  command and control networks of, 5, 27, 79–85, 196, 203, 205, 218, 225, 232, 240–41, 274, 280
  contractors of, 84, 127, 152–53, 179, 182–83, 191, 228
  cooperation with CIA and DoD on space debris, 71–77
  cooperation with DoD in command and control and space surveillance, 26–29, 77–78
  creation of, 2–3, 8, 14
  development and growth of, 8–9, 14–16
  geodetic satellite programs of, 95–102, 255–60
  goals, principles, and responsibilities of, 2–3, 13
  human spaceflight programs of, 14–16, 47–58, 120–21, 123, 153–68
  intelligence, U.S. Congress and, 40–43
  intelligence, White House and, 40–43
  as intelligence analyst, 22–23, 58–67
  intelligence conflicts with CIA, 43–47
  as intelligence consumer, 5, 16–22, 37–58
  leadership of, 3, 7, 8–9, 13–15, 18, 34–35, 48, 190–91
  network support to DoD vehicles, 28–29, 79–85
  oceanographic satellite programs of, 260–67
  participation in cover stories, 23–26, 70–71
  personnel of, 35, 81, 113–14, 160–61
  remote sensing aircraft of, 109, 131–32
  remote sensing programs of, 103–20, 123–31, 175–78, 267–75
  weather satellite programs of, 12, 29–33, 69–70, 85–95, 246–55, 276
  See also specific missions and programs
National Aeronautics and Space Council, 3, 19, 73–74
National Council on Marine Resources and Engineering Development, 122
National Geodetic Satellite Program (NGSP)
  Beacon Explorer I, II, and III, 98
  designation of, 69–70
  establishment of, 98, 256
  Geodetic Earth Orbiting Satellites I, II, and III, 99
  Passive Geodetic Earth Orbiting Satellite, 99
  programs of, 95–102
National Intelligence Estimates (NIEs), 18, 20–22, 37–39, 58, 63
National Oceanic and Atmospheric Administration (NOAA), 94–95, 120, 130, 245, 249, 250, 265–67, 271, 275
  geodetic satellites and, 256, 259
  oceanographic satellites and, 260, 265–67
  weather satellites of, 235, 237, 247–48, 253–54
National Oceanic Satellite System, 266–67
National Oceanographic Satellite System, 246, 260, 265
National Operational Meteorological Satellite System, 9, 281
  establishment of, 69, 87, 106, 246, 278–79
  requirements for, 87–90
National Photographic Interpretation Center, 120, 125, 130, 153, 156, 158, 268
"National Policy on the Security of Meteorological Satellite Information," 250–51

Space debris, analysis and collection of, 71–77
Space Detection and Tracking System, 29
Space Electronics Corporation, 190, 197
Spaceflight Tracking and Data Network, 196, 203, 215, 240
Space Intelligence Panel, 45–46, 54–55, 63–64
Space Launch Policy Working Group, 224–25
Space Partnership Council, 280
Space Policy Committee, 272
Space program, Soviet Union, 1
   competition with, 3, 13, 36, 47–58, 67
   human space flights of, 37, 40, 50, 54, 59, 77
   intelligence on, 3, 13, 19–22, 37–58, 61–67, 69
   lunar probes of, 18, 43
   manned circumlunar program of, 49–56
   manned lunar landing program of, 7, 17, 38–40, 57–59
   secrecy regarding, 16
   weather satellites of, 59–60
Space program, U.S., 1–2. *See also* National Aeronautics and Space Administration; *specific missions and programs*
The Space Shuttle Decision: NASA's Search for a Reusable Launch Vehicle and History of the Space Shuttle, vol. 2, *Development of the Shuttle, 1972-1981* (Heppenheimer), 7
"Space Shuttle Implications on Future Military Space Activity," 197
The Space Station: A Personal Journey (Mark), 8
Space surveillance, 26–29, 77–78
Space Task Group
   establishment of, 191
   members of, 191–92
   reports of, 192–93
The Space Telescope: A Study of NASA, Science, Technology, and Politics (Smith, R.), 6
Space Test Program, 198, 209, 234, 238

Space Tracking and Data Acquisition Network (STADAN), stations of and support to DoD vehicles, 79–83, 728
Space Transportation System (Shuttle), 8, 279–80
   astronauts on, 212–13, 238
   Carter, J., on, 205–17
   Challenger accident and effects on DoD, 234–40
   and CIA, 196, 200, 211, 213
   costs and funding of, 194–96, 199–201, 211–12, 226, 237, 242–43
   development of, 4, 182, 191–99
   DoD and performance specifications for, 189, 192–94
   DoD examines use of, 195–99
   DoD payloads on, 223, 229–33, 237–38, 240–43
   DoD payload transition to, 198, 207–11, 215–16, 226–28
   DoD political support for, 205–13
   DoD review of civilian experiments on, 203–5, 231
   early flights of, 178–79, 220–26
   experiments on, 221–25, 241–43
   flight schedules of, 200, 226, 234, 237–38
   launch sites for, 196
   NASA-DoD conflicts over continued use of ELVs, 226–29
   NSC and, 220–21, 229
   pricing of, 200–203
   problems with, 205–13, 219–26
   security for, 200–203, 223, 232, 240–42
   STS-1, 220
   STS-2, 222, 230
   STS-4, 223–24
   STS-6, 225–26, 231
   STS-9, 222
   STS-10, 231
   STS-29, 240
   STS-32, 240
   STS-38, 241
   STS-39, 241
   STS-41C, 229–31

JAMES E. DAVID is a retired curator in the Space History Division at the Smithsonian National Air and Space Museum. He has written numerous articles on the intersection of the U.S. civilian and national security space programs and the classification and declassification of government records.